STONE CIRCLES

STONE CIRCLES

A Modern Builder's
Guide to the Megalithic Revival

ROB ROY

CHELSEA GREEN PUBLISHING COMPANY

White River Junction, Vermont

Totnes, England

Dedicated, with love, to

George Roy, Alan Roy, and Ann Roy Carlin

Book design by Andrea Gray.

On safety: Moving large stones is inherently dangerous. While this book
tries to promote safe work habits, what is safe for one builder under
certain circumstances may not be safe for you under different circumstances. Due to the variability of local conditions, materials, skills, site, and
so forth, Chelsea Green Publishing Company and the author assume no
responsibility for personal injury, property damage, or loss from actions
inspired by the information in this book. Consult professionals when
learning new methods, and please be careful.

Printed in the United States of America.
First printing October 1999.

03 02 01 00 99   1 2 3 4 5

Printed on acid-free, recycled paper.

Library of Congress Cataloging-in-Publication Data
Roy, Robert L.
        Stone circles: a modern builder's guide to the megalithic revival
    / Rob Roy.
            p. cm.
        Includes bibliographical references and index.
        ISBN 1-890132-03-9 (alk. paper)
        1. Stone circles—Design and construction—History—20th century.
    2. Megalithic monuments—Design and construction—History—20th century.
    3. Spirituality in art. I. Title.
    NB1450.R69   1999   99-33259
    721'.0441—dc21

Chelsea Green Publishing Company
Post Office Box 428
White River Junction, VT 05001
(800) 639-4099
www.chelseagreen.com

Contents

Acknowledgments

This book could not have been written without the help and collaboration of many people, including friends, relatives, and people I haven't even met, except through their kindness in answering my many telephone questions. First, I must thank my partners in this long megalithic journey, my wife, Jaki, and son Darin, who have had to endure the loss of valuable family time during nearly two years of work on this book. Next, megathanks to my "Club Meg" friends, particularly Drs. Stephen and Robin Larsen; Rev. William H. Cohea Jr.; Edward Prynn; Ivan McBeth; Cliff Osenton; Ron and Margaret Curtis; Ed Heath; and John "Trigger" Tregurtha of The Australian Standing Stones Committee, who all provided manuscript corrections and criticisms, as well as written material and photographs far beyond the call of duty. *There wouldn't be a book without you.*

During our megalithic journey, Jaki, Darin, and I also benefited from the friendship and hospitality of many others, including Enrique Noguera of the Early Sites Preservation Society, Jaki's brothers and sisters in Devon and Cornwall, Julie Britton, Maggie Osenton, Fred Lindkvist and Brad Colby of Columcille, Inc., Barbara Heath, Harold Makepeace, Ivan and Angie Hicks, David Brandau and Lynne Hall, Ella Anstruther, Dominic and Abigail Ropner, Orren Whiddon of Four Quarters, Gaudry and Pauline Normand of the Crystal Sanctuary, Milan and Jeanette Warford, Andy and Val Scrase, Reva Seybolt, Brian Sullivan, Roy Dutch, and especially Glynis Kent, who also coined the title phrase "the Megalithic Revival." Chapter 15 would be a mere shadow of itself were it not for material provided by Prof. Judy Young, Helen and Brian Fosgate, Elberton Granite Association, Inc., Casimiro Bargas in Costa Rica, Brian Williams and Cheryl Power at Powys County Observatory in Wales, Mark Hough-ton Brown, Dr. David A. Summers, Lydia Pettis, Paul E. Phaneuf of Friends of Carhenge, and a very helpful librarian in Masham, North Yorkshire, whose name I failed to record. For artwork, valuable contacts, and information, I must also thank Chuck Pettis, Larry Mayer, Marty

Cain, Steph's Studio of Alliance, Adam Jonas Horowitz, Jay Markel, Patrick MacManaway, and Sig Lonegren.

Professional thanks to authors John Martineau and John Michell for leading me to new stone circles; and to author and leading stone circle authority Dr. Aubrey Burl, for cheerfully enduring several phone conversations over the past year or two, and for providing me with valuable leads and facts. Selena Fox of the Crystal Sanctuary helped in so many ways: providing contacts, pictures, and not one but two excellent literary contributions to the book itself. Many personal friends helped with experiments and research, including Russ Keenan, Alex Brett, Larry and Ed Garrow, George Barber, Eileen Allen, John Carlson, Scott Carrino, Brendan Kelleher, Dave and Karen Hornell and my neighbors on Murtagh Hill who helped build the Earthwood circle. Thanks to my sister Ann Carlin, brother George Roy, and my son Rohan Roy for research assistance. Thanks to all who have read the book in its manuscript form, including editors Alan Berolzheimer and Rachael Cohen (who also kept me on track), Anne McClellan, and Al Pratt and Erika Barber (who, by the time of publication, will be married at a stone circle built with some help, I hope, from the manuscript copy of this book).

Thanks to designer Andrea Gray for improving the organization of the book with her tidy layout, Production Manager Martha Twombly (who built a stone circle next to her desk), Hannah Silverstein for editorial and other assistance beyond the call, and, finally, Editor-in-Chief Jim Schley and Publisher Stephen Morris at Chelsea Green, for their support and faith in this project and for their understanding as I went beyond deadline.

And apologies to anyone I inadvertently left out of this list.

Introduction

This is a book about stones, big stones, circles of big stones—and how to build them. Equally, it is a book about people, very special people, the people in fact who build stone circles.

To learn how to design and build a stone circle, it is not an absolute requirement to know about the ancient stone temples, or even to know what other people are doing in the field today. But if there is wisdom in the old adage "Don't reinvent the wheel," I submit that it makes no more sense to reinvent the circle.

This is my tenth book, a milestone of sorts, and it's been far more work—and lots more fun—than any of the others. It has taken me into living dreamworlds, my own and those of others, ancient and modern, known and unknown, with magic yet to come. And you, the reader, are invited along for the journey. But first some important introductions need to be made, much like the cast of characters at the beginning of a Tolstoy novel. After all, what is an introduction for?

First I must introduce a word, a word that some of you will already know, and others will not. But it is a word I will use quite a bit—in fact, I'll need it soon—so here it is:

megalithic (meg-a-lith'-ic), adj. [from the Greek *megas* (great) and *lithos* (a rock or stone)] Consisting of huge stones; applied to ancient structures and monuments constructed of large, often unhewn, stones.

Second, as the three chapters of part 1: A Personal Odyssey tell of my personal involvement with rings of stone for the past thirty-odd years, I had better introduce myself, briefly. I suffer from the unlikely but catchy name of Rob Roy, and have recently completed my first half-century on this planet. I am director and chief instructor at Earthwood Building School in West Chazy, New York, where students are taught to build their own cordwood masonry and underground homes. But long before I became interested in alternative house-building techniques, I was caught up with the wonder of megalithic buildings such as the famous Stonehenge temple and Maes Howe, a magnificent stone-slab burial chamber in the Orkney Islands.

Because it is a story, a true story, the book is written chronologically. Chapter 1 begins with my own introduction to Stonehenge, but also tells a bit about humankind's hazy knowledge of a variety of ancient stone circles, built in prehistory. When I first visited Stonehenge, Avebury, and the other great circles, the same question kept coming to mind. *How in the world did the ancients wrestle the multi-ton stones into such extraordinary positions?* While we may never know the complete answer with certainty, I will at least describe how the monuments *might* have been built. In 1987 I began my search for these answers using a hands-on technique; I built my first true stone circle, described in chapter 2. Ten years later, using methods more akin to those of an investigative reporter, I sought the knowledge, views, and opinions of some of the world's greatest experts on modern megalithic building. On that journey, I found some unexpected links with the ancient past.

Another equally important question came to me, usually from friends who wonder if I've left all sanity behind: *Why? Why were the megalithic monuments constructed, and, for Pete's sake, why are you building stone circles?* The *why* question has proven to be equally challenging as the *how* question. In this book, I'll offer several answers, from experts, from the stone people who are building today, and from myself.

Next, I must introduce my family, because they are characters in this true account of a journey we all have shared, a *megalithic journey*. Jaki has been my wife for the best half of my half century. She has taken an active interest in the ancient and modern stones, and has joined me on stone-seeking expeditions to England, Scotland, Brittany, Quebec, and Pennsylvania. As you will see, we were fortunate to meet some of the warmest and most interesting characters imaginable—stone people all—on both sides of the Great Pond.

Sons Rohan and Darin have been part of the journey from the start. At the age of twelve, Rohan (now twenty-two) accompanied me on a winter sunrise observation at a megalithic stone-slab chamber in southern New York and helped to build the Earthwood Stone Circle. His brother Darin, now thirteen, accompanied Jaki and me on the European leg of our megalithic journey, showing great interest and aptitude. In England, Darin learned how to lift a 6-ton stone with a 10-foot fence post. Each has taken delight in building small stone circles on beaches, and in sand piles, and some of their artistic efforts will find their way into chapter 12.

As the chapters unfold, a group of extraordinary people will weave through the story to create a human tapestry that perfectly complements the ethereal beauty of the stone sites themselves. Here are some of the fascinating people you will meet:

Doctors Stephen and Robin Larsen are codirectors of the Center for Symbolic Studies near New Paltz, New York. Personal friends for several years, Steve and Robin commissioned me to direct the construction of a magical stone circle at the Center in August 1997, a story recounted in chapter 3. The authorized biographers of Joseph Campbell, their knowledge of world myth, Celtic and otherwise, is extraordinary, and they have developed a keen interest in things megalithic.

Part 2: The Stone People introduces a memorable collection of modern stone builders, whose roots are planted firmly in the past.

William H. Cohea Jr., a retired minister, was inspired in a dream while visiting the mystical island of Iona many years ago. He has designed and constructed Columcille in Pennsylvania, a spiritual retreat consisting of 17 acres of magical megaliths. Bill's story, which also tells of his own spiritual journey, is recounted in chapter 4.

Edward Prynn is one of the truly great eccentrics left in Cornwall, an English county that has produced more than its share. Archdruid Ed has created a living museum of megalithic artistry. His garden contains, in my view, the world's most spectacular rock collection, a beautiful display of full-scale monuments including a major stone circle, an eleven-ton rocking stone, a three-legged dolmen with an twenty-ton capstone, and two "holed" stones, among others. Only Ed could have pulled it all off in a country where planning permission to build *anything* in the countryside can be extremely difficult to obtain. You'll meet Ed and his friend Glynis Kent in chapter 5.

Ivan McBeth is another druid building stone circles in the south of England. A gentle giant of a man, Ivan shares the story of the creation of the Swan Circle, near Glastonbury, in chapter 6, and is the prime mover of the Dragon Circle in Surrey, described in chapter 10.

Cliff Osenton is a modern stone mover who does things the old-fashioned way. He moves mighty stones using "equipment" that would have been available to builders five thousand years ago. Cliff shuns backhoes, loaders, cranes, tractors, and winches. In an on-going effort to solve one of the great mysteries of the neolithic (new stone) age, Cliff continues to experiment quietly in rock quarries in the English Midlands, and at megalithic sites all around Britain. His fascinating story fills chapter 7.

Chapter 8 is the saga of Ron and Margaret Curtis, who live and work at Callanish on the Isle of Lewis, Scotland. They are regarded by many, including myself, as the most knowledgable experts on the great Callanish group of megaliths, including several stones circles and other sites. In 1977, Margaret rediscovered a fallen stone beneath the peat at the famous main circle at Callanish, and, sixteen years later she discovered an entire stone circle lost in the mists of time. Ron and Margaret, working with two friends and only simple tools, re-erected with great accuracy a very important and particularly beautiful three-ton standing stone at an outlying site. Along with Cliff Osenton, I see Ron and Margaret as holding keys that connect modern stone builders with the ancients.

Ed Heath is an award-winning American sculptor who now resides permanently in Quebec. Over the past several years, he has created some of the most astounding modern megalithic monuments on the planet, including huge circles, dolmens, and Inuit *inuksuit*, or totem stacks of stones. Although Ed lives with his wife, Barbara, less than 100 miles from Earthwood, we met them for the first time in November of 1997. The story of Ed's incredible work is told in chapter 9.

Along the way, we'll visit other stone builders and ancient and modern stone circle sites in England, Scotland, France, Canada, Costa Rica, Australia, and the United States.

Scores of books have been written about stone circles and other megalithic sites, dozens about Stonehenge alone. The bibliography lists just a few. Why do we need another? Well, until now, none of them have discussed *modern* stone circles, which are becoming more widespread. And none of them have told *how to actually design and build a stone circle, or why*. A few books, notably *Stonehenge* by R. J. C. Atkinson, speculate on how Stonehenge *might* have been built, but as scholarly as they are, they do not recount the actual construction of stone circles, as does the book in your hands, particularly part 3: How to Build a Stone Circle.

Although several of the earlier chapters detail the construction of particular stone circles, part 3 tells you, step by step, how to plan and build your own, from minilithic to megalithic. Chapter 11 discusses stone circle design, including such important considerations as siting, astronomical alignments, geometry, units of measure, geomancy, and architectural harmony. In chapter 12, we'll look at models and small circles, including a delightful garden variety mini-Stonehenge just a few miles from the real thing. Chapter 13 tells how to build a major stone circle using heavy equipment, or, if you prefer, chapter 14 explains how to do it with methods that might have been used by people five thousand years ago. Chapter 15 is a tour guide to some of the stone circles that have been built during the 20th century, including a full-scale replica of Stonehenge built between 1918 and 1930, the spectacular Australian Standing Stones, and several other wonderful private and public circles built in the past twenty-five years.

Some people believe that they know the answers to the large quiet questions posed by the ancient stone circles. They believe that they are somehow tuned in to the subtle meanings, purposes, even methods of design and construction. One or two stone circle aficionados I have recently encountered really believe—or try to give the impression—that they know how a circle *must* be built, and why. I make no such claims, and am wary of absolute truth. My experience in life is that truth is both personal (it varies from person to person) and transient (it varies from time to time in one's life). The dogmatists to whom I refer are not to be found among the heroes of this book, all of whom will consider wisdom and new ideas from any source.

We stone builders but mimic the ancient circles; few of us claim to know their full depth of meaning, their nuances, their subtleties. But we are learning. Archaeologists, archaeo-astronomers, heavy-lift engineers, dowsers, and backyard stone builders may all contribute to refining our knowledge of the ancient circles. In the meantime, I do not claim that this book is the complete and final answer even to modern stone circle construction. My goal has always been to bring within one volume as many considerations as possible into the discussion. Perhaps readers with a strong background in "sacred geometry" will be disappointed with my treatment of their field. Likewise, heavy-lift engineers may find

the building discussion to lack some technique or principal of which they are aware. Ditto archaeo-astronomers, sculptors, landscape gardeners, archaeologists, et cetera; all will find certain sections lacking. Fortunately, there are whole books devoted to these subjects, many listed in the bibliography. As I cannot go into depth in all the related fields, I have settled on a twofold intent: first, to introduce the reader to an exciting rebirth of interest in stone circles, a *megalithic revival,* if you like; and, second, to provide sufficient information for the reader to build a darned good stone circle, large or small. In short, it is the kind of book I wish had been available to me in the early days of my stone-building career.

Stone Circles: A Modern Builder's Guide to the Megalithic Revival is a story that predates written history, and is open-ended into the future. Megalithic construction has been with us at least seven thousand years, and, in the late 20th century, renewed interest has inspired a rate of construction that rivals the "golden era" that occurred between 3000 and 1500 B.C., when roughly 1,500 stone circles were built. I'm constantly hearing of more circles being planned, constructed, and completed. While crop circles are fun and can be created in an hour or so with practice, they are as sidewalk chalk art next to a good stone circle. And, as we will see in chapter after chapter, a ring of megaliths can be every bit as useful and spiritually invigorating as a church or temple . . . and at least as much fun! Go ye and do likewise.

On Tons and Tonnes, Colors and Colours, Backhoes and Diggers:

Writing a book for a transatlantic market has its problems. Yanks and Brits, are, after all, "two peoples divided by a common language," according to Churchill. So here's what I'm going to do, to help my friends on both sides of the pond to make sense of this tome.

Tons or tonnes? To compound the spelling problem, these terms refer to different quantities. Thankfully, a pound of weight is the same in all English-speaking countries; but there are all kinds of tons, so be wary. British publications might use either ton or tonne, but, unless specified, the term refers to the British or *long* ton of 2,240 pounds. In America, however, we use the *short* ton of 2,000 pounds. The devaluation of the ton in America has helped stone builders; it is exactly 12 percent easier to move a 10-ton stone here! In this book *ton* is used for the Ameri-

can ton and *tonne* for the British ton. American tonnage figures predominate, the main exception being when I quote from British sources. An easy conversion, incidentally, is that 8 tonnes is almost exactly 9 tons. Having clarified that, watch out (in other books, not this one) for the *metric ton*—also called *tonne, tonneau,* or *millier*—which refers to 1,000 kilograms, or 2,204.62 pounds!

Color or colour? American spellings predominate, except when I quote directly from a British publication.

Backhoe or digger? Building and equipment terms vary between countries. In fairness, and to maintain the correct voice, I use the British terms when speaking of British projects, and American terms when describing American projects. The translation, in both cases, will follow immediately in parentheses; a little awkward, perhaps, but it does the job.

PART 1

STONE CIRCLES
A PERSONAL ODYSSEY

— CHAPTER 1 —

A Neolithic Love Affair

When you visit an ancient stone circle, you are standing
in a space-time hologram of 2000 B.C. You're getting
an echo of the past, just out of phase with today
because the heavens have changed.

—JOHN MARTINEAU, STONE CIRCLE DESIGNER

In October of 1966, at the age of 19, I arrived at Southampton on the *SS United States,* famous as the fastest ocean liner ever built. It had only recently been revealed that the ship could travel at speeds up to fifty-six miles per hour, incredible for something weighing 80,000 tons and as long as three football fields. I was soon to learn that "Stone Age" people were capable of engineering feats at least as impressive, given their lack of basic tools and materials that we take for granted today.

The liner arrived after dark, and I immediately took a taxi to the Old Bell Inn, Salisbury, just a few miles from Stonehenge. Within an hour I learned to drink "lager and lime" and discovered that central heating had not yet been introduced to the British accommodation scene.

STONEHENGE

The next morning, I hired a bicycle and pedaled the ten miles of winding back road to Stonehenge, traveling though picturesque English villages such as the Woodfords (Little, Middle, and Upper) and Wilsford. No matter the mode of transport, the first impression of Stonehenge is almost always the same: that it is small. This illusion has been reported by visitors for centuries, and it may stem from the fact that the stones are first spotted from at least a mile away, from which vantage point they seem very small indeed.

I made my final approach from the north, just as visitors do today. The present pedestrian tunnel under the old A360 road (now the A344) to Devizes did not exist, nor the food kiosks, toilets, bookstore, and offices. One simply left one's bike or car in the small car park and walked over to the monument at leisure, stopping to pay a shilling to the custodian manning a small black booth. Admission charges to Stonehenge began in 1900. One of the large upright stones fell over in that year, prompting the owner to fence the monument and to charge an entry fee. When Stonehenge was auctioned off for £6000 in 1915, the auction notice said, "A charge is made for seeing Stonehenge and the net receipts average about £360 per annum." At six pence per person, this would indicate roughly

14,400 paying visitors a year. Today, Stonehenge is visited annually by three quarters of a million people, most of these in the summer, when 5,000 a day is the average, and the admittance fee is £2, about $3.20.

Stonehenge. Here I was at last, and feeling the kind of thrill experienced by a romantic and naive teenager. The *stone* part was obvious, as the megaliths became taller and taller with each approaching step. But what was *henge*? As I crossed the circular bank and ditch—about 350 feet in diameter—I didn't realize that I was also crossing the very feature that gave this stone circle part of its name. *Henge* now refers to all circular ditch and mound monuments, whether they enclose stone circles or not. In the 12th century, one Henry of Huntingdon described *Stanenges* (the place of the stone gallows or hanging stones) as one of Britain's four wonders. In 1932 Sir Thomas Kendrick gave the name "henge monuments" to circular ditch and mound sites, Class I (like Stonehenge) having a single entrance and Class II having opposing entrances, like the great stone circle henge at Avebury, just twenty-seven road miles north of Stonehenge.

As I entered the circle itself, and the true size of the individual stones became evident, I began to feel like Alice, just after nibbling from the small-making side of the mushroom. One Wiltshire family during the 19th century had a long-standing tradition designed to prevent a certain kind of natural disappointment, "that on this extended plain at such a distance it appears nothing, and by the time you are at it all astonishment ceases." The trick was to bring guests by coach right into the monument itself, with the window blinds closed for the final approach. "Stop, down with the blinds," was called out in a majestic tone, and the visitors found the stones "suddenly and all at once on the eye" and "not familiarized by a graduated approximation." (Chippendale, *Stonehenge Complete,* p. 142–143, citing *The Times*, London, 17th November, 1883.)

Alas, such a trick of perception is hardly available now. While I was extremely impressed by my first visit, it is with sadness that I return to Stonehenge today and listen to the disappointment of tourists, kept 33 feet away from the outermost stones by a low rope barrier. The tourists, of course, comment that the stones seem much smaller than expected. My family and I experienced this disappointment during a visit in 1985, when the barrier was a horrible 90 feet from the stones. Five years later, armed with a special

The author's first visit to Stonehenge, 1966.

Special Access to Stonehenge

To apply for a special access permit to view Stonehenge before and after regular hours, write to English Heritage (First Floor, Abbey Buildings, Abbey Square, Amesbury, Wiltshire, England, U.K., SP4 7ES) several weeks in advance of your planned visit. You will be asked the nature of your visit, and acceptance is not guaranteed, and there are times of the year when special access is not given, particularly at the summer and winter solstices, the vernal and autumnal equinoxes, and any time that special work is going on inside the circle. The special access charge in 1998 was twelve pounds (£12) per person, half price for children aged five to fifteen, payable by check in pounds Sterling at least two working days in advance of the visit. If you are serious about stone circles, the higher admission charge is worth it. There are two pages of rules and restrictions as part of the permit package, but none of them are unreasonable. A one-hour time limit is imposed.

Our family has visited Stonehenge twice under a special access permit, and thoroughly enjoyed the experience both times. The guards will leave you alone, if you like, or even show you around and point out some of the special features you might not otherwise know.

There is encouraging late-breaking news as we go to press. English Heritage and the National Trust (which owns over two square miles in the immediate area) are spearheading a comprehensive plan that will return Stonehenge and the many nearby barrows and other ancient structures back to an environment approximating that of 2000 B.C. Part of the plan is to close and remove part of the A344, which runs within sixty feet of the outer ditch at Stonehenge, as well as to put a mile or so of the A303 trunk road underground.

When this plan is finally actualized, which could take several years, it will again be possible to approach Stone-henge by foot from across Salisbury Plain, as in days of old, and without distraction from modern traffic. All of the present-day car parks and underground buildings will disappear, in favor of a visitor's center 2½ miles from the stones. Once again, the *inner sanctum* will be accessible, and without the need for a special access permit. In the meantime, the permit is the best way for serious stone people to visit the monument.

During our initial visit to Stonehenge in 1995, we could only visit with the stones from a distance.

During the nineties we twice visited Stonehenge by way of a special access permit. (Photo by Ray, one of the night guards.)

access permit to the *inner sanctum,* we were able to experience the full power and wonder of this greatest of all stone circles.

But, back in 1966, anyone could pay a shilling, fourteen U.S. cents at the time, and walk in. The floor of the monument was gravel in those days, instead of the lush grassy carpet of today. The unfortunate but necessary years of restricted access were still in the future.

Although my memories are more than half a lifetime in the past, I remember, aided by pictures I took on the day, that the sky was clear and blue, and the stones were large and a light greeny-gray. There was no hint of the orange color reported for freshly hewn sarcen stones, or the "blue" color that gives the smaller bluestones their name. Four millennia of weathering; erosion by millions of hands, feet, and, yes, hammers; and not a little lichen growth on the surface, gave the monoliths their regularity of hue.

The previous paragraph just about wraps up all the specific details I can remember from my first youthful visit to Stonehenge. On a deeper level, the power and the beauty and the mystery of these great stones touched off my lifelong love affair with megaliths, although, at the time, this "love" may have been more akin to a youthful infatuation. In years to come, the infatuation would pass through what author John Michell calls *megalithomania,* and finally settle into the kind of love I feel for the stones today, a love based on respect, knowledge, wonder, aesthetics, and—increasingly—a difficult to define spiritual outlook.

The capsule portrait of Stonehenge that follows is based on information I've learned since, through many subsequent visits and through research. I've tried to concentrate on some of the highlights of the voluminous Stonehenge history, particularly where it might be pertinent to the designing and building of a modern circle. For more thorough discussions, see category 1, "Stonehenge," in the bibliography.

Sarcen Stones

The sarcens are the massive stones that stand out in all the pictures. Their name is a rustic diminutive of the word *saracen*, the name of the Muslim "infidels" encountered by Crusaders in the Middle Ages. Saracens were foreigners, just as these stones were almost totally foreign to this part of the Salisbury Plain. Geologically, the sarcens are an extremely tough sedimentary sandstone, harder than granite. The layer of sand formed during the tertiary period, up to seventy million years ago, and covered the predominate chalk strata found throughout this part of England. The sand layer was only a few feet thick, which lends a uniform dimension to many of the sarcens used for building both at Stonehenge and at Avebury. It is said that when freshly quarried, the sarcen is somewhat easier to work, as compared to the obdurate surface that results from years of exposure. In any case, other stones will not break it: flint shatters and granite is too soft. Even had iron been available—the major sarcen rings were built during the early bronze age—it would have helped little, as Victorian day-trippers found when they discovered how much labor was involved in breaking off even the tiniest souvenir.

The sarcens used at Stonehenge came from the Marlborough Downs, a good 18 miles distant as the sober crow flies, farther if the gentlest route is taken. Even if roughly shaped on the Downs prior to shipment, some of the monoliths moved to Salisbury Plain would have weighed 45 tons. When first built, the outer sarcen ring of giants consisted of 30 uprights, supporting, in turn, 30 lintels, horizontal stones that joined the uprights to form a continuous circle about a hundred feet in diameter. Each upright stands about 13½ feet above the ground, is roughly 7 feet wide, and is 3½ to 4 feet thick. With roughly four feet of the uprights buried in their socket holes, their average weight is about 29 tons. The lintels are about 10½ feet in length, 3½ feet wide and 2½ feet deep, and weigh five to six tons.

Although we cannot know with certainty that the outer circle was ever fully completed—so many stones, particularly on the west side, are missing—the collective opinion among experts assumes that it probably was. Certainly, the socket holes are accounted for, and it is known that Stonehenge served for some centuries as a convenient local quarry from which to obtain good building stone.

When completed, the outer ring must have been awe-inspiring. Each lintel was shaped with a radius curve, inside and out, so that the ring comprised two almost perfect concentric circles. The walkway thus formed would have been about 3½ feet wide and 16 feet off the ground, and all evidence is that the curvature and height of this ring is incredibly

accurate, possibly within an inch or two or true. And, to assure that this accuracy would be retained, the architects and builders employed two remarkable construction techniques. First, each lintel was keyed to its two supporting uprights by a mortise and tenon joint, the mortise (circular depression) carved into the underside of the lintel, and the tenon (projecting knob) left revealed on the top surface of each upright. In addition, the lintels were keyed to each other with another kind of joint most often seen in wooden structures: the tongue and groove. A carved tongue at the end of one lintel was fitted snugly into the corresponding groove of an adjacent lintel. When we remember that a sarcen stone four thousand years ago could only be shaped by pounding it with another sarcen maul, the creation of this detailed jointing is more than impressive, and rivals the skill displayed in the close-fitting Incan stone jointing in South America, and the precision stone fitting at the Giza pyramids.

Sacred Measure?

One wonders why such care and time-consuming labor was considered so important by these ancient builders. John Michell, writing in *The New View Over Atlantis*, offers a tantalizing answer.

> The curved inner faces of the lintels formed a circle, and so did the polished inner faces of the upright stones supporting it. The diameter of that circle has been precisely measured by modern surveyors, confirming the figure which W.M. Flinders Petrie obtained in 1880 of 97.325 ft., with a margin of error of 0.06 ft. It has been commonly accepted that that length was intended to represent 100 Roman feet, which is the same as 96 Greek feet. The values of these units . . . mean that the inner diameter of the Stonehenge circle was 97.32096 feet. (p. 128)

Using a reasonable and logical progression of thought, Michell then arrives at a lintel width of

3.4757485 feet, which he admits "may seem a long and meaningless number for this important measure." It turns out that this *long* number is not entirely *meaningless*, being one six-millionth of Earth's polar radius. Multiplying six million times Michell's average width of the lintel yields a figure of 3949.7142 miles, which, he says, is "within a few feet" of results from modern satellite surveys.

Taken alone, this numerical circumstance could be easily dismissed as a coincidence, but there's more. Michell goes on to demonstrate important correlations between many of the ancient units of measure, and by an interesting progression, he is able to show that the "long number" representing the lintel width of the Stonehenge circle can be precisely related to measures that Isaac Newton found in the plan for the Temple at Jerusalem. Michell believes that the lintel width is, in fact, the *Jewish* or *sacred* rod.

While most present-day archaeologists would scoff at such a connection, it should not be dismissed too quickly. Michell tells us that Hugh Harleston Jr., during survey work at the ritual city of Teotihuacan in Mexico, identified a common unit of measure that he called the Standard Teotihuacan Unit (STU), which happens to be the same as the sacred rod, the width of the Stonehenge lintel, and one six-millionth part of the polar radius. Stonehenge, then, if Michell's figures hold up to further research and scrutiny, could very well be a repository of sacred measure, something that many researchers also ascribe to the Great Pyramid at Giza.

As a counterpoint, the careful and methodical Alexander Thom, a retired professor of engineering, found a completely different unit of measure used at many stone circles and megalithic monuments throughout Britanny and the British Isles, a unit he calls the megalithic yard (MY), which is roughly 2.72 feet. Michell, although he acknowledges Thom's work, does not correlate the megalithic yard with the other ancient measures. And Thom's work has had some recognition among mainstream archaeologists. In chapter 11 we'll look

into the question of units of measure in greater detail.

All of this was unknown to me as I wandered in wonder among the standing and fallen stones of Stonehenge in 1966. My main concern was, "How did stone-age people build this thing?" And probably, in the back of my mind, the first seeds of another question were beginning to germinate: "How could *I* build something like this?"

The question of *why* Stonehenge, or, indeed, any stone circle or ring, was built, also did not trouble me too much at the time. Michell suggests that the building was a repository for sacred measures, and acknowledges that many astronomical alignments may be incorporated into the structure, a theory which, in various forms, is held by many, including Thom, author Gerald Hawkins *(Stonehenge Decoded),* and the astronomer Fred Hoyle, just to name a few. The combination of mathematics, measure, and astronomy suggests that Stonehenge was some sort of scientific instrument or observatory. And, in truth, many other stone circles incorporate wonderful alignments and geometries in their design. As one amazing example, in chapter 8, we'll look at the Callanish stone ring alignments in more detail, through the eyes of Ron and Margaret Curtis.

Rings or Circles?

In the previous paragraph, I found it necessary to introduce the term *stone ring* alongside of stone *circle*. A word of explanation is in order. Not all rings of stone are, in fact, circular. Stonehenge is circular, and precisely so, as is the great Ring of Brodgar in Orkney and so many others. Callanish is not a true circle, but it is a very beautiful ring set within intersecting avenues of standing stones. Avebury appears to incorporate two stone circles within a great ring of stones that is by no means round. We'll return to stone ring design, circular or otherwise, in chapter 11.

Personally, I like the term *stone circles* better than *stone rings,* and I'll use the term for monuments ancient and modern that are not true circles, as many others have done before me. For example, Aubrey Burl's definitive book on the subject is called *Stone Circles of the British Isles,* even though true circles represent only about two-thirds of those that he has catalogued. In fact, after Thom, Burl has charted at least five different shapes of stone circles: circles, flattened circles, ellipses, egg shapes, and compound rings (rare). Burl tells us, "Of the 900+ megalithic rings known in the British Isles, the general proportions are 600 circles, 150 flattened circles, 100 ellipses, and 50 eggs. Circles are the easiest to lay out, requiring only a central peg and a length of rope or thonging. It is not surprising that throughout the whole 'megalithic' period most rings were circular." (Burl, *Stone Circles of the British Isles,* p. 43.)

What Were They Used For?

Stonehenge, Avebury, Callanish, Brodgar, Stenness . . . so many of the great circles are commonly called *temples,* a designation that is in no way far-fetched. When I think of the uses of today's churches and temples, uses that have not changed a great deal in recorded history, it seems more than likely that the ancient circles would have served similar purposes, including: social meeting place, place of celebration, and ceremonial place for various rites of passage such as baptism, coming of age, matrimony, and death ritual. In short, they functioned much like churches, past and present. It is no wonder that the early Christians often built their churches on stone circle sites, and for much the same reasons that they incorporated ancient holidays: People were meeting at these places anyway at certain times of the year. The easiest way to ease the transition from one religion to another is to maintain as much continuity as possible. This system was also used by Spanish missionaries in Central and South America, resulting in some very interesting hybrids, such as combinations of Christian with Mayan ritual, which persist to this day.

Throughout the world, temples ancient and modern make use of astronomical alignments, so religion and science are certainly not mutually exclusive uses of the sites. During the Middle Ages, churches were built with easterly entrances, sometimes (according to some writers) with an orientation toward sunrise on the day of the saint to whom the church is dedicated. Sometimes the churches simply followed the same orientation as the stone circles they replaced. This provides continuity, but also has the practical advantage of using the existing approach to the site.

Another clue to the ancient use is to find out what modern stone circles are used for, although strict archaeologists, concerned only with facts, might dismiss this idea as readily as they dismiss the efforts of some researchers to try to find the answer through psychic readings, a method, incidentally, used by David D. Zink in his book *The Ancient Stones Speak.* Ivan McBeth's Swan Circle is used as a venue for the great Glastonbury Festival each year. Ed Prynn's Seven Sisters is sometimes used as a druid temple. The Columcille monuments of Bill Cohea provide a venue for quiet spiritual contemplation. The Stone Mountain Stone Circle is used to celebrate the various natural festivals in the calendar year. The Earthwood circle is where we hold our bonfire socials at the building school, our annual Fifth of November bonfire, and mark the seasonal astronomical alignments in a celebratory fashion. All of the reasons that contemporary stone circles are used could be compatible with the reasons why the ancient circles were built.

The Trilithons

The most striking feature of Stonehenge is the remains of the five ancient *trilithons,* a word meaning "three stones" that was coined by the 18th-century antiquarian William Stukeley. Originally, five of these giant trilithons stood in a horseshoe-shaped arrangement within the circle formed by the thirty uprights already described. All fifteen pieces still exist on site: Three of the trilithons are complete, and the other two consist of a single standing stone, its fallen brother, and the capstone or lintel lying nearby. Stone 56, considered by many to be the most beautiful standing stone in the British Isles, is a fully dressed (shaped) monolith standing 22 feet high, with an additional 8 feet buried in the ground. Its weight has been estimated at 40 British tonnes, about 45 American tons. (It is not quite the tallest standing stone in Britain, a distinction belonging to the 25'9"-tall Rudston Monolith at All Saints Churchyard in Rudston, Humberside.)

Although Stone 56 is deeply socketed, which explains its longevity, the fallen Stone 55's socket was only four feet deep, and the stone has been recumbent for centuries. According to R. J. C. Atkinson, "The date of the collapse of this trilithon is unknown, but must have been earlier than A.D. 1574, since an illustration of that date shows stone 56 as already leaning" (*Stonehenge,* p. 41).

One of the trilithons is known to have fallen on January 3, 1797. The shock was felt by plowmen a half mile away as "a considerable concussion, or jarring of the ground." This trilithon was restored in 1958 under the supervision of Atkinson, the sockets filled with concrete.

The trilithons follow the central axis of Stonehenge, which aligns with the rising sun on the summer solstice—the longest day—and the setting sun on the winter solstice. The pair of trilithons closest to the entrance on the northeast part of the circle are (or were) about 20 feet high to the top of their lintels, the next opposing pair are 21½ feet high, and the great trilithon, which frames the winter setting sun, was originally 24 feet high, its three component stones weighing almost 100 tons. Maybe it helps to think of a trilithon three stories high and weighing as much as 62½ new Volkswagen bugs.

The Heel Stone and its Mate

Long before the major sarcen circle and trilithons were built, the Heel Stone stood just outside the ditch to help mark the summer sunrise. If people remember one "fact" about Stonehenge, it is that the sun rises over the Heel Stone on the longest day. But a rescue dig performed in 1979, when the Heel Stone area was being assaulted by the laying of a new telephone cable, shows, as many experts had suspected, that even this widely accepted "fact" was not true. The dig revealed that another stone had existed twelve feet to the left of the Heel Stone, as viewed by an observer standing at the center of the temple. This makes a lot of design sense. If you want to observe the sun rising on the horizon, the last thing you want is a hulking 38-ton stone right in the way. Rather, you'd place two stones side by side, like a gunsite, so that the sun would rise between them. While Atkinson (who was not in charge of this particular dig) takes issue with the coexistence of two "gunsite" stones, I'm inclined to disagree for practical alignment reasons. Incurable romantic that I am, I would even speculate that a clever designer would place the two stones one sun's diameter apart from each other, as viewed from the center.

Nevertheless, pictures continue to show the sun rising over the Heel Stone, an illusion that is possible to photograph just a few minutes after sunrise, if the photographer is a little loose on his or her interpretation of the center point. The true effect is better experienced on the shortest day looking at the sunset from any point along the center of the ancient earthen causeway that approaches Stonehenge from the northeast, known as the Avenue. The sunset is beautifully framed by the entrance uprights on the most complete section of the sarcen circle.

Some writers have surmised that Heel Stone should be spelled Hele Stone, a link to Helios, the Greek sun god, but this is not generally accepted. Equally doubtful is the connection of hele to the Anglo-Saxon verb *helan,* meaning *to conceal,* but applied to the stone because it conceals the rising of the sun on the longest day. The more likely derivation of the name—and the one most commonly quoted by authorities—is based upon the ancient legend given in the sidebar.

Earlier Phases of Construction

When people envision Stonehenge, they most commonly think of the giant sarcen stones already described, but the sarcen building was actually a late phase of construction. The temple was begun about eight hundred years earlier, around 3300 B.C., according to the latest carbon dating adjustments made in 1997. This first phase of construction produced the henge, the roughly 300-foot-diameter ditch and bank, as well as fifty-six curious *Aubrey holes* just inside the bank, first recorded by antiquarian John

The Legend of the Friar's Heel

This version of the legend is taken from *Stonehenge, Today and Yesterday,* by Frank Stevens (His Majesty's Stationery Office, London, 1924).

The devil, so the story goes, determined one day to undertake some great and stupendous work, for the like of which he is famous throughout the world. In this devil we can still discern the Scandinavian "giant" legend, which in later Christian times became "devil" legends. The work had to be great, puzzling, and amazing to all beholders, for, as the Wiltshire story-teller adds, "he had let an exciseman slip through his fingers." In the course of his wanderings up and down the earth he had noticed some large stones in the garden of an old crone in Ireland; and he determined, therefore, to transport them to the stoneless waste of Salisbury Plain as being the most unlikely spot in which to find such things. There yet remained the old woman's permission to be obtained before he could commence his labour. His request was at first met with a flat negative, but eventually the devil so played upon her cupidity, by the assurance that she could have as much money as she could count and add up while he was engaged in the work of removal, that she readily gave her consent. As usual the devil had the best of the bargain, for he, knowing her powers of arithmetic to be but scanty, handed her a number of pieces of money, whose value was fourpence halfpenny and twopence three farthings. The dame had barely managed to add the first two coins together when the devil called upon her to stop, and looking round she saw the stones were all removed, and had been tied with a withe (twisted twig) band into a neat bundle which was slung upon his shoulder. Away flew the devil towards Salisbury Plain, but as he sped onwards the withe cut deep into his shoulder, so heavy were the stones. He endured it as long as he could, but just towards the end of the journey, while passing over the valley of the Avon, he winced, and readjusted his burden; in so doing one of the stones fell down and plunged into the river at Bulford, where it remains to the present day, as witness to the veracity of this legend.

Right glad to be rid of his burden when he reached the Plain, the devil made haste to set up the stones, and so delighted was he with the result of his first efforts, and with the progress he was making, that he cried aloud with glee, "Now I'll puzzle all men, for no one knows, nor ever will know, how these stones have come here." Unluckily, this blind boast was overheard by a holy friar walking near, who straightway replied in right Wiltshire fashion, "That's more than thee can tell"; and then realising who the builder was, turned and fled for his life. Enraged at his discovery by the friar, and perceiving that his scheme had failed, the devil, who had just taken up a stone to poise it upon its two uprights, hurled it at the holy man, and struck him on the uplifted heel as he made haste to run. The friar's sanctity was evidently greater than his personal courage, for it was the stone and not the friar which suffered the most from the impact. Even today the huge impress of the friar's heel is to be seen upon the stone. At this juncture the sun rose, and the devil had perforce to relinquish his task. This accounts for the present scattered appearance of the stones.

Aubrey in the 17th century. The summer sunrise alignment was already included in the design by the selection of the entrance. Some authorities, including Atkinson, report that the Heel Stone and its mate, date from the earliest period, and Aubrey Burl suggests that another pair of stones would have framed the rising midsummer sun, perhaps four hundred years after the Heel Stone and its partner.

Certainly, by 2000 B.C., several large unshaped sarcens had been positioned at the entrance to Stonehenge, as well as the four so-called station stones, set in the form of a rectangle just inside the circular bank. The long axis of the station stone rectangle runs perpendicular to the entrance axis, which faces the midsummer sunrise. Therefore, the short ends of the rectangle also mark the midsummer sunrise (as well as the midwinter sunset), while the long axis appears to mark the major extremes of the moonset. A sidebar in chapter 8 compares the solar and lunar cycles in detail; for now we only need to know that it takes the moon 18.61 years to go through a full cycle from extreme northerly moonrise back to extreme northerly moonrise. And how long would it have taken the ancients to discover and verify this 18.61-year lunar cycle? The archaeological evidence points to a mind-boggling tenacity! Post holes where the Stonehenge avenue crosses the circular ditch indicate where observation sticks were positioned for six complete lunar cycles, close to a hundred years of observations. Lifespans are known to have been much shorter in Britain four thousand years ago, and an adult observer would have been unlikely to see the same position in the cycle twice in a lifetime. Therefore, the task of recording observations to determine lunar alignments with a degree of certainty would have been passed down from generation to generation. This singleness of purpose typifies the approach taken by the ancients to so many aspects of the design and construction of Stonehenge and other great circles. These were monuments or temples for the ages and they were neither planned nor built over-

night. Learning of the posthole avenues greatly augmented the wonder and respect components of my love affair with stone circles—and, I suppose, with their mysterious creators.

It is not universally accepted that Stonehenge was a lunar observatory, although the respected stone circle authority Aubrey Burl, in his *A Guide to the Stone Circles of Britain, Ireland and Brittany,* argues strongly in favor of the idea. (The *Guide,* by the way, is indispensible for any serious tour of stone circles.) My own view is that evidence is convincing for both the solar and the lunar major observations. The reader who wishes to learn more about these alignments at Stonehenge and other stone circles will find useful references in the bibliography.

The Bluestones

After the earliest building period, but before the great sarcen circle and trilithons were erected, there was a flurry of stone-building activity that made use of the enigmatic *bluestones,* known to have originated in the Preseli mountains of Wales. During this phase some sort of monument was built with fifty or so of these stones. (Eighty-two stones would have been required to complete the pattern, but the archaeological evidence suggests that the work was never finished.) Most of the bluestones were around eight feet in total length and weighed between two and five tons. All of the stones are not the same, geologically; there are examples of spotted dolerite, rhyolite, and volcanic ash. Atkinson identified unspotted dolerite "bluestones" in the 1950s and outcroppings in the Preseli mountains to match.

While no one argues where the stones came from, the debate continues about whether they were transported from the outcropping in Wales, 135 miles from Stonehenge, by glacier or by the hand of man. Atkinson argues fervently for human transport, and even helped organize a well-publicized experiment in July 1954. One oft-reprinted photograph shows four school boys punting a concrete replica blue-stone weighing 1.75 tons in a composite boat made of two hulls tied together with wooden cross-members. Atkinson designed a simple sledge made of squared 6-inch timbers, about 9 feet long and 4 feet wide, and thirty-two senior schoolboys were able to haul the same concrete "stone" up a four-degree slope, "though it is doubtful whether they could have continued this effort for long." By using wooden rollers under the runners of the sledge, the schoolboy team was reduced to fourteen, plus an ancillary crew of ten to keep moving the rollers forward (Atkinson, *Stonehenge,* pp. 113–15). The so-called Altar Stone, at seven tons, is the heaviest of the stones that came from Wales. Based on his experiments, Atkinson calculated that 110 men would have been required to transport the Altar Stone to Stonehenge.

Present-day experiments, described in this book, indicate that far fewer people are required than Atkinson calculated, perhaps a third as many. However, many other experts believe that the ice sheets brought the Welsh stones to what is now Salisbury Plain, with no contribution of human effort. This debate about how the stones got to site could go on for a long time, and I must admit that I am swayed back and forth by the reasoned arguments from each side. A good recapitulation of the debate is recounted by Cynthia Page in the first chapter of *Secrets of Lost Empires* (Sterling, 1997).

Final Phases of Construction

The final era of construction at Stonehenge commenced with the building of the sarcen circle and the five great trilithons, already discussed. During the last stone-building phase, according to Atkinson, the bluestones were reused in two configurations: a circle of standing stones between the sarcen circle and the sarcen trilithons, and an inner horseshoe- or oval-shaped ring within the sarcen horseshoe setting of trilithons. Examination of the dressed bluestones of this innermost ring reveals evidence of mortises and tenons, suggesting that it originally

had incorporated two or three mini-trilithons. The last stones were probably set sometime between 1550 B.C. and 1100 B.C. Atkinson postulates a subsequent period when the Avenue was extended around to the right (south) and toward the River Avon. While the temple itself was not altered, this work shows that Stonehenge was still revered and probably in use as late as 1000 B.C., and maybe later. Indisputably, the site was in use, in its various manifestations, for two thousand years or more. Although some writers speculate that it was abandoned five hundred to one thousand years before the Celts and their druid priests arrived upon the scene, it is possible that the druids saw some activity at Stonehenge, and they might have been able to question the indigenous British people about the temple.

OTHER STONE CIRCLES

Two years after my first visit to Stonehenge, I moved to the Highlands of Scotland, where I resided for most of the next seven years. I had become truly hooked by Stonehenge, so during this time I visited many stone circles in Scotland and England. A schoolteacher rented a room from me, waiting for the time when he could get a teaching job back in his beloved Orkney Islands. Angus and I became friends and I would occasionally visit him in the islands during summer vacation. Orkney, with its wide open vistas, friendly and intelligent people, good pubs, and wonderful concentration of megalithic and other archaeological sites, became one of my favorite places to visit.

I met Jaki in September 1972, and we were married two months later. She has accompanied me on most of my megalithic journeys since that time.

Orkney

I've visited four relatively small areas with great concentrations of spectacular megalithic and other prehistoric sites: Cornwall, the Carnac region of

Brittany, Wiltshire, and Orkney, but nowhere are the really great sites packed together quite as tightly as on Orkney Mainland, the largest island. And nowhere are the sites better preserved, thanks to Orkney's island setting off the north tip of Scotland. I will mention only four of the sites by name, although there are dozens of others in close proximity. Orkney's history of spectacular building spans from neolithic times, through the broch builders of five hundred B.C. and the Viking period (lasting five hundred years), up to modern times; I refer to the massive Churchill Barriers built during the Second World War, a series of causeways made of thousands of 20-ton interlocking concrete pieces shaped like children's jacks. Connecting Mainland to Burray and South Ronaldsay, the barriers turned the inland sea of Scapa Flow into a sheltered bay, concealing the British fleet.

Stenness and Brodgar

Two spectacular henge-type circles exist within a mile of each other on Orkney Mainland, the Standing Stones of Stenness and the Ring of Brodgar. The rings are also among the oldest in the British Isles. The earliest carbon dates from the bottom of the ditch at Stenness are just over five thousand years old, a date that could be pushed five hundred years earlier by the carbon dating adjustments of 1997.

To me, Stenness and Brodgar are two of the most impressive ancient circles, partly because of their positions at each end of a causeway across Loch Harray, partly because of Orkney's huge sky and open vistas, and partly because of the wonderful natural geometrical shapes of the stones. Some writers have described them shaped like playing cards but there are trapezoids, rectangles, triangles, diamonds, and irregular polyhedrons aplenty.

The geology of Orkney Mainland is perfect for megalithic work. The overlying strata is a thick layer of red and brown laminated sandstone, with excellent bedding planes for easy stone splitting. Many of the old farmhouses and barns in the area are roofed with large slabs, some of them three feet wide by four feet long and over three inches thick: 500-pound shingles! Other slabs are used for stall dividers in barns, fence posts, and even long straight walls to enclose stock paddocks. The stone age, it could be fairly said, persisted well into the 19th century in Orkney.

Stenness, smaller and older than Brodgar, and called by some "The Temple of the Moon," originally consisted of twelve tall relatively thin stones, of which only four striking examples remain; two of those were re-erected in 1906. The two imposing southernmost stones have been standing as sentinals on the Orcadian landscape for over five thousand years.

In 1814, a now notorious tenant farmer felled one of the stones and broke the adjacent stone to pieces. The fallen stone was re-erected in 1906, and, during the work, the distinctively crooked northerly stone was discovered and also re-erected. The temple was also taken into public guardianship at the same time, to prevent further damage. The same farmer also destroyed the famous Stone of Odin, nearby to Stenness, an act considered by Orcadians to have been an even greater loss because of its long-standing—and still active—romantic associations. The Stone of Odin was about eight feet high and featured a hole five feet above ground. Before its destruction, it was customary to leave some offering at the stone, such as a little food, a rag, or even a small stone. And, for time out of memory, any agreement punctuated by clasping hands through the hole was considered sacred. Young couples would "plight their troth" in this way. The story may end up with a happy ending. In early 1999 Orcadian photographer Charles Tait told me that the stone's original socket has been found, and that drawings exist of the original stone. The resourceful Orcadians are planning to erect an accurate replacement stone at the original location.

Other holed stones throughout Britain served the same purpose, and the custom has persisted up

Aerial view of the Ring of Brodgar. (Charles Tait photo.)

to the present day, such as at the Men-an-Tol stone in Cornwall, described in chapter 5. On our megalithic journey in the fall of 1997, Jaki and I would consistently come upon offerings of flowers, cloth, and even poems at the centers of stone circles as well as at special stones or ancient megalithic chambers.

The completed Stenness circle would have been 104 feet in diameter, about a third of the diameter of its neighbor, Brodgar, a mile to the northwest. Brodgar's diameter is third largest in the British Isles, out of roughly a thousand circles. The circle encloses an area of 90,790 square feet, according to Aubrey Burl, and he surmises that 1,500 people could fit inside of it, even if half the space is given over to the ceremonies. (Burl, *Guide,* p. 145–46.)

Brodgar originally consisted of sixty stones, equally spaced every six degrees of arc around the perimeter. Estimates vary as to the number of survivors, from 27 all the way to 36. My own count in 1995 was around 30, but the difficulty of defining a surviving stone was impressed upon me. There are perfect full standing stones, stones broken in half, stumps, recumbent stones, and fragments. No wonder so many legends persist throughout Britain that it is impossible to count the stones at a particular site, although it surely must have been easier when the circle was freshly completed.

The diameter of the Brodgar stone circle is numerically interesting, almost exactly 125 of Alexander Thom's megalithic yards. Curiously, this is the same diameter as the two inner circles at the great temple of Avebury (discussed below), and the same diameter as the stone circle that surrounds Ireland's greatest megalithic monument, the passage grave at

The Standing Stones of Stenness. The author's son, Darin, lends a sense of scale to the tallest stone.

Newgrange, County Meath. It strikes me as more than coincidence that three of the greatest ancient stone circle sites all incorporate a dimension of 125 MY.

The stones average 7 feet high, according to the thorough Burl, "with soaring pillars at two cardinal points, a 12'6" tall stone at the south, an even greater and heavier at the west, fully 15'3" in height." Like the taller stones at Stenness, the Brodgar menhirs are narrow compared with those found in most other circles, so that a stone much lighter in weight will still yield great height and, from certain angles, impressive silhouettes against the sky. The tallest stone at Stenness, for example, is 18'9" tall, 4'7" wide and only 10 inches thick, a total weight approximating 7 tons.

Although the Orkney stone builders benefited from easily cleaved stones that were close at hand,

they were not afraid of work. The ditches of the henges at both Stenness and Brodgar were hewn out of solid sandstone, not earth. Brodgar's ditch, about 1,000 feet in circumference, is, in places, 30 feet wide and 10 feet deep. "Following the excavations of 1973," says Burl, "it was computed that 12,000 tons of rock had been quarried, pried, and hoisted from the ditch, 80,000 hours of labor." (Burl, *Guide,* p. 147.) This suggests that a man could quarry and remove a tonne (a long ton in this case, of 2,240 pounds) of sandstone in an average time of 6 hours and 40 minutes, a pretty tough pace to keep up, long ton or short.

Sunset at Maes Howe

Within a mile of the Stenness stones is Maes Howe, considered by many to be the finest chambered tomb in Western Europe. The 5,000-year-old stone slab tomb was the site of one of my most memorable megalithic experiences, so I just have to tell you about it.

Around 1970, just before Christmas, I had occasion to visit my friend Angus in Orkney. Late one afternoon, around five, we pulled into the Queen's Hotel on Kirkwall harbor, as was our custom, for the "tea-time session." Midway through our first pints, three boisterous lads, acquaintances of Angus, came rollicking into the pub, their demeanor in advance of the hour, as the bar had only just opened. The three had just returned from Maes Howe, where they had waited inside the central chamber to catch the last rays of the setting sun on this rare clear midwinter's day. At sunset, at the winter solstice, the setting sun passes through a long passageway, described below, and illuminates the central chamber. A quantity of McEwan's Export Ale, apparently, had helped to pass the waiting time. Their description of the sunlight breaking into the chamber and lighting it up was intoxicating, and I wished that I could have been with them, if only I'd known.

The next day, Angus and I set out in a driving rain in the direction of Maes Howe, nine miles due west of Kirkwall, but the weather was formidable and we drove back to the Queen's, disappointed, and drowned our sorrows at the tea-time session. I had to catch my flight back to the Scottish mainland the next day, so I missed any further chance of solstice observation. I resolved that, someday, I'd be back.

Several years later, Jaki and I were back in Scotland visiting her parents for the Christmas holidays. (In Roman times, the winter solstice occurred on December 25, and it is almost certain that Christians commandeered that ancient holiday as Christmas. Therefore, our holiday visit could be considered a continuation of a pattern dating back more than five thousand years!) Armed with Jaki's father's camera, I made my way to Orkney to visit Angus and to catch the setting sun at Maes Howe. I had a three-day window of opportunity before I had to rejoin my family for Christmas.

The first day was dry but overcast all the way to the edge of the extensive Orcadian horizon. I took the early afternoon bus from Kirkwall, and met Charles Woods, the caretaker for Maes Howe, who lived in a cottage just across the road from the tomb. Charles, a reddish-blonde, full-bearded, and amicable fellow of about my own age, had the key to the tomb and he gave me a special tour of the monument. But, alas, there was to be no sunset this day. Nor the next; I met Charles again, and hung around the old mill, now a tearoom and craft shop near his cottage. It rained on the second day and I took the bus back to the Queen's Hotel, damp as well as disappointed.

The third afternoon looked no better, and it didn't seem even worth the while to get on the bus and waste the whole afternoon on a fool's errand. But it was my last chance, maybe forever, and I traveled again to Maes Howe. The weather there was no better—heavy overcast—and Charles and I commisserated on our bad fortune. We said cheerio and I went over to the tearoom to wait for the return bus. Through a window, I noticed a narrow but bright band of sky along the southwest horizon. Orkney Mainland has a long and generally unobstructed view almost everywhere on the island, and I figured that the sun would have to pass through this bright band on the way to its bed. Grabbing only the camera, I ran across the road and made for the beginning of the path to the tomb's entrance.

Charles Woods had made the same observation from his cottage. Simultaneously, we met at the gate, said not a word as we fumbled our way through, and sprinted the final eighty yards to the locked tomb entrance, which Charles, bathed now in full sunshine, unlocked with swift dexterity. I readied the camera for the light conditions. As the iron gate was swung open, the sun began to perform its yearly magic on the stones.

The five thousand year-old original entrance to Maes Howe features a 32-foot-long passageway, averaging 4'3" high and 3'3" wide. Each time I enter the tomb, I think of one of the runic messages left in the stones by the Vikings: "Many a haughty woman has stooped to enter here." After a few feet, a little relief for the back is provided by a 4-inch step down. The second half of the entrance chamber displays one of the great curiosities of Maes Howe, a design feature unmatched, to my knowledge, anywhere else in Western Europe. Four giant slabs of red sandstone form the passage: one on the floor, one forming each side, and one as a ceiling. At least two of the slabs are now broken, but the west-side slab is completely intact and measures 18'3" long, 4'3" wide, and about 6" thick, measurements that yield a weight of almost three tons.

Because the slabs were installed deliberately for the winter sunset, their walls are very nearly parallel to the sun's final rays. But the sun does not set straight down, as it does near the equator. In extreme latitudes, it sets on a shallow angle. The effect is that the sunlight moves quickly along the west stone. You can watch its progress, which takes, as I recall, about a minute to move along the 18-foot wall. Then the sunlight's progress is slowed by an extending 4-inch doorjamb, probably used to seal the tomb, some say from the inside, during ceremonial observations. After perhaps another minute, the first gleam of direct sunlight makes its way past the doorjamb and strikes the back wall of the chamber. The effect is like turning on a light in a darkened room; our eyes, accustomed to low light, could immediately see the ancient room clearly, including the four standing stones that help support the corbelled arch roof and the entrances to three ri-

One of four large standing stones that help support the corbelled arch roof at Maes Howe, Orkney.

The Stone Age village of Skara Brae, Orkney Mainland.

parian burial chambers. It is hard to describe the emotion that comes from experiencing the dramatic results of such careful planning, actualized more than five thousand years ago by peoples unknown. Think of the word *grandfather,* with 250 "greats" preceeding it. Suffice it to say that—leaving family events aside—this was one of the greatest thrills of my life. (See photo in color section.)

Skara Brae

Just five miles west of the Ring of Brodgar is one of the most perfectly preserved villages of the neolithic (new stone) age. During the winter of 1850, a tremendous storm buffeted the Bay of Skaill on the west coast of Orkney Mainland. The grass was stripped from a high dune known as Skara Brae, revealing the ruins of ancient dwellings, and, im-

portantly, the midden (refuse) heaps between buildings. Professor Gordon Childe, who excavated from 1928 to 1930, thought that the village may have been inundated by a sandstorm and suddenly abandoned—some authorities disagree—which explains the amazing state of preservation, not only of the houses, but of their stone furniture as well: beds, dressers, hearths, water tanks, grinding stones, and other domestic artifacts. The midden heaps tell us a great deal about the diet of the people, the seafood and terrestrial animal life they ate, the grains they grew, even the trinkets they made and used, such as dice, pins, and jewelry. A full discussion of what we know—and don't know—about neolithic life is beyond the scope and intent of this book, but the important point is that a village contemporary with the construction of the nearby stone circle sites exists almost intact. No visit to Orkney is complete without seeing Skara Brae. Nor should the excellent Tankerness Museum in Kirkwall be missed; it gives a close look at neolithic life, and houses many of the most important artifacts found at the various archaeological sites of Orkney.

Callanish

While living in Scotland, and again in later visits, Jaki and I would occasionally make our way to the Isle of Lewis, largest of the Outer Hebrides, mainly to see the ancient stone circle at Callanish. Called by some the Stonehenge of the North, the main circle at Callanish (Callanish I) is as different from Stonehenge as different can be. The Stonehenge megaliths are carefully shaped, much greater in size, and often support huge lintels. The Callanish stones show no evidence that they were shaped, the circle is much smaller, and the stones, while tall, do not have nearly the girth of their cousins to the south. And yet, of the two, Callanish is somehow the more evocative, the more mysterious, the more compelling. I attribute the special atmosphere of the place to three circumstances: the dramatic positioning of the circle on a promontory overlooking Loch Roag, the changeable Lewis weather that can alter the character of the site in a matter of minutes, and the wonderful natural shape of the stones themselves, which, in certain conditions, look like ancient cowled figures walking across the hilly landscape. Visually, Callanish is my favorite circle, and when it came time to design the Earthwood Circle, Callanish provided the inspiration.

During a visit in 1985, I accidentally met Margaret Ponting, who had done so much study and work at Callanish with her first husband, Gerald Ponting. Twelve years later, Jaki, Darin, and I visited Margaret and Ron Curtis, her second husband, and the story of our fascinating three days with the couple is recounted in chapter 8. So further descriptions and discussion about Callanish—and Margaret and Ron's amazing work there—must wait.

Avebury

I have visited the Avebury rings in Wiltshire many times with my family during the past twenty years, and it remains one of our favorite stone circle sites. Avebury also played an important part in our meeting Ivan McBeth, the hero of chapter 6, and, thus, in the writing of this book.

The 17th-century antiquarian John Aubrey might have best described the scope of work at Avebury when he said that the temple "did as much excell Stonheng as a Cathedral does a Parish church." And yet many people are unfamiliar with the site, twenty-seven road miles almost due north of its more famous neighbor.

Built of the same sarcen stones as Stonehenge, but having the advantage of being much closer to their source, Avebury consists of a gigantic henge earthwork, about a mile in its outer circumference, enclosing the largest known stone circle. Within the henge and its accompanying ditch appear approximately a hundred unshaped stones, averaging ten to thirteen feet tall, originally forming an irregular circle of about 1,100 feet across or roughly 3,500 feet in perimeter. This great circle, in turn, enclosed two perfectly round inner circles, each 340 feet in diameter, equivalent to 125 of Alexander Thom's megalithic yards.

The outer circles and henge are so large that they enclose the central portion of Avebury village, complete with inn, post office, several shops, and a few private residences. Encroaching on the henge from the west side is the remainder of the village, including Saint James Church dating back to well before the Norman Conquest of 1066.

What the visitor sees today is a bare shadow of the original monument, which featured more than 600 large standing stones, including three miles of avenues to the south and west, each lined with monoliths. During horrendous periods of destruction in the 14th century and again from the 17th through the early 19th centuries, most of the stones

The Avebury stone circle is enclosed by a ditch-and-mound henge of almost a mile in circumference. The circle encompasses a substantial part of Avebury village.

were felled and/or destroyed. Luckily, the 14th-century stone executioners, driven by religious fervor more than greed, were not so skilled at dismembering the bodies as their later counterparts (who used Avebury as a quarry for building stone), and buried them instead. Around 1934, virtually the entire monument was bought by Alexander Keiller, heir to a Scottish marmalade fortune. Keiller became an expert archaeologist and discovered—and re-erected—the great stones that had rested under the turf since the early 1300s.

One interesting discovery may give a clue as to why the razing of the stones ceased for a few hundred years. Modern excavation revealed a man's skeleton, wedged between a nearly 15-ton stone and a hole that had been prepared to bury it in the early 14th century. In the definitive work on the great circle, Aubrey Burl describes the scene:

His rotted leather pouch held three silver coins, dated about 1320–25, confirming the time when the stones were being buried. His iron scissors and a lance or probe lying beside him suggest he was an itinerant barber-surgeon traveling from one market-fair to another with his knives and leeches, happening to be at Avebury when stone-felling was in progress. Joining in, he was crushed and his body was buried in the rubble thrown back over the stone. Nor was this the end of his misfortune. His skeleton was given to the College of Surgeons in London and was destroyed in an air raid during the Second World War. (Burl, *Prehistoric Avebury*, p. 39.)

Speculation is that the surgeon's death might have dissuaded the superstitious villagers from further destruction, at least until the late 1600s, when the

21

stones were seen as a convenient quarry. For a century and a half thereafter, massive destruction took place at the temple and the two great avenues that emanated from it.

Stone Giants

As this book explores the building of stone circles, some facts about the size of the Avebury stones must be included here. Pairs of gigantic sarcen megaliths originally stood at each of the four entrances to the great circle. Two survive at the south entrance, stones "each big enough to be house walls, metres-thick slabs that are amongst the grossest, most backbreaking megaliths ever heaved upright in prehistoric Britain. One facing the causeway, with a natural ledge like a seat, the background for a million family snapshots, is known as the Devil's Chair" (Burl, *Prehistoric Avebury,* p. 181).

At the north entrance, only one of the original gate stones remains. Called the Swindon Stone—it marks the road to the city of Swindon—the diamond-shaped stone is estimated at 67 tons. Burl, citing two different sources, also tells of a fallen stone, broken up for building material, which would have contained 1,295 cubic feet, just over 100 American tons. In 1694, another stone from the north ring yielded "twenty cartloads" of building stone for the construction of the Catherine Wheel Inn, itself now destroyed. Many of the remaining village buildings within the great circle are constructed, in part, of sarcens smashed to bits by "Stonekiller" Tom Robinson and his cronies around the turn of the 18th century.

I will not presume to tell the reader how to move or erect 100-ton standing stones, using methods ancient or modern. By any method, work of this

The Devil's Chair, Avebury.

Of Obelisks and Megaliths

Even greater stones were shaped, moved, and stood up in Egypt at approximately the same time as the construction at Avebury. Called *obelisks,* the monuments were commissioned by pharaohs as an exercise in self-aggrandizement. Tall and elegantly shaped—like the Washington Monument, which was patterned after them—truly gigantic obelisks were erected in the second and third millenniums B.C., and they stood as sentinels for thousands of years until Europeans, beginning with the Romans, began to ship them back home as souvenirs. The largest, the Lateran, hauled away by the Romans, stands in Piazza San Giovanni in Rome. It is 105 feet high and weighs 505 tons. The unfinished Obelisk at Aswan, which was abandoned in the quarry when a disastrous fault in the granite showed up after a year of shaping, would have weighed 1,300 tons at completion. It would have been the largest single piece of stone ever moved in the history of engineering, but it is quite possible that even the great engineers of Egypt had exceeded the limits with this monolith. It remains in the quarry today, after four thousand years, with the marks of the tools and the measuring lines still in place.

The megaliths at Baalbeck in the mountains of Lebanon feature what may be the largest cut stone blocks ever moved. Graham Hancock, writing in his well-researched and compelling best-seller *Fingerprints of the Gods* (Crown, 1995), writes: "Long predating Roman and Greek structures on the site, the three (blocks) that make up the so-called 'Trilithion' are as tall as five-story buildings and weigh over 600 tons each. A fourth megalith is almost 80 feet in length and weighs 1.100 tons. Amazingly these giant blocks were cut, perfectly shaped and somehow transported to Baalbeck from a quarry several miles away." *Eleven hundred tons!* I visited Baalbeck at the age of nineteen, just two months after first seeing Stonehenge, but I had not yet learned to appreciate such things. My complete diary entry for December 8, 1966, reads: "Picked up Iraq visa. Barry and I took a service taxi to Baalbeck. Went to movie. Stayed at American University (Beirut)."

The Great Pyramid at Giza, which I'd visited a few days earlier, is probably the greatest engineering feat in the history of humankind. Composed of 2.3 million blocks weighing an average of 2.5 tons (tens of thousands of them weigh 15 tons or more), the total project is mind-boggling, and makes Avebury and Stonehenge seem like child's play. The King's Chamber alone has walls built "of 100 separate blocks weighing around 70 tons each," according to Hancock, "and the ceiling is spanned by 9 further blocks each weighing 50 tons" (Hancock, *Fingerprints of the Gods,* plate 44 caption). Furthermore, these blocks are squarely shaped and fitted against each other with remarkable accuracy, problems that did not concern the Avebury builders.

The civilizations that produced the obelisks, the pyramids, and the magnificent temples at Baalbeck were more highly developed than the neolithic peoples of ancient Britain and France. And yet it remains an open question whether we could duplicate the finest megalith-moving feats of any of these civilizations today.

A final example are the ancient standing stones of Brittany, called *menhirs.* In 1998, Jaki, Darin and I visited the greatest of these, some in the 60- to 100-ton range still standing after five thousand years, and one, the Grand Menhir Brisé, reputed to have fallen around 1700, weighing about 350 tons.

The author, age nineteen, playing Samson at Baalbeck, Lebanon.

The gargantuan Grand Menhir Brisé at Locmariaquer, Brittany.

Kerloas, in Brittany, near Plouarzel. Aubrey Burl calls this the largest menhir in Western Europe. It is 31 feet tall with probably another 10 feet of the stone below grade, in which case the total weight is about 100 tons.

Standing stones are commonplace in Brittany. This one, almost certainly ancient, is in a front yard at Plouharnel, near Carnac.

kind must be considered prohibitively hazardous, and, therefore, foolhardy. Today, cranes capable of lifting even 50 tons are far and few between, and would be extremely expensive to hire. A certain medieval barber wandering through Avebury found out that even knocking megaliths down can cost one's life.

If the reader wishes to be further dumbfounded by ancient engineering feats, and enjoy some highly interesting and well-thought-out speculation about them, I highly recommend Graham Hancock's *Fingerprints of the Gods,* as well as his new book, beautifully photographed by his wife Santha Faiia, *Heaven's Mirror* (Crown, 1998).

Strange Doings at Castlerigg

Castlerigg is a special stone circle for me for three reasons: One is obvious to anyone who visits, one I learned from Burl's excellent guide, and the third is personal to our family.

First, Castlerigg is situated in the most perfect location imaginable for a stone circle. Surrounded by the Cumbrian Mountains in England's beautiful but touristy Lake District, the site commands a 360-degree view, perfect for solar alignments.

Second, the ring is old, very old. Aubrey Burl says that Castlerigg may be one of the oldest stones circles in Europe, suggesting a construction date of 3200 B.C. (Burl, *Guide,* p. 43).

Third . . . well, I have to tell you another tale from our megalithic journey.

John and Robina Jacobsen are old friends from our Scottish days. Jaki and Robina had nursed together at the local hospital, and John was an engineer for the county. Not long after we moved to New York, the Jacobsens settled in Bermuda for several years, and we stayed in touch and they visited us during the design stage of the Earthwood house. With his engineer's cap in place, John showed me the need for special buttresses within the curved walls of Earthwood, in order to resist the lateral load of five hundred tons of earth to be placed against the home's northern hemisphere. I only

mention this to establish the man for the careful and learned engineer that he is.

After the Bermuda stint, the Jacobsens decided to return to the Newcastle area, home of their youth, and settled into the village of Wylam, near Hadrian's Wall. We stopped to visit them on Saturday, October 7, 1995, on our way from Scotland to England's West Country. The weekend started in good form, with John and I sampling several different real ales at a local pub while Jaki and Robina caught up on old times. Robina prepared a fish casserole, probably the tastiest meal we had on our month-long visit to the U.K. Great camaraderie with old friends!

Sunday was a beautiful clear autumn day, but was clouded for Robina by news of the death of a favorite aunt. It was a very awkward situation for all of us, but John and Robina said that we ought to carry through with our plan to travel to Castlerigg, an hour's drive west of Wylam.

Getting there was half the fun. Jaki and Darin joined John in his metallic green open-topped Morgan sports car, while Robina and I followed along in our hired car. At Langley Castle, we switched about and I joined John for a hair-raising hairpin journey across the Yorkshire and Cumberland Fells, and into Penrith. After we'd come down off the moor, I mentioned to John—the man of science—that we'd recently learned from Ed Prynn (featured in chapter 5) to dowse for the energy given off by stone rings. "Well, then," said the engineer, "we'll need some equipment!" John pulled over at the next tree-lined lay-by, and we hopped out of the Morgan in search of forked hardwood sticks. We made three dowsing rods from dead branches lying on the ground, one only a few inches long.

It was my first time to Castlerigg, though John had been once before, many years earlier. We were dutifully impressed with its setting, a flat-topped knoll in a treeless pasture, surrounded by some of the highest mountains in the Cumbrian Range: Skiddaw (3,054 feet) four miles to the Northwest, and Helvellyn (3,116 feet) six miles to the Southeast.

Several other near–3,000-foot mountains were in view on this blue sky day, in every direction. Some of the stones seemed to have been deliberately shaped so that their tops superimpose with the mountains behind them.

The stones themselves, few of which exceed four and a half feet in height, are not as impressive as the natural site upon which they stand, or sit. Two of the bulkier stones, at 5'6" and 5'8" tall, respectively, mark the main entrance to the circle, which is oriented due north. The tallest stone, at 8'3", is actually not a part of the circle itself, but part of a mysterious trapezoidal stone configuration adjacent to the inner eastern part of the ring.

We came not to measure the stones, but to dowse them, a pastime that held great interest for Darin, then almost ten years old. He had developed quite

a skill at determining the edge of stone circles, blindfolded, with a divining rod. Again, at Castlerigg, he proved to be top witch in a contest. But other results are, perhaps, of even greater interest. John, who can fairly be described as an open-minded skeptic, never failed to find the edge of the circle with our homemade divining rods. He would approach the circle from without, eyes closed, holding the forked stick as we'd shown him. At the edge of the circle, the point of the stick plunged downward, and he would open his eyes, laughing to find it pointing to the exact perimeter. Jaki, at this time, had still not learned how to hold the stick properly—something she would learn at Ed Prynn's place five days later—so she had no success at all. But the most astounding reaction of all came from Robina, who was holding the smallest divining rod in her substantial

Engineer John Jacobsen finds the exact perimeter of the Castlerigg stone circle with dowsing "equipment" found in a roadside lay-by.

hands. Once, on her approach to the circle from the north entrance, and just at the perimeter of the circle, the little divining rod suddenly leapt from her hands, as if propelled by a small explosive charge. We laughed uproariously, Robina for the surprise, the rest of us at her reaction to the leaping stick.

Some of the other visitors may have thought our group had gone completely bonkers, but I did notice a couple of others engaged in similar study. We returned to our cars and motored south a few miles to the King's Head, a country pub with a sun-bathed beer garden. Robina had brought a wonderful ploughman's lunch basket of rolls, Stilton cheese, and apples, which washed down well with glasses of Jennings, the excellent local ale. The weather, the outing, the company, all had been perfect and our parting was teary-eyed. Robina told us that it had been a joyous day for her, just what she needed to properly celebrate her love and memories of her dear departed aunt. Castlerigg, I think, played its magical part.

Stanton Drew

A description of what we see at Stanton Drew, Somerset, does not hint at the wonder that we *don't* see. Visibly, the site consists of three stone circles. The central ring, largest in diameter at 372 feet, encloses 2½ acres, more area than any other circle in Britain with the exception of the outer ring at Avebury. The big stones are in the 97-foot-diameter northeast ring, some of them, by my estimate, weighing 20 tons or more. The badly ruined south-southwest circle is on private land and is 145 feet in diameter. On a perfect alignment with the northeast circle and the central circle (and almost a thousand feet farther on to the southwest) is a cove of three stones arranged like three sides of a shed, but unroofed. A similar configuration existed at the center of the the northern inner circle at Avebury, and elsewhere. This cove actually sits today in the beer garden of the local pub, called—what else?—The Druid's Arms. Unfortunately we were there at

5:30 P.M., but the pub did not open until 6:30, so I cannot offer a report from the interior.

We visited Stanton Drew on October 8, 1997. Except for us, the place was empty of sentient life save a single very friendly sheep, who was happy to be petted, much to Darin's delight. There was no sign of the important archaeological activity that, just a month earlier, led to the discovery of a great new wooden-posted temple at the site. Our local upstate New York newspaper a month later carried the story with the provocative headline, "Temple Dwarfs Stonehenge." More details appeared within ten days of the first announcement. The English Heritage discovery team used a highly sensitive magnetometer to find the existence of perhaps four hundred now-filled postholes, each about three feet deep and three feet in diameter, and spaced about three feet apart. Speculation is that the holes supported twenty-five to thirty foot-long oak pillars, nine concentric rings in all, with an overall diameter of 95 meters (about 312 feet).

Experts agree that the building was a temple, similar to seven other known wooden-posted henge monuments, including Woodhenge. The difference at Stanton Drew is the enormity of the structure, at least four times larger than any previously discovered timber-posted temple. Experts *disagree* about whether or not the structure was ever roofed. What seems certain is that the wooden temples preceded the stone temples, often on the same site, as at Stonehenge. The largest stone circle at Stanton Drew is built within the same henge and uses the same center as the wooden temple. The obvious conclusion may not always be the correct one, but it would seem that when the oak posts rotted at the ground, the local people thought to replace them with something more permanent: megaliths. The wooden monument at Stan-ton Drew is tentatively dated at around 3000 B.C., whereas the stone circles seem to have been built about 2700 B.C. Geoffrey Wainwright, chief archaeologist for English Heritage, considers the discovery to be among the top twenty archaeological finds this century.

In 1998, yet another closely related discovery literally revealed itself in the shifting sands of Holme next the Sea, a coastal village about twenty-five miles north of King's Lynn, Norfolk. An actual ring of fifty-four large oak tree trunks has been preserved for at least four thousand years because they have been buried for millennia under wet sand and brine. Archaeologists are in a race against time to learn all they can about this ancient wood henge before it disintegrates from exposure to air or gets covered up again for another four thousand years.

In 1999, an ancient "inverted stone circle" was discovered in Miami Beach, Florida, of all places, during excavation for a highrise condominium. The circle consists of holes carved into the underlying strata of limestone, the sockets taking on shapes of animals. Work was stopped, thankfully, while emergency excavations were made. The hope now is that the circle will be preserved, perhaps as a tourist attraction at the development. I am constantly amazed by the exciting new archaeological discoveries that continue to be made—even whole stone circles, as we will see in chapter 8. It certainly is an exciting time in the field.

Other Ancient Stone Circles

The ancient circles described above are among the "first-stringers," the all-stars of megalithic stone circle construction. There are hundreds of others in Great Britain and Ireland worth visiting, out of the thousand or more sites that can still be identified as stone circles. But the book in your hand was never intended to be a guide to ancient circles, although I have included (as chapter 15) the first comprehensive Guide to Modern Stone Circles. Use the "Guide Books" section of the bibliography to track down the ancient rings.

I have visited a couple of dozen ancient circles, and cannot remember being disappointed by any of them. The stones of each circle have their own distinct character, a function, no doubt, of the geological strata from which they were sprung. When I look at new stone circles I often find myself saying, "This stone reminds me of Avebury." Or Callanish, or Castlerigg, or wherever. One of the outliers at our Earthwood circle, where an observer stands to catch the summer sunrise, could definitely find a home at Stenness.

In addition to Orkney, already mentioned, one of the best places for a concentrated megalithic journey is the county of Cornwall, where some of my favorite stones reside, including the Merry Maidens; the Hurlers; Duloe Circle, made from huge white quartz mehhirs; and the friendly yet mysterious Bosca-wen-Un. Cornwall also offers spectacular dolmens, the unique *fogous* (underground chambers), the Carn Euny neolithic village, and Men-an-Tol, the famous holed stone, actually part of a ruined stone circle. Cooke's *Journey to the Stones* is a beautiful and thorough guide to Cornish megaliths, with many suggested walking tours. And don't forget to take in Ed Prynn's amazing Seven Sisters Temple, described in chapter 5.

Burl's *Guide* includes stone rings in Ireland and Brittany. His *Megalithic Brittany* adds wonderful menhirs, stone rows, and spectacular megalithic chambers. Balfour's *Megalithic Mysteries* includes stone rings and other megalithic monuments in Europe and Africa.

Egypt is best known for the pyramids and temples dating from its golden age at the time of the pharaohs from the third to the first millennium B.C., but at a much earlier time, around 5000 B.C., megalithic monuments were built that may be the precursers to stone circles. Working in the Sahara desert in southern Egypt, researchers over a period of several years in the 1990s have discovered a complex of astronomical alignments and other megaliths spread over an area of 1.8 by .75 miles. Announced in the April 2, 1998, issue of the journal *Nature,* J. McKim Malville of the University of Colorado at Boulder describes a "calendar circle" of stones, a twelve-foot-wide arrangement of slabs about eighteen inches long, most of them recum-

The unusual central stone at Boscawen-Un, Cornwall. Note offerings under the stone left by previous visitors. The author has not matured a great deal since visiting Baalbeck thirty years earlier. (Jaki Roy, photo.)

bent. Two pairs of uprights face each other across the circle and mark the sunrise on the longest day.

This Nabta Playa discovery will no doubt rekindle the "diffusionist" debate among researchers. Did the idea of astronomically aligned megaliths originate in the Middle East and spread through Europe? Or did the idea spring up independently around Europe and, indeed, the rest of the world? Other parts of the world have seen stone circles on their landscapes since ancient times: India, North Africa, Japan, Korea, Gambia, even North America (see chapter 16). Over the years, my own view has shifted from a solid belief in diffusion to accepting that widely separated peoples might well have come up with similar approaches to the astronomical and temporal questions they sought to answer; such approaches would have been logical and natural responses to the environment, their observations of it, and the available materials they had to execute their ideas. Ancient humankind had a brain capacity equivalent to our own, but little in the way of

specialized tools and materials: no iron, plastic, fuel-burning mechanisms, electronics, and so forth. One of the few ways people could exercise their intellect was in building megalithic structures, and they developed the skill so highly that we have difficulty duplicating their work today.

The attractions of stone circles, ancient or modern, differ from one to the next. Some, such as Castlerigg and Callanish, command spectacular sites. Others, such as Stonehenge and Avebury, instill a sense of awe and wonder. Some are friendly, such as Boscawen-Un; others spooky, such as the Rollright Stones, described in chapter 7. Some are simply beautiful: Callanish and Stenness.

In the ensuing chapters, you'll meet the present-day stone builders, and see their creations, starting in the very next chapter with our own Earthwood circle.

And when it comes time for you, dear reader, to build your own stone circle, large or small, rest assured that it too, will be one of a kind.

– CHAPTER 2 –

The Earthwood Stone Circle

*As life may have formed in the primordial soup, the Earthwood
Stone Circle began to form in the chowder of my mind.*

—Rob Roy

*W*hy build a stone circle? I am asked. The answer
varies from person to person, but the one
unifying thread that I have found in every case is
compulsion. The use or purpose of the circle is often
secondary to the oft-repeated refrain, "I was compelled to build it." My own case follows this pattern,
but the compulsion was not all-consuming; it simmered low on the back burner until circumstances
were right. And, as it simmered, the flavor improved.
A certain aesthetic as well as a sense of purpose entered the mix. It was as if I had to think of some
justification for the stone circle, that to yield to compulsion alone was some kind of weakness. I no longer
feel this way. In the past few years, I've observed that,
sometimes, the reason for the stone circle doesn't
become manifest until after it's built.

THE STONE CIRCLE AT LOG END

My first stone circle came as a response to a very
real and practical problem. In 1975, Jaki and I moved
from Scotland to West Chazy, New York. That same
year, we built Log End Cottage, a small home of

cordwood masonry, whereby short logs are layed transversely in the wall, much as a rank of firewood is
stacked. The experience led to my first book, about
cordwood masonry. Originally, we'd hoped to put
an earth roof on the cottage, but we decided that
the 45° roof pitch was too steep for earth.

In 1977 we built a second house at the homestead, an earth shelter called Log End Cave, with a
gentle two in twelve (2:12 or 10°) pitch. We planned
the site carefully, oriented the house with a south-
facing exposure to maximize the solar gain in winter, and, with a large front-end loader, proceeded to
dig the foundation.

The excavation yielded dozens of large boulders, some roundish, some having parallel faces.
While most of the excavated soil and earth would
be useful in creating the earth berming on the sides
of the Cave, the large boulders, some weighing a
ton, were a problem to get rid of. Many of them,
particularly the ones with parallel surfaces, were
useful in creating the retaining walls that framed
the south elevation of the home. Creating those
walls was my first serious experience building with

large stones. The strategy I used is worth recounting, as it has been successfully repeated on several megalithic projects with which I've been involved.

Catalog the Stones

The first important step on any building project, but particularly for megalithic construction, is to know the materials you have to work with. Intimately. With large stones, this requires a two-step process. First, place the stones in a convenient and easy-to-reach depot. Do not "pile them up"; rather, store them with a little space between each, so that you can easily measure all of the dimensions, and visually determine which are the attractive—or unattractive—sides. Second, catalog *all* of the stones, even those of questionable character. Number each one with chalk or crayon, and list all of the important characteristics of each stone: dimensions (height, length, and width); color; and special features, such as unusual shape or if it is a good building stone. The chart on page 38 is the one I used to catalog the stones of the Earthwood Circle.

At the Cave, the stones were scattered in two or three locations, and some were hard to see clearly because they were piled together. I've been more careful since then. Nonetheless, I was able to gain a close familiarity with the stones, and had a pool of perhaps forty to choose from. Heavy equipment hire is expensive, so time saved by knowing your stones translates into major money saved during construction. (My early projects all involved heavy equipment, usually backhoes, and these are the techniques described first. Later in my megalithic experience, I became interested in moving and erecting large stones without machinery, described later in this chapter, and in chapters 7, 10, and 14.)

Having worked three years for a stone mason in Scotland, I brought a somewhat experienced building eye to the retaining wall project. However, I hadn't a clue about how to physically move the heavy stones into place. Here the heavy equipment operator's long experience was invaluable. He would tell me to tie a chain around a stone, and I would stand there looking at the chain like it was an unknown artifact, never before seen by man. The operator shook his head, hopped off the machine, and showed me how to use the chain. He must have wondered what kind of nerd *(with his fancy clipboard)* he was going to be working with the rest of the day. (See the sidebar "Chaining Stones.")

The retaining walls were not quite as impressive as those at Maes Howe, but I was very pleased with them, and they have happily stood the test of more than twenty years. The operator was impressed by how well it went together, how I seemed to always know which stone would fit in the next position. The secret was on the clipboard. By cataloging the stones carefully, I knew each one on a first-name basis.

The Cave, incidentally, was a great success, and it precipitated another how-to-build-it book (*Underground Houses*, Sterling, 1979, revised and updated in 1994).

A Circle of Circumstances

After building the retaining walls in front of the Cave, a dozen large stones were left over. Most of these had rounded corners, or only one flat side; not much good for building a wall, but perfectly adequate to provide a reactionary load to the human posterior. I'd always fancied having my own stone circle, and the stones had to be put somewhere, so I decided on a little firepit surrounded by large stones; sitting stones, really, as opposed to the standing variety. At the same time, I was able to make use of a large rocky outcrop on the future lawn just to the east of the Cave. As it was impossible to remove and difficult to cover, I decided that the outcrop would make an excellent centerpiece to the stone circle, a place to build a fire. We even made a stone barbecue pit upon it to support a grill for cooking.

The circle consisted of eleven stones weighing from 350 to 1,000 pounds arranged in a ring around the bedrock outcrop. Two other pieces of the ledge were at finished grade level and we plunked a sitting stone on each. Stones were delivered easily to the site by the front-end loader part of the backhoe, and tipped into place. Final adjustment was made by gently pushing with the machine or levering by hand with an iron bar. All inner surfaces of the stones were kept six feet from the arbitrary center point on the outcrop. The two stones on bedrock were steadied by shimming with small flat

Chaining Stones

The primary concern working with stones is safety. Respect for suddenly shifting heavy stones is paramount. Working around heavy equipment must be considered hazardous duty. Wear a hardhat, heavy leather boots—steel toes are a good idea—and leather or heavy rubber masonry gloves. Work alone with the operator. It's all you can do to keep yourself out of harm's way, without having to worry about your spouse, weekend guest, kid, or dog.

Actually, fastening a chain around a stone is quite simple, providing that your chain has a hook or pinned fastener at one or both ends. One end could have an iron ring attached, as long as it is big enough for the other end (hook) to pass through. I have worked with Plain Chain (no jewelry at either end) and it is possible to fasten chain both to stones and to equipment by just tying it like rope. Tying chain is easy to learn and to do, and will be necessary sometimes even on chains with hooks, if slack needs to be taken up. Generally, the chain is fastened first to the stone, and then to the machine. Just pass the chain under the stone and fasten the hook to one of the links. If there is no space to pass the stone through, the operator may have to lift one end of the stone with the machine to assist you. Never put your hand or any other delicate thing under a large stone! Pass the chain through with a long stick. Next, turn the fastening point (the chain knot or the place where the hook is connected) so that the backhoe or loader will be lifting the stone from the top, as you envision its placement in the wall. Refasten the hook, if necessary, to snug up the chain loop against the stone. In most cases, you should try to reach for a link that will maximize chain tightness. A chain end with a ring instead of a hook is more difficult to snug up, but it will work. Be particularly wary of a shifting stone if the chain is not snug. Stand well clear when the stone is lifted, regardless of how well you've fastened the chain. Accidents happen.

Now fasten the other end of the chain to the machinery: bulldozer blade, front-end loader, backhoe bucket, whatever. Often, the equipment will have designated fastening mechanisms, such as a link hook or hole drilled in the plate steel of the bucket, designed to receive chain. When these are not available, the chain may have to be tied around the bucket, blade, or boom with a half hitch or square knot. Seek the operator's guidance. He or she will probably have to show you one or two techniques early in the game. They know their equipment and how to work with chain. Never tie chain around hydraulic equipment such as hoses or shiny metal piston tubes.

Tying chain in a knot worried me at first. *Won't it slip?* Actually, as tension is put on the chain, the knot automatically tightens. Later, releasing the chain is easy. When the chain is made slack, the knot is easy to untie. But, yes, it can slip. Always assume that it will.

When building a wall, a stone must be carried in the horizontal position in which it will be laid, so the chain will have to be fastened at or very near the balance point of the stone. See also "The Megalithic Hitch" in chapter 14. When the stone has been positioned satisfactorily in the wall, the chain must be removed. When tension is released, it is easy to unhook (or untie) the chain. The tricky part is removing the chain. Usually the machine can simply pull the chain out, but if the full weight of the stone is on it, such a pull might adversely move the stone. In this case, use a long metal bar or wooden post to lift one end of the stone so that the chain is easily removed.

The circle at Log End.

stones. The inner diameter of the circle was about twelve feet, small as stone circles go, about the same as the smaller circles in Gambia and Ireland. The circle comfortably sat seventeen people—if additional cushioning was provided—and enclosed about 120 square feet.

The circle saw many years of use as a meeting place for our hilltop community of homesteaders, a venue for summertime potlucks, and a site for the very first campfire socials at what was to become Earthwood Building School. One July night we were treated to an excellent display of the Northern Lights, which quite a few of our students had never before experienced. Years later, the new owners of Log End did some major landscaping and used some of the larger and better stones for parts of additional stone walls. But they had become accus-

tomed to the joys of a stone circle for summer evening outdoor fires, so they moved a few of the stones about forty feet and formed a new smaller circle, thus following a pattern that has been repeated many times at ancient stone circles, even Stonehenge.

THE EARTHWOOD CIRCLE

By 1980, writing and the building school had become my full-time occupations. And I had caught the building bug: hook, line, and sinker. Although we needed another home as much as our new son needed a louder voice, we began (in 1981) construction of the Earthwood house in an abandoned gravel pit. Earthwood is a round two-story earth-sheltered cordwood masonry house. It was, and is, an ongoing experiment in integrated design, not simply shelter. It accommodates food production and preservation, energy production and conservation, recreation, and home industry (including Earthwood Building School). Readers interested in these concepts are directed to my *Complete Book of Cordwood Masonry Housebuilding* (Sterling, 1992), which describes the construction in detail, or *Mortgage-Free!* (Chelsea Green, 1997), which elaborates on integrated design concepts.

Taking a leaf from the notebook of fellow underground author and compatriot Malcolm (Mac) Wells, we scoured the land for the worst possible piece of land, and found it just a half mile away, already scoured for us by years of gravel extraction. The gravel run was only about five feet deep, so when the quarriers reached the clay layer below, they dug laterally until two acres had been completely laid waste, looking very much like pictures sent back from Mars. Mac's basic concept goes something like this. Any house that we can design and build is an imposition on nature. We can't improve upon wild lands. However, the earth-sheltered or underground house has the potential for the *least* negative impact. The footprint of such a house can be returned as nearly as possible to its original pristine state. But people, particularly Americans I'm sorry to say, look for the best and most beautiful pieces of land upon which to build their homes, and they don't even have the grace to cover them back up with earth again when the structure is complete. This adds insult to injury. Mac's idea, instead, is to take land that has already been laid to waste by humankind and bring it back to living, green, oxygenating habitat once again, not only for people, but for all of the diverse creatures that are equally important to the ecosphere.

So we built Earthwood on the edge of a two-acre gravel pit. It's all Mac's fault, but we thank him for it, and so does the planet. Very little of the gravel pit remains, having regenerated itself, with some help from us, I might add modestly. The site plan called for a house and outbuildings on the arc of a great circle. A stone circle, it seemed, would complement the site plan, make use of "waste" materials indigenous to the site, and improve the landscape. Stones that were too large to haul away for road building were simply left behind at the gravel pit, many conveniently pushed to one side. My stone circle dream was coming closer to reality.

The Geology of Murtagh Hill

Two indigenous rock strata make up the geology of Murtagh Hill on the edge of the Adirondack Mountains in northern New York. The underlying strata is called anorthosite. Composed largely of soda-lime feldspar, it can be thought of as a quartzless granite. Anorthosite is found in only three known locations: here in the Adirondacks, Glen Torridon in Scotland, and on the moon. It is said to be among the hardest rocks on the planet, but, in fairness, lots of other rocks make the same claim, including the sarcen sandstone of Marlborough Down. The anorthosite is about two billion years old, making it one of the oldest rock strata as well.

The overlying strata is equally interesting. Four hundred million years ago, a very large inland sea covered the area where we now live. Early fossils such as trilobites and nautiloids are found in this layer. We have some pieces from the top layers of sandstone that are the fossil remains of the rippled beaches along the shallow bottom. The resulting stone layer is known as Potsdam Sandstone, and it is generally red, although yellow, purple, and other colors can be found. Unlike the anorthosite, an igneous rock, the Postdam Sandstone cleaves very well along its sedimentary bedding plane. It is almost identical to the Orkney red sandstone used at Brodgar, Stenness, and Maes Howe.

Murtagh Hill borders a fault line and has been thrust upward about 1,500 feet over millions of years, exposing the anorthosite bedrock. The sandstone is also exposed here and there on the top of the Hill, including on our property. The more recent Champlain Sea, which dates to glacial times, was probably responsible for the various sand, gravel, and clay layers encountered at the Earthwood site.

As life may have formed in the primordial soup, the Earthwood Stone Circle began to form in the chowder of my mind. The problem was that there didn't seem to be enough good standing stones on site. Today, friends sometimes accuse me of being "a few stones short of a circle," or something like that, but, in 1987, that was exactly the case. Here, serendipity played a big part to push the fledgling project forward. Enter George Barber, a close friend and land surveyor. At the request of the several families who came together in 1975 to form the Murtagh Hill community, George had researched properties on the Hill, taking the chain of title right back to the first Murtaghs from Ireland who settled in the 1800s. George became a fellow traveler and he eventually filed a deed in his daughter's name for a property he discovered about a mile and a half from Earthwood, which encompassed a small abandoned anorthosite quarry. Curious, I went to visit the old quarry, which, although just fifty feet from the edge of the road, was hard to see for the thick growth of trees. There, lying in their original quarried positions, were several of the most perfect potential standing stones imaginable. (They were, of course, recumbent stones when I saw them, and had been sleeping there for sixty to seventy years.) I had the kind of rush that I imagine Thor Heyerdahl must have felt at Easter Island when he first saw the huge incomplete stone heads still in their quarry. I raced around the quarry and found one wonderful stone after another, and realized that I would need to catalog the treasure to see if there was a stone circle here.

I rushed home for a pad, pencil, and tape measure, returned, and took stock of the quarry in earnest. I found at least seven potential standing stones with lengths of up to seven feet. Within days, I approached George's daughter and asked her if she would be willing to part with the stones. She told me that I could have whatever I could carry away!

Designing the Earthwood Circle

Now, with a pool of good stones to form the nucleus, I could seriously begin to design a circle. Ideally, the circle would have twelve standing stones. I chose the number for two reasons: First, because it would be easy to lay out—there would be "corner" stones at the four cardinal compass points—and second, because the circle was being built in North America, and I had heard that the Native American medicine wheels often had twelve radial spokes in their geometry. This theory turned out to be wrong. However, Aubrey Burl discovered something of equal interest and import, which shows that sometimes it is possible to be on the right track for all the wrong reasons. In *The Stone Circles of the British Isles,* Burl provides a statistical abstract of the number of stones in 316 stone circles. Of all the possibilities, the number twelve is the clear winner with thirty-two examples. The next most popular number, four (with twenty-nine examples), hardly counts, as these "circles" are actually a special class of monument called *four-posters,* almost all of which are found in northeast Scotland.

And why was twelve the most popular stone circle number? No one has the foggiest idea. But if I were to venture a guess, it would be that twelve was chosen for much the same reasons that I chose it: The number is easy to work with (having several factors), has elegant geometrical symmetries when placed in a circle (see Ivan McBeth's herb garden circle in chapter 6), and has the obvious potential for true north-south and east-west alignments.

By this time, I had at least heard of Alexander Thom's compelling theory, based on years of meticulous surveying of megalithic monuments, that the ancient stone ring builders all used a common unit of measure. Thom calls this unit the *megalithic yard* (MY), equivalent to 2.72 feet (or 0.829 meters, for those of you caught up in the unfortunate metric system). I decided to incorporate the unit at the Earthwood Circle, just on the off-chance that it might be right. And, who knows, it's use here might provide good fun and debate among stone sleuths five hundred years from now. (Aubrey Burl tells me that he has no objection to modern-day stone circles—in fact, he finds them "quite delightful"—*as long as no attempt is made to pass them off as genuine.* So don't worry, Dr. Burl, I was just kidding about throwing a monkeywrench at future archaeologists.) Besides, when I build round buildings—and I've built a lot of them—I always use the measuring unit of currency, to wit: feet and inches in New York. If I were to build a round building in France, I would use the meter, despite my distaste for the unit (and I'm of French descent on both sides of the family). If I'm building a stone circle, why *not* use the megalithic yard? (Chapter 11, "Designing a Stone Circle" includes an expanded discussion of the megalithic yard.)

A circle has just one unique point, its center. Early in the design stage, it must be decided what, if anything, is going to occupy the center point. At Earthwood, the decision was not difficult. Jaki and I liked being able to sit around a firepit in the evenings, and found that the circle of stones made for good camaraderie. A central standing stone, then, as at Boscawen-Un, was out of the question. And a firepit, unlike a stone, would not obscure alignments taken through the center of the circle. (If the designer has two special features in mind for the central part of the ring, he or she should consider an ellipse instead of a circle; it has two focal points.)

We also liked the sitting stone feature that we had enjoyed at the Log End Circle. It seemed sensible to have twelve of these as well, so that there would be a standing stone behind each seat, giving a sense of enclosure within a natural outdoor space. The sky would form the ceiling, and the twelve gaps between standing stones would be portals to enter and exit the circle. Logically, the firepit could have twelve small stones as well, with spaces between for easy flow of combustion air.

Choosing a unit of measure is one thing, but choosing the size of the circle is an altogether dif-

ferent problem. The Earthwood Circle seemed to have a mind of its own, and practically designed itself; certain mathematical and spatial relationships evolved almost automatically.

Given that we wanted to incorporate the megalithic yard, and given that we wanted a sitting circle and a firepit, what would be practical dimensions for the sitting area? The Log End Circle had a 12-foot inside diameter, but the firepit was just 3 feet across. Small fires were dictated by the size of the fire enclosure, and sitting about six feet from the center seemed to be about right for the comfort and enjoyment of the parishioners. At Earthwood, we wanted to be able to build much larger fires, particularly for our annual Guy Fawkes Night bonfire in November.

Many a design problem is answered by asking the simple little question, "What if . . . ?" I learned this trick from writer John Wyndham, who liked to say that the ideas for his successful science fiction stories usually started from just such a question. *What if* the firepit was 2 MY across? (That's 5'5¼" for those not yet thinking in terms of megalithic yards.) This seemed about right for the larger fires we wanted. The next question was one of practicality: What distance should the sitting stones be from the fire to promote comfort? The stone circle at Log End proved a valuable model. A 6-foot radius was the right distance from a small fire. *What if* we go with a radius of three megalithic yards (8'2") for a bigger fire? Again, the proportions seemed about right.

A pattern was developing. One MY to the inner surface of the twelve little firepit stones, 3 MY to the inner surface of the sitting stones. *What if* we go with a 5-MY radius to the standing stones? (When building stone circles, it is most convenient to measure from the center point to the *inner surface* of the stones. Measurements are easily lost during construction if an attempt is made to use the *center* of the stones.) By this final "what if," we create the pleasing geometrical progression of one-three-five with regard to radius, or two-six-ten on diameter. I placed football-sized stones at roughly the right places on the flat gravel pit surface to see what the space felt like. It seemed to be a good size for a gathering of from twenty to sixty people, our likely range of guests. In feet, the inner diameter of the standing stones would be 27.2. (The average diameter for all stone circles in the British Isles, according to Burl, is just under 60 feet.) Using πr^2 for the area, we'd enclose 581 square feet, a fair-sized area, and comparable to some of the smaller ancient circles. Our chosen diameter, therefore, felt comfortable in and of itself, and was in keeping with the other round buildings in the architectural scheme at Earthwood. Finally, the size and number of stones available seemed to be in pleasing aesthetic proportion to each other, although this did not become clear until I drew an accurate scale plan.

The location of the circle was almost self-determining. It couldn't go in front of the house, because the tall two-story home would spoil the true north alignment. Also, I wanted to minimize the visual impact of a lot of buildings, even round buildings, close to the stone circle. The best way to isolate the circle, then, was to continue with the eastern arc of buildings, on beyond the little round sauna, already built. We actually didn't have a lot of space to work with, as one large flat area was being reserved for a future tennis court. A plan of how the stone circle relates to the other buildings and the clearing is shown below. The centers of all five buildings fall on the arc of a great circle. The closest standing stone is still over twenty feet from the edge of the sauna, and most of the circle has a natural green forest background.

With the basic plan of concentric circles in mind, details had to be worked out, and the only way to

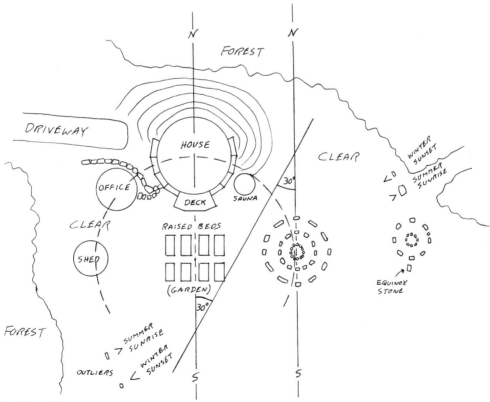

Site plan at Earthwood.

do this was to make an accurate inventory of all available stones and work out how to make best use of them on a plan. The following chart is the actual catalog of standing stones that I used to design the Earthwood Stone Circle. I made a similar chart for the sitting stones, not shown here.

The first seven stones listed had been found and measured at the old quarry. All were first quality and eventually installed in the plan. The other eleven stones were found at the gravel pit where we were building and at other nearby properties belonging to neighbors. I spent quite a long time over a period of weeks to find, measure, and catalog all the available stones on Murtagh Hill. The only anorthosite was from the quarry. Most of the other stones were Potsdam Sandstone. Two of the stones,

CATALOG OF POTENTIAL STANDING STONES— MURTAGH HILL

Note #	Estim. Total Length	Estim. Final Height	Color	Location	Notes	Rating	Rank	Clock Location in Circle
1	85"	70"	Gray/G	Quarry	Good base	100	1=	12 (N)
2	79½"	64"	"	"	Top-heavy, great point, bad base	100	1=	9 (W)
3	78"	60"	"	"	Big heavy triangle, fair base, must have socket	100	1=	3 (E)
4	77"	75"	"	"	Tallest final height, great flat base	100	1=	6 (S)
5	52"	50"	"	"	Bunny Stone, ears align with summer sunrise	100	1=	2
6	69"	60"	"	"	Good base, good shape, *difficult access*	95	6=	8
7	59"	?	"	"	Can't see base *in situ*	90	8=	4
8	61"	45"	Gray	Near GS's driveway	*Poor base,* heelstone.	78	17	Outlier
9	63"	51"	Red	Near wind-plant	"Big Red," good stone, poor base. Jam it in ground	95	6=	1
10	62"	49"	Red	Near Office	Beautiful red, flat and thin; too bad to move it	88	13	11
11	54"	42"	Red	In wall	Nice shape, short. Must remove from wall	87	14=	alt. 11
12	43"	35"	Purple	Stone pile	Short but very pretty, may be tough to stand up	85	16	?
13	57½"	45"	Red	Susan's	Nice shape	89	11	5
14	51"	41"	Red	At dome	Wide, accessible	87	14=	alt. 11
15	50"	38"	Gray	R's drive	Fat gray menhir, possible alternate heelstone	73	18	Outlier?
16	72"	57"	"Tan"	Allen's	Pleasing shape	90	10	10
17	73"	58"	Red	Don's	Intriguing shape	92	8=	7
18	59"	44"	Red	"	Thin	88.5	12	?

numbers 8 and 15, might have been glacial erratics, "foreign" stones, as it were. I would choose the twelve best for the standing stones and the other six would provide a pool from which to select needed outliers later on, foresight and backsight stones, for example, for solstice sunrises and sunsets.

Next, I sought and received permission from my neighbors (four different ones) to take stones from their properties. Best to stay out of trouble as well as to avoid crippling disappointment at crunch time. Assume nothing. If you don't know the people very well, it's a good idea to get a written release for the stone. This precaution should also be taken if the property is in danger of being sold, or if there are any other unstable circumstances connected with the owners. Be sure to let him/her/them know that you will clean up any mess associated with the removal of the stone. Offer to put *that* in writing.

Now, with the stones secure if not actually on site, the fun part commences: drawing the plan. Although it's fun, it's important. I tell students at our building school that the most valuable tool they own is the eraser on their pencil. It's a lot easier to change numbers on a plan than it is to change your mind and move two-ton stones around during construction.

Note that the catalog chart shows several characteristics of each stone, some objective, some subjective. Obviously, we are concerned about the height (length) of the stones; we must plan in three dimensions, not two. But the important column for determining final stone height is not the first, "Total Length" (a measured figure), but the second, "Estimated Final Height." The figures in this column are "guestimates," but I have found that such guestimates are well worth making. I've stood up a lot of stones, and the only ones where my EFH has varied much from my guess is when the socket hole was inadvertently dug too deep. This happened on one of the stones at Earthwood, the one called Susan's Stone, which could easily be six inches higher than it actually is. How do I come up with these esoteric EFH numbers? First of all, I visually examine the shape of the stone, and try to get a sense of how its weight changes along its vertical axis. Usually, we want to maximize the height, human nature as it is. This means determining *which end is up*. Generally, the heavy end gets set in the ground and the slender pointy bit goes up. This maximizes height because there is a more massive base in the ground to stabilize the stone.

Sometimes, exceptions must be made. Several of the standing stones at Stonehenge have (or had) their small ends in the ground, because the builders needed full-width tops to receive the lintels. This caused some of the stones to tumble down prematurely; they might have lasted only one or two millennia, instead of three or four. At Stone Mountain (next chapter), we deliberately put the East Stone in "upside down" because of the wonderful cowled figure that appeared at the larger end. You can break rules as long as you do something to make up any structural deficiencies that might arise as a result. Extra care might need to be taken with the packing stones in the socket (foundation hole), for example, or a little discreet concrete might need to be placed in the socket. But I am getting ahead of the story.

I like to have some white chalk in hand as I inspect a potential monolith. I look the stone over in its recumbent position—I've never seen one just standing around in nature—and try to visualize it upright, as well as to judge where ground level will be, based upon the shape of the stone. Shape determines center of gravity. A large heavy end pulls the center of gravity in that direction. The Rainbow Stone outlier at Earthwood stands over seven feet above ground, although only about fifteen inches of below-grade base anchors it in its socket hole. Why so little underground? Well, the base of the Rainbow Stone suddenly splays out from a thickness of about five inches to a comparatively hefty stump of twelve inches thick. So, although only about 15 percent of the overall length is below grade, probably 30 percent of the stone's mass is jammed in the socket, which is plenty. See also the section called "Sockets" in chapter 13. After making these judgments as best I can—call it seat-of-the-pants reckoning—I mark what I think will be ground level on the stone with the white chalk. I measure the part to go in the ground, subtract it from the overall length of the stone, and record the difference in the Estimated Final Height column.

The next column, "Color," is obvious. Color was important at Earthwood, because there were two radically different colors of stone available. This is unusual. Most ancient and modern circles I have seen are constructed of the same kind of stone, and are, therefore, the same color.

"Location" is really important. You don't want to spend any extra time on moving day trying to locate a stone. Also, any access difficulties must be worked out ahead of time. Note any such problems in this column.

"Notes." This is my catch-all column, where I include special features of shape or any other interesting considerations or observations.

The "Rating" column reflects my analytical approach to structural problems; indeed, all sorts of problems. My own habit is to use a 0 to 100 scale, although, in the final analysis, I usually end up using just the 70 and up portion of the scale. My thinking runs like this. Any stone rated over 90 is excellent, a score in the 80s is good, and a 70s score is only fair. Anything below 70 is poor and doesn't usually make it onto the chart unless I'm desperate. Below 60 is a failing grade. This scale works for me, but you may decide on a one to ten scale. If you watch a lot of figure skating or gymnastics, you might go from 1.0 to a perfect mark of 6.0. The scale doesn't matter as long as it makes sense to you and there is room enough for fine tuning and tie-breaking. And what judgments do I use to make these arbitrary determinations? Well, that's where the fun comes in. What man or woman fails to make value judgments of the opposite sex

based on little more than a two-dimensional glossy photo? Judging stones is even more fun; you can touch them and look at them from different angles without fear of getting involved in a lawsuit, although some people might look at you a little strangely. If you've decided to build a megalithic stone circle in your backyard, you're probably not the kind to get too worried about a few strange looks now and again.

Said without apology, size *is* important. But shape, color, appearance, and personality are equally important. (If we judged people this way, or if other people knew we judged people this way, we'd be in trouble. With stones, there's no problem. We can fantasize, *and* actualize.)

All of these factors roll around in the judgment center of my brain, and out pops a number. Five of the quarry stones were shoo-ins, quintessential standing stones. Although their qualities varied in character, none could be left out of the outer ring. I gave them nominal 100s, and established one end of the scale. The purpose of this exercise, after all, is stone selection, something like football team try-outs. Another important benefit that came from this extensive cataloging, ogling, and judging is that I became very well acquainted with each and every stone. In fact, one could say that I knew them, literally, on a first-name basis: Big Red, the Bunny Stone, Susan's Stone, the Cat's Back, et cetera. All of this helps in design, and, ultimately, in placement, because special problems or considerations can be identified and charted.

I believe that the research, value judging, and systematic charting described above greatly contributes to the success of the project, in terms of architectural quality as well as in ease and speed of construction (which can translate into economy of construction when the hourly cost of heavy equipment is considered). Many other stone ring builders have taken a similar approach. My father used to say: "A smart man learns from his mistakes. A wise man learns from the mistakes of others." I would add that you can learn from the successes of others as well.

Drawing the plan is like filling in a crossword puzzle. You start with the easiest or obvious answers. The others become easier as the pool of good stones dwindles and the circle starts taking shape.

Circles are easy to draw. Yes, you can use plates and other kitchenware, but it's hard to relate such equipment to the center, and it's never in the right scale. Buy an inexpensive drawing compass, adjustable to any scale. Using the megalithic yard does not complicate the tasks of designing and constructing very much. Just use the conversion table below. I've even included meters, for those of you hung up on that artificial, antiseptic, totally ahuman unit of measure. You see, I try to keep an open mind on these things.

MEGALITHIC YARDS (MY) TO FEET AND METERS

MY	Feet	Feet & Inches	Meters
1	2.72	2'8⅝"	.829
2	5.44	5'5¼"	1.658
2.5 (MR)★	6.80	6'9⅝"	2.073
3	8.16	8'2"	2.487
4	10.88	10'10½"	3.316
5	13.60	13'7¼"	4.145
6	16.32	16'3⅞"	4.974
7	19.04	19'0½"	5.803
8	21.76	21'9⅛"	6.632
9	24.48	24'5¾"	7.461
10	27.20	27'2⅝"	8.290
20	54.40	54'4¾"	16.580
30	81.60	81'7¼"	24.870
40	108.80	108'9⅝"	33.160
50	136.00	136'0"	41.450

★ MR = Alexander Thom's "megalithic rod" or 2.5 MY, often found at stone circles.
Note: Feet and inches column based on decimal conversion, rounded to nearest ⅛".

Using the table to convert the desired megalithic units to feet, I then drew my rough plan on a piece of plain white letter-sized paper, using a scale of ¼" = one foot, and keeping true north at the top of the page. Intersecting the diagonals of the page gave me its center point, which was important because the planned circle of roughly 30 feet (7½" in scale) barely fits on an 8½"-wide sheet. Lightly, in pencil, I drew three concentric circles with radiuses of 1, 3, and 5 MY. Next, I drew a north-south line through the center, and then an east-west line. The twelve points where these lines intersected the circles marked the center points of the inner faces of twelve of the stones: the four station or directional stones (N, E, S, and W) in the standing circle, the four primary sitting stones, and four of the little firepit stones (roughly the size of footballs). Next, with my drawing compass, I trisected each of the quadrants described by the intersecting lines and the outer circle. This takes just a little bit of trial and error to get it right, or you can use a protractor and mark every 30° around the circle. (12 x 30° = 360°, the full circle.) The compass method is easier, but it's good to know your way around the circle in degrees. (The Ring of Brodgar, originally, had 60 standing stones, spaced every 6° around the circle.) I completed my basic location plan by lightly drawing straight lines from each edge point to the center. Now there were thirty-six intersections marking the center of the inner faces of the thirty-six stones. I darkened the intersections with the point of a pen.

On the plan, I first drew in the four station stones, which happened to be the the ones cataloged as numbers 1, 2, 3, and 4, all tall "anorthostats." The idea was that they would be the tallest stones, and have a unifying color and texture. That left me three more quarry stones; the remaining five would have to come from farther down my list. There were four excellent red Potsdam Sandstone slabs, all within a quarter mile of the site, so I decided to place them symmetrically on either side of the north-south axis, contributing a balance of color and shape to the

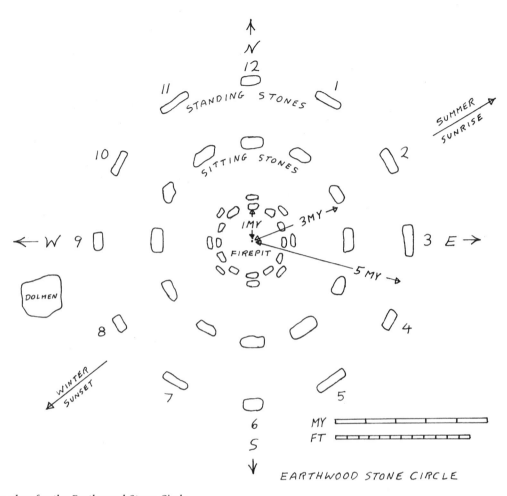

Working plan for the Earthwood Stone Circle.

circle. The last four standing stones, one more for each of the four quadrants, would come from the three remaining quarry stones and Bob Allen's stone, a tall well-shaped foreign stone, possibly a glacial erratic, possibly a Potsdam stone from a little distance away.

Bob Allen lived about a quarter mile farther on up the Hill. I had noticed the stone next to his driveway one day while I was sniffing out local menhirs. I told Bob what I was doing and asked him if he'd be willing to part with the stone. "I don't know," he said. "That stone saved my life." "How so?" I asked. "Well, I was shoveling the driveway last winter, been working about an hour, ex-

hausted, thought I was gonna die. Made it to that stone and sat down. It saved my life." I told Bob he could come and visit with it anytime, he agreed to donate it to the circle, and it has been Bob's Stone ever since. I'll tell you about some of the other special stones as we come to them during construction.

Once a few concepts were worked out, the standing stones really found their own places in the circle. The catalog was the key.

I did a similar catalog for the sitting stones, quarter- to half-ton boulders that were mostly just scattered around the gravel pit. One fine flat two-passenger sitting stone came from the old quarry. There were actually twenty or so to choose from, and the leftovers were used for another stone circle, the Children's Circle, fifty feet farther to the east.

The twelve footballs for the fire circle were selected from a pool of twice that number. Again, the remainder would define the firepit at the Children's Circle.

Astronomical Alignments at Earthwood

Accurate astronomical alignments require a long sight line. Two closely spaced lumpy stones do not lend themselves to the construction of a precision instrument. The Heel Stone and its now missing partner were about 256 feet from the center of Stonehenge. Alexander Thom reckoned that the Grand Menhir Brisé (Great Broken Standing Stone) at Locmariaquer in Britanny served as the foresight for lunar and solar alignments, with the backsights several miles distant. The Curtises showed me alignments at Callanish where notches or bumps on the hillsides miles away were used as foresights. Now *that's* accuracy!

Despite popular association of stone circles with astronomical alignments, Aubrey Burl says, "Had prehistoric man's prime intention been to design an astronomical monument it is unlikely that he would have constructed a circle" (Burl, *Stone Circles of the British Isles,* p. 53). It's true. Lots of better configurations can be imagined: stone rows, for example, or even simple pairs of stones. Long passage graves work well and are particularly dramatic during the actual event. But the ancients knew all this as well as we do, and this may be why they sometimes incorporated *outliers* as satellites to their stone circles. An outlier is simply a remote stone, not part of the circle. Although it is true that the purpose of some outliers near ancient rings remains a mystery, many are known to have astronomical significance, the Heel Stone at Stonehenge being the most obvious and best-known example. Not only do outliers greatly improve accuracy of alignments, but they can be added to the site after the circle is built, a compelling design advantage for someone like me who had not yet performed the necessary observa-

tions to determine sunrise and sunset alignments. I decided that the unused potential standing stones from my list would be brought to the site anyway, and stockpiled for later use as outliers, perhaps several years down the road.

One alignment, however, could be determined with accuracy any clear night of the year—or any day with the right equipment—and that alignment is the true north-south axis. While lumpy stones close together make poor sightlines, Polaris, the North Star, is a very accurate point in the sky, and it was possible to position it at the exact center of the North Stone, as seen from the exact center of the South Stone. I went to my surveyor friend George Barber for help in aligning the circle to true north. There followed a long and enjoyable discussion, punctuated by more than one glass of beer, during which he suggested several ways of finding north, some of which are enumerated in the sidebar toward the end of chapter 11.

Incidentally, there is another good reason to have four station stones at the four accurate compass points. The east-west axis automatically aligns with the rising and setting suns on the two equinox days: about March 21 and September 23, and this configuration had some importance to the ancients. These are the two days in the year when the sun spends twelve hours above the horizon and twelve hours below it. *Equinox,* literally, means "equal night." Of course, this depends on the site being blessed with a true horizon. Building in a valley, or in a clearing surrounded by forest—the case at Earthwood—spoils this easy alignment. The ideal site would be a low-lying sandy island in midocean, although importing stones could be troublesome. Salisbury Plain is a pretty darned good site, and bringing stones in was *certainly* troublesome.

I take some pleasure in my visitors' amazement when, on the shortest day, the sun sets brilliantly right over the point of the Sunset Stone. "How in the world did you figure that out?" they ask, incredulously. I bask for a few moments in the glory, but then have to admit that any two ten-year-olds could accomplish the same thing with a pair of wooden poles. This expostulation transforms me from genius to modern-day dunce, as the sharpest of the visitors says, "Aha, but the ancients didn't have a calendar to tell them the shortest day. They had to figure it out!" This is true. But, even so, the project becomes just a little more complex. Observations are taken for, perhaps, a month at the obvious dark time of the year. The two sticks are adjusted each day to mark the sunsets, and it is observed that the alignment continues to swing to the south until one day it stops—called the *standstill*—and stays in the same position for a few days. (Literally, *solstice* means "sun stand still.") Then the sunsets start occurring a little bit farther north each day as the days get longer. The standstill point is a pretty good indication of the shortest day. Truth be told, the alignment works with no apparent loss of accuracy for two or three days either side of the true solstice. If an exact day was required for ceremonial purposes, this is most easily determined by counting the days (sticks) needed for the sun to return to a known point, say, five degrees or so north of the solstice event. In future, the ceremony can be held that number of days after the sun rises over the indicator stick. This degree of accuracy might take a year or two longer to determine, but it's still simple stuff and requires no mathematics.

Determining the moon's extreme rising and setting positions, however, is another magnitude of difficulty altogether, as it takes 18.61 years for the moon to return to its major standstill. And it would take an observer another 18.61 years to verify the cycle, 37.22 years in all. According to the archaeological study of life expectancy in neolithic times— a 40th birthday was seldom experienced—very few people would have lived long enough to verify the moon's cycle with two adult observations. Many authorities agree that the postholes at the northeast entrance of Stonehenge are evidence of verification of the lunar cycle over a period of six observa-

tions, or roughly ninety-three years. People had a different sense of time, a different kind of patience, four thousand years ago.

A more detailed look at the vagaries of the lunar cycle appears in chapter 8 and more information on alignments appears in chapter 11.

The Afterthought Stones

The plan was cast, set in stone, you might say, although I did have a reserve menhir—the 13th-ranked stone on the list—in case one of the stones broke during construction. But something came up that slightly altered the plan. While looking for standing stones, I came across a stone wall near an outcropping of Potsdam Sandstone. Lying on top of the wall were four or five beautifully shaped flat slabs of the same red stone. They had been there some time, as their edges were eroded into gentle curves. Each stone had its own special personality, but all would have passed as miniature standing stones that belonged in a garden-variety stone circle. I began to look farther along the wall, and even moved a few stones to facilitate my search. Within a half-hour I had found another half dozen of these beautiful stones, all two to three feet in length. They were light enough to load in the wheelbarrow and take back to the site, where I stored them in a sandy area well out of harm's way. The alteration that resulted from finding these stones was to set up a ring of eight "miniliths" just outside the innermost ring of twelve firepit stones. They would mark the eight primary points of the compass rose (N, NE, E, SE, etc.) and provide visual interest to celebrants during bonfires. I marked them on the plan.

BUILDING THE EARTHWOOD CIRCLE

The plan on page 42 is retraced from my 1987 original, tidied up just a little for publication. The plan and a handwritten copy of the catalog chart were my constant companions during construction.

Laying out the Circle

Re-enter George Barber, megalithic surveyor extraordinaire. This wasn't the first time I had called on George with a problem of alignment. Six years earlier, I had asked him to lay out the foundation of the Earthwood house on a true north-south line. The intent was that the thirty-two radial roof rafters of the round building would correspond precisely with the thirty-two points of a ship's compass card. "How close do you need to be?" George asked. "Can you get me to within half a degree?" I asked. "I can get you to within half a minute," he replied. Using his wonderful old German transit and an accurate and up-to-date correction for magnetic declination, George laid out the home with great precision and the house now serves as an amazingly accurate directional finder for the night sky. If you know the azimuth (degrees from true north) of a particular planet, comet, or other celestial object, you can go to the appropriate rafter and find it easily. To get the stone circle alignment, George simply transposed the house alignment fifty feet or so to the east, by using equal alternate interior angles. (See page 37.) Not everyone has a true north alignment already handy on site, so, again, the sidebar in chapter 11 lists several other methods of finding north. Almost everyone (in northern latitudes, at least) has access to Polaris, the North Star. It's the bright one at the end of the handle on the Little Dipper (Ursa Minor). It's real accurate nowadays, although, curiously, it did not serve as the North Star in neolithic times. There was no accurate north star at the height of megalithic construction.

I chose a center for the stone circle that I judged to be on the arc of a great circle passing through the centers of the other three round buildings already on site at the time. This decision was intended solely to achieve aesthetic balance with the existing architecture. We drove in a strong 2 x 4 stake at the center, and a little finishing nail (no flat head) defined the exact center. I left the nail sticking out a half inch so that I could clip one end of my tape measure to it.

George stationed his transit exactly over the nail, using a plumb bob to get this just right. Once he was set up, and the transit bubbles indicated level in each direction, laying out the circle was quick and easy work. Beginning with north, George would focus the telescope crosshairs on the 5 MY mark and I would drive a 20-penny (4-inch) nail into the ground at that exact point. A piece of red surveyor's tape tied around the nail made it easy to spot, even from the backhoe operator's elevated seat. George moved his azimuth to 30° and we installed another positioning nail, and so on, every 30° around the circle. We repeated the procedure for the sitting stone circle, until twenty-four red-tagged nails marked the position of the major stones. The little firepit guys would be easy to do later with a yardstick aligned on the major stones, as would the eight little afterthought standing stones.

Until someone produces a tape measure in megalithic yards, it is best to use the tape of the realm, available from the local hardware store, ironmonger, or quincaillerie. Just use the conversion table on page 41. For setting radial distance from the center, I like a 50-foot flexible tape that can hook on to a finishing nail. To measure stones, a more rigid 12-foot or 5-meter tape is useful, or even a wooden folding rule.

Building the Inner Circles

In a concentric circle design, it's wise to start with the inner rings and to build the outer ring last. This order of events maximizes working space.

A couple of weeks before doing the standing stones, friend and neighbor Peter Allen helped me to build the sitting stone ring. Peter had access to a small backhoe and pushed the sitting stones into place, using the loader bucket most of the time. Fine adjustments could be made with the digging bucket. No sockets were required for any of the stones, although I sometimes dug a small depression for a stone if it was higher than its fellows. The stones were all on site or very close by, except one that Peter plucked with his backhoe from the old quarry on his way up the Hill. The sitting stones were in the 500- to 1,000-pound range, and very fine adjustment was easy to make with a five-foot iron levering bar. (See "Walking the Stones," in chapter 7.)

Nine leftover stones became the Children's Circle: eight stones octagonally placed in a circle six megalithic yards across, and an extra smaller stone slipped in as a recumbent on the east side. This smaller stone is a lovely clean diamond-shaped piece of red sandstone. It was added to the circle for our nearly two-year-old to sit on, like a throne, during children's gatherings. It is still known as Darin's Stone, the only one named in the Children's Circle.

Incidentally, I took the same care to catalog the sitting stones as I had taken with the standing stones, so I didn't waste any of Peter's time when he graciously came to help.

At my leisure, I arranged the innermost ring of twelve small stones, each exactly 1 MY from the center, and the eight little red standing stones. This work was pleasant and relaxing, and the result was a pretty little firepit with standing stones that would grace any backyard or garden.

BUILDING THE OUTER CIRCLE

This was my first real stone circle, with large standing stones, sockets, important alignments, and all. Log End was a toy by comparison, a whetting of the appetite, Stone Circles 101. Earthwood was

Earthwood Stone Circle in winter. The Children's Circle appears through the snow at the upper left.

Stone Circles 202. I would need help in the form of experienced people on the ground and on the backhoe, which was the piece of equipment I considered the best choice for the project. For my ground crew, I knew I need look no further than friend of my youth, Russell J. Keenan. We'd grown up together in Massachusetts and had kept in touch over the years. In 1987, Russ was involved in the restoration of an 18th-century water-powered sawmill near our hometown, and I'd helped him move some very large stones in the sluiceway of the mill. His father, then in his late seventies and repository of copious amounts of mechanical experience and Yankee ingenuity, helped us to move stones of two or three tons easily with a long iron bar. Russ was cut of the same cloth and had plenty of experience moving large stones with heavy equipment, and without it. He was the one person that I wanted on site, and I was more than pleased when he agreed to come north for a few days to help. Building a stone circle is just the sort of thing that appeals to a guy like Russ.

If Russ was the ideal foreman, than Ed Garrow was the perfect equipment operator. Ed had helped with me on all facets of the heavy equipment work at the Earthwood house, from digging the well to building the final megalithic retaining wall. Always a pleasure to work with, Ed was ever willing to share of his many years in heavy construction. By 1987, Ed was 72 and now worked for his sons, instead of the other way around. As I write in 1999, Ed is 84 and still operating heavy equipment for Ed Garrow and Sons. One of his sons proudly tells me that "no one knows more about moving heavy stones than my father." But I already knew that in 1987, and I made sure that Ed would be the operator for the stone circle construction.

Russ arrived on a Friday afternoon in late October. Ed would arrive the next morning with his backhoe on a lowboy, all towed by a dumptruck. (A lowboy is a low open-sided trailer used for hauling heavy equipment, and is known as a low loader

The Bunny Stone, as found in the grass near the quarry. The loader boys prepare to move her.

in Britain.) The truck and trailer would also serve to bring the stones up the Hill from the quarry.

We intercepted the equipment as it arrived at the quarry and I directed Ed to the seven standing stones that I'd chosen. Some of them still lay tight together like freshly cut brownies in a baking pan. The Bunny Stone lay in the grass a few yards from the edge of the quarry. Most likely, it had not moved since it hopped out of the small pit some sixty to seventy years earlier.

The stones all weighed between one and two tons, and Ed managed to fit them all on the lowboy. By 10 A.M., the quarry stones were on site, lined up neatly and easily accessible fifty feet outside the circle. Some of the other cataloged stones were indigenous to the gravel pit itself and didn't need to be moved until we were ready to plunk them into place. We also left the six stones on neighboring property in situ until we needed them. I knew them well enough, and exactly where they were located, so I figured there was no sense chaining them and moving them twice. The farthest, Bob's Stone, was only a quarter mile away, and the others were much closer.

Stones with Sockets

Nine of the stones would require sockets, and my chart indicated how deep each socket would have to be. The modus operandi for these was established pretty early in the game. I would guide Ed to one of the twelve standing stone locations and point out the red-flagged nail that marked the center of the inner surface of the stone to be installed. The backhoe boom and bucket were like powerful extensions of Ed's own anatomy, and he could cut as close to the nail as a man could with a hand shovel. I would draw the footprint of the stone on the gravel to give Ed an idea of size and tell him how deep the socket was supposed to be. Sometimes he'd come down off the machine to look at the stone itself. Never did he question the selection, orientation, or socket depth that I'd chosen. We trusted each other's judgment implicitly. He always listened carefully, and would either simply nod or make a suggestion that improved upon my idea. The single most important quality in an equipment operator is the ability to understand exactly what is in the mind of the designer/builder. This is a two-way street, of course, and it could just as accurately be said that the most important quality of the designer is to be able to impart his or her ideas into the mind of the operator. Ed and I have a symbiotic relationship this way, and I hope that the reader will be as fortunate in finding such a perceptive, skilled, and patient operator.

People with the ability to operate the equipment themselves can save a lot of money, because it is less expensive to hire the machine without the operator. You might even have your own equipment and further reduce your costs. And a substantial stone can be moved and raised with the front-end loader of a farm tractor, too, so keep Farmer Jones down the road in mind as a valuable resource.

I have a 4-foot metal rule that I like to use for this kind of work, although a common yardstick or folding rule will do. I constantly checked the depth and dimensions of the hole and let Ed know where more of the undisturbed earth needed to be removed.

Russ knows how to handle a chain from long experience and uncommon common sense. One method he uses effectively for lifting and lowering long stones vertically into position is a little trick he calls the "megalithic hitch," but it will be more useful as a reference in chapter 13, so you'll find it on page 273.

With Ed driving, Russ tying, and me directing, we soon got into a smooth flow. None of us ever stood idle. Ed talked slowly and moved slowly, but always carefully and methodically. Never was a move wasted and Russ and I had all we could do to keep up our end on the ground.

Jammers (Packers)

After a stone was lowered into its socket, Ed would hold it in place, still chained, while we dropped packing stones around the monolith. I have always called these supporting pieces *jamming stones,* or simply *jammers,* but British archaeologists refer to them as *packing stones* when discussing ancient megaliths. Take yer pick. I'm sure I'll intermix the two terms in ensuing chapters: jammers in the U.S., packers in the U.K.

We were fortunate to have an almost endless selection of jammers close at hand. The gravel pit at Earthwood is littered with them, and I had a pile left over when we built the masonry stove at the center of the house. *Make sure that the standing stone is in the right place before you begin to pack it in tight.* Step back and eyeball the thing. Is it facing the center? Is it plumb? Is it too far right or left? *Is it upside down?* (Hey, it can happen.)

My theory on jammers (and it has worked well over the years) is to put the largest one that you can in a particular space. One large jammer between standing stone and the edge of its socket will support the monolith better than two or more smaller ones. Put a jammer on each side of the upright, before attempting to tighten up the backfill with a sledgehammer, iron bar, or wooden post. Driving the first jammer in with the heavy sledge—a jammer hammer?—can easily shift the base of the stone out of position; then you have the nasty problem of removing the jammer so that the stone's base can be properly repositioned. Leaving the error in place is sloppy and you won't be pleased with the result. After the standing stone is surrounded with substantial jammers, you can tighten everything up

wonderfully by driving in wedge-shaped stones—we call them wedgies—and a regular hand-held hammer can be quite useful for this purpose.

How big are the jammers? It depends on the socket hole. The more oversized the hole, the larger the jammers will have to be. Neatness counts; it lessens the work, and makes for stronger support. I've used jammers the size of basketballs and bigger. When a socket hole is snug, I've had trouble getting baseballs or cricket balls into the space. I speak of game balls only to give a sense of size. Spheres are not really the best shape for jammers. Square, flat, or wedge-shaped stones all work better. Pack right up to grade level for best support.

When Ron Curtis took us up on the moors of Lewis above Callanish and showed us some ruined

The author tightens the Cat's Back with stone jammers and a sledgehammer. Ed Garrow (left) and Russ Keenan (right) look on. (Jaki Roy, photo.)

circles, the packing stones were very clearly in evidence. I found it interesting that the old-timers seemed to favor wedge-shaped stones, sometimes quite large, as packers.

After tightening the stone with jammers, we'd shovel earth and gravel into the hole. Russ used a garden hose to water the material into the hole, and then tamped the ground with a sledgehammer. The water drives the material into all the voids. When the socket hole dries out, the material in it is nearly as stable as concrete.

Freestanding Stones

Three of the stones demanded to be erected at their full height, no socket. The Bunny Stone is an extraordinary natural sculpture, and the first time I saw it lying in the thick grass near the quarry, I recognized the sitting rabbit carved into the anorthosite, probably by an explosion around 1920. Stone had been removed from this quarry for over twenty years, but a study of the deeds and the surrounding land suggests that the owners probably moved their quarrying operation to a larger property nearby, with a better quality of stone. A field of odd-shaped stones, like the Bunny, were probably culls, not good enough to be hauled into the village and made into gravestones.

The base of the Bunny, a rectangle about 15" by 24", was flat, but angled about 15° from the perpendicular face of the stone. The center of gravity was fairly low with respect to the full height of the rabbit to the top of her ears, most of the weight being concentrated in her sitting posterior. I marked the bunny's "footprint" on the undisturbed gravel

With the backhoe bucket, Ed lowers the Bunny Stone onto her base.

and earth, and carefully removed one to three inches of the surface, in an attempt to approximate the 15° angle of the bottom plane of the stone. Using Russ's "megalithic hitch," Ed simply lowered the Bunny Stone onto her base. Later, I built up an inch or two of gravel around the base, creating a sloping skirt away from the base of the stone. We were all impressed with how solidly the one-ton stone sat there, and how much more she looked like a bunny in her upright position.

I'd like to say "end of Bunny Stone story," but I'm afraid it is not. A few years later, at springtime, frost heaving tipped the stone outward in the direction of the slope, and she fell over. We had fun one night at a bonfire standing her back up again, by hand, using just levers and support timbers. When the wayward rodent went over for the second time, a year or two later, I thought of the legend of the Eagle Stone, a carved Pictish standing stone in the village of Strathpeffer in Scotland, not far from where I'd lived. A famous Highland soothsayer, known as the Brahan Seer, whose predictions have proven to be fairly good since his death in the 15th century, said that when the Eagle Stone falls over for the third time, the Cromarty Firth will rise in elevation and flood out Strathpeffer, five miles from the sea. Well, the stone has fallen over twice since the Seer's time, and was re-erected both times. Although it seems to be stabilized now (cemented onto bedrock), some locals joke that the stone should not have been set up again after the second fall!

No such legend threatened Earthwood with destruction, but I was determined that the Bunny Stone would not fall again. I dug a foundation hole three feet deep and dropped a huge boulder into the pit, a stone comparable in size to the sitting stones of the inner ring. The foundation stone came within a few inches of the surface. I cleaned its top and coated it with Acryl-60, a bonding agent made by the Thoro Corporation. (Other masonry bonding agents are available on both sides of the Atlan-

The Eagle Stone, Strathpeffer, Scotland, is a good example of an ancient carved Pictish stone. It is thought that the Picts carved these stones around the 7th century A.D. to record marriage patterns. In this case a clan with an eagle symbol is joined with a clan that has a rather more abstract symbol. Pictish stones appeared in Scotland two to three thousand years after the uncarved standing stones found in circles and alignment groupings. However, some of these carved stones are spectacular and well worth a visit in their own right. See the bibliography for an excellent guide to Pictish stones.

tic.) Next, I poured plenty of stiff hand-mixed concrete into the hole and troweled it flat, but with a slope to match the angle of the Bunny's bottom. Now I had a foundation three feet deep, which should be proof against further frost heaving. On a day when a few able bodies were on-site, I painted both the concrete foundation and the underside of the Bunny Stone with more Acryl-60. Then I mixed up a batch of stiff mortar—two parts coarse sand, one part Portland cement—and buttered the foundation with an inch of this "mud." We stood the stone up again, for the third time, and I tooled the oozing mud back into place with a pointing knife. The Bunny has since shown no inclination to hop off her base.

The Shark's Tooth—it looked just like a petrified shark's tooth in my son's rock collection—also stands on his base. (The reader may wonder how I assign gender to the various stones. All I can say is that some of them are just plain obvious. See also chapter 3.) But even though this long pointy stone is much taller than his Bunny mate two stones away, he has a base that is perpendicular to the vertical plane of the stone, so there is no natural tendency to tilt. Nevertheless, frost heaving can still happen, and I have had to drive in wedges once or twice when the Tooth showed signs of leaning in the springtime. Drainage is key to any hope of success with a freestanding stone. I have discovered a little trick that seems to have worked on my unsocketed stones, so far, at least. I created a sloped skirt away from the base of the stone in all directions. Lacking clay at Earthwood, I then installed 6-mil black polyethylene over the skirt, extending outward a foot or so in all directions. Finally, I covered the plastic with sods of grass that eventually knit together over the plastic and meld with the rest of the grass that now covers the site. The plastic draws rain and surface water away from the stone's base. No water, no freezing. No freezing, no heaving. At least that's the theory and it seems to be working so far. At Stone Mountain (chapter 3), we used a concrete skirt covered with clay to accomplish the same goal of shedding water away from the stone.

Building the Old-Fashioned Way

The South Stone was the third one that I had in mind to stand without benefit of a socket and jammers, but three other considerations were important. First, this was the largest stone, weighing, by calculation, 3,785 pounds, nearly two tons. (Calculating the weight of stones is explained in a sidebar in chapter 11, where I use our South Stone as an example.) It had a large flat squarish base, however, and I was determined to realize its full potential height of 6'6" or so above grade. Second, I wanted to bury a time capsule in the stone circle, and figured that under the largest stone would probably be the next best place to put it, after the center. (I was concerned that many years of bonfires would destroy any but a deeply buried time capsule in the center, and I didn't want to make it too hard to find.) Third, I wanted to stand up the large stone without benefit of modern equipment, using only materials that might have been available to the ancients.

Ed Garrow had finished all he could do by noontime on Sunday. Eleven stones were standing solidly in their places. "Are you sure you don't want me to put the other one up?" he drawled, indicating the big stone lying sixty feet from its pad. "It'd only take a couple of minutes." I thanked him for the offer, but declined, and watched as he drove away in his pickup truck. His son would come and collect the backhoe first thing Monday morning. Russ and I looked at each other. We were floating free; our lifeline had just driven away.

Lunch was hurried. I'd arranged for a few friends on the Hill to come over and help—Megalithic Construction Company—and I had to be fully prepared. I also had a letter to write.

The time capsule would contain only flat goods that could lie horizontally between two sheets of aluminum offset printing plates from the local newspaper. The plates are used once and the paper sells them for scrap metal, or, individually, to people off the street. I use them for all sorts of things, and even shingled a shed roof with them in 1975, a roof that holds up to this day. The entire contents of the time capsule can be seen opposite: the local daily newspaper, the current *Newsweek* (all about the October 1987 stock market crash), two of my ear-

Contents of the time capsule, placed under the South Stone.

lier books, an Uncle Scrooge comic book for kids of the future, a 1987 penny, and a letter to the finders with the background story of the stone circle. The last line is *Please put the stone back up.*

I laid all of these flat goods between two printing plates, and sealed the entire package with two large pieces of waterproofing membrane that I use with earth roof construction, W. R. Grace Bituthene. It sticks marvelously to aluminum and to itself and I overlapped the edge of the printing plates by three inches. The toughest part was purging the capsule of extra air. We didn't want an air bubble under the two-ton stone. Sealed against moisture, the flat "capsule" was placed about three inches below the surface, and then covered with a mixture of number one (½") stone and crusher dust. After compaction with a wooden tamper, this material made a very

stable foundation pad, but accommodated the irregularities of the base of the stone.

Incidentally, in case future archaeologists did excavate at the center, I buried another note in a sealed bottle, under a flat stone, with basic information similar to the other letter, but also telling of the time capsule's location. Maybe this bottle will survive, maybe not. It's like making a backup file on a computer disk.

The Hill folk arrived raring to go on this extraordinarily clear and comfortable first day of November. Conditions could not have been better, nor spirits higher. These were some of the same people who, in a few days time, would return for the Guy Fawkes Night bonfire. A local newspaper reporter stood by to record the event for his readers. Crazy people are newsworthy.

We had two separate tasks: move the recumbent stone to its foundation and, then, stand it up—without it falling over again. The first problem was to get the stone off the ground so that we could put wooden rollers under it. Russ lifted the stone with a fulcrum and a five-foot iron bar. ("Foul!" shouts Cliff Osenton from way over in chapter 7, reading this account at his home in Banbury, England. "No iron in the bronze age!" We knew that, but it could have been done with a long wooden lever just as well, as Cliff showed us ten years later. The sun sets rapidly in northern New York on November 1st, Cliff, and the race against darkness, we knew, would be critical. We rationalized the iron bar, and, later, the concrete blocks, as poetic license, for their jobs could have been done just as easily with similar aids crafted from heavy timbers.)

Once Russ lifted one end of the stone ever so slightly, son Rohan, then eleven, could drive in a wooden wedge. Now we could get a good purchase with a long wooden lever and lift what would be the bottom end of the stone clear of the ground. (The mechanics of leverage is explored in more detail in chapters 7 and 13.) In a very few minutes, we had a series of five or six maple rollers under the stone, with others in reserve. Each roller was about five inches in diameter and four feet long.

Our crew of eight experimented with three means of propelling the stone along the ground: manhandling, pulling with a rope, and levering. Manhandling is the first thought that comes to mind. Eight people ought to be able to push this stone along the ground on rollers. And we could, but it was a struggle. We didn't have room to work, and it was difficult to grab the stone. The guys pushing in the back, and groaning, were doing most of the work. Larger-diameter rollers would have been better. Second thought: let's pull it with a rope. (Expert opinions differ as to whether or not the original Megalithic Construction Company had ropes in their toolbox, but I have no doubt that they did.) We tied a stout manilla rope around the stone

lengthwise and put four people on each side. Pulling the stone was even more of a struggle than manhandling; it was very difficult to get the stone going, and we had to combine a couple of our strongest pushers along with the pullers to make any progress. Finally, we figured out that our old friend the lever could make this job a lot easier. Levers could be used in at least two ways, we found. (A) They could be used to greatly increase pushing strength from the rear. Simply plant the lever hard into the ground at the rear of the stone, using quite a steep angle. Then push against the lower rear edge of the stone, which is, after all, only five inches clear of the ground. With a six foot lever (about the longest most people can effectively use by this method), quite a mechanical advantage can be gained. The stone moves easily, but slowly. (B) Two people with levers, we found, can lift the leading edge, where the worst friction takes place. Pushers or a person with a lever to the rear, as per (A) above, can easily move the stone. By this method, the stone lunges forward, and then the levers need to be repositioned.

All three methods require a crew of one or two people whose job it is to take the spent rollers from the rear and place them as lead rollers at the front. We discovered early in the game the importance of smooth round rollers and a carefully prepared roadway. A bump on the roller or a stone sticking out of the ground can each slow progress considerably. Early on, we found out that the rollers worked better if they rolled on a trackway of two parallel flat timbers.

By a combination of the methods described, we managed to move the stone to its base in a little over an hour. Sixty feet per hour. That's a mile in eighty-eight hours, or eight eleven-hour workdays. And this was a two-ton stone. Pyramid builders would have to have placed a precisely shaped *five-ton* block in position every thirty seconds to have built the great pyramid in twenty years. Moving 100-ton menhirs, like we saw standing in the Brittany countryside, seemed, well, impossible. Never

Russ lifts one end of the two-ton South Stone with an iron bar.

The leading edge of the South Stone is raised with a long wooden lever and several rollers are installed.

mind the 350-ton Grand Menhir Brisé, which, whether or not it was ever stood up, was certainly dragged to the site, and from at least a mile away. We were starting to gain a real respect for the megalithic builders.

Taking charge, Russ said, "We've got to lower the leading edge to the ground without the whole damned thing going forward. We'll need to remove the first roller." With levers, which we were now manipulating with more speed and skill, it was easy to remove the lead roller, and then the second. Finally the lower leading edge could be pivoted on the middle roller and lowered to exactly the right position on the little foundation pad. The pad was made of dampened and compacted "number ones and dust," to use the quarry parlance.

The top end of the stone was very easy to raise—for the first foot or so. Everyone was kept busy from then on at a number of tasks, and we could see that the sun was within a half hour of setting. I knew that we wouldn't get a long twilight at this time of year. With Russ and myself barking orders, we built a cribbing of heavy timbers and, yes, concrete blocks to support the stone. Another crib was built to provide a place to put the fulcrum. A couple of helpers with sharpened wooden poles provided resistance to the stone's leading edge, to restrain its tendency to slide forward as the top end was lifted. If memory serves, the front edge did slip once, costing us valuable time as the stone was maneuvered back to the right place.

Progress slowed as an even greater structure of blocks, logs, and stones had to be built to support the great weight. The sun had now set, and the light faded by the minute. The stone was at a 40° angle, and was actually getting easier to lift as its center of gravity shifted more and more toward the north. But it was still too heavy for eight of us to manhandle into the upright position. I knew we were beaten and wasn't about to risk someone getting hurt for the sake of proving something. Discretion is, indeed, the better part of valor. We secured the

support pillar, I thanked the crew, and we called it a day. One or two stayed for well-earned beers, most went home to their suppers. But we *had* proven something. We proved that the ancients were a lot better at moving heavy stones than we were.

The sun again rose clear the next morning. We'd certainly been lucky with the weather. But there was no way that we could raise any help on a Monday morning. Russ and I discussed the idea of completing the job by ourselves, with Jaki assisting. We both knew from experience that by moving slowly and methodically, we could move heavy stones with great control. But we knew that Ron Garrow would be along to collect the backhoe at 9 A.M. "When it's gone, it's gone," observed Russ. "We mess up and we're out of luck." We decided on the prudent course and, with Garrow's chain, Russ tied his hitch a quarter of the way down from the top. Ron arrived about nine, in a hurry to get to the next job, but he took the time to help us lift the South Stone onto its pad. The backhoe boom could not quite lift this stone, as it had the others, and we had to use the front-end bucket instead, which had greater lifting power. With signals, we helped Ron get the stone in place. The tension on the chain seemed to cut the air, and Ron began to transfer the stone's load to the earth. The stone compressed the pad, and we signaled for Ron to release the tension on the chain. Immediately, the stone began to tilt toward the center of the circle, and we waved our hands wildly for Ron to regain control. The stone would have fallen over had he completely let go. The base of the stone had a very slight angle to it, perhaps 5°, and although we'd tried to correct for that angle in the creation of the pad, something was wrong. Maybe our allowance was a little off, or compaction was not uniform, or the darned time capsule was destabilizing the foundation farther down. Ron lifted the stone clear of the pad and within three minutes that seemed like ten, Russ and I reshaped the pad according to the impression made by two tons of stone. On the second attempt,

THE EARTHWOOD STONE CIRCLE

This is far as we could raise the stone before darkness stopped work.

The South Stone is installed with the loader, and we gained a lot of respect for the ancient stone builders.

the stone stood proud and perfectly straight. It was, well, solid as a rock.

Russ and I said our thanks and good-byes to Ron and looked at each other. We suddenly realized that had we succeeded in standing up this stone on the previous evening, we might not have been able to stop it from tumbling over and, perhaps, smashing onto the sitting circle. As there was a hairline fault around the waist of the stone, such a fall would probably have broken it into two neat one-ton blocks. We breathed a sigh of relief. The early darkness of the night before might well have saved a nasty time, and the stone itself.

Setting the South Stone was—and remains—a critical moment in my stone building experience. We had been careful, and cautious, and made the right decisions. But perhaps we'd been lucky, too. I gained a greater respect for heavy stones that day, and the ancients who built with them.

For longevity and complete safety—during construction and down the road—standing stones should be set in sockets and tightly jammed. That's the reality. To maximize their height and appearance, three of the stones at Earthwood are freestanding. They are solid now, but all have given us trouble at one time or another. Only the Bunny Stone, with its 15° angled base, ever actually fell over, and I don't think it will again, at least in my lifetime. The South Stone leaned six inches out of plumb one spring after the thaw, and I had to prop it with poles. Later, by levering against a heavy dumptruck parked nearby, I managed to straighten it, pack it with the appropriate stone shims, and create the plastic-lined drainage mound previously discussed. None of these stones have caused further problems, but I keep a close eye on them each spring. The safe way to make full use of a particularly wonderful stone is to build a solid concrete or stone foundation under it down to frost depth, and then either concrete the stone into place, or pin it with stainless steel pins, as was done at Jon Isherwood's stone circle in Plattsburgh, New York, and at the

Guy Fawkes Night bonfire at the Earthwood Stone Circle. (Frank Brasacchio, photo.)

huge granite Georgia Guidestones, both shown is chapter 15. Do not rely on unsocketed tall stones to remain standing without careful safeguards. I have a responsibility to advise against the use of free-standing stones, because I cannot be sure that everyone will take adequate precautions. Consider this to be a disclaimer.

USING THE EARTHWOOD CIRCLE

To finish this chapter on a more upbeat note, I am happy to report that the Earthwood Circle has been a success beyond my wildest expectations. Building the circle cost $500 in heavy equipment contracting, and it's the best $500 I've ever spent. Certainly, no other similar expenditure has given me nearly

as much pleasure. The fun began right at the design stage, and it has never left. When we finally got the South Stone in place, the circle itself—not counting grassing the area—was finished. Still to come would be the outliers. This is an ongoing project, and I'm going to save the discussion of the Earthwood outliers for chapter 11, when I speak of stone circle design in more detail, as well as astronomical alignments.

Guy Fawkes Night is celebrated throughout Great Britain with bonfires on November 5, the anniversary of the Gunpowder Plot, an attempt in 1605 by Guy Fawkes and some coconspirators to blow up the Houses of Parliament. The plot failed. Fawkes and his buddies were caught out, and most of the conspirators were killed. Fawkes himself was hanged in 1606, and the "guy" you see hanging on country porches in America on Halloween is a direct descendant. But bonfires at this time of the year date back centuries before 1605, maybe for millennia. This is the time of the ancient Celtic New Year of Samhain, the beginning of the dark quarter, and one of the four station holidays of the ancient year. Although we didn't know it at the time, some stone circles show evidence of solar alignments on Samhain. We celebrate it because it reminds us of our British background. Jaki is English and remembers Bonfire Night from her youth. ("Please to Remember the Fifth of November, Gunpowder, Treason and Plot."—Traditional.) I had lived long enough in Scotland to gain permanent resident status, and my Scottish friends observed Bonfire Night because they thought that Fawkes was on the right track even if the plot *was* found out. Besides, its a good excuse for a booze-up.

The Fifth of November, 1987, was our thirteenth consecutive Bonfire Night, but the first at the stone circle, completed just days earlier. You need a big fire to drive away the November cold in New York's North Country, but something feels right about it and I sometimes wonder if thousands of years of

Samhain—and pre-Samhain—bonfires have left something in the deep recesses of our racial memory that we latch onto at this time of year. Each year's celebration is as different as the weather; there's nothing religious or nonreligious about it. Mostly, it's a celebration of fellowship with good friends, and November 1999 will mark our twenty-fifth year.

In addition to Bonfire Night, the circle is used for social gatherings at the building school, and has been a great playground for the kids. Sometimes I just go out to the circle to sit and think. Someday, I hope, my ashes will be scattered around the site, so that I won't miss future Bonfire Nights.

Additional pictures of the Earthwood circle can be found in the color section.

One morning, during a cordwood masonry workshop at Earthwood, I looked out the bathroom window, about sunrise, and watched, fascinated, as one of our students greeted the rising sun at the stone circle with the wonderful movements of his Tai Chi exercise regimen. I watched, transfixed, little suspecting that years later this very same student would ask me to build a stone circle for him.

CHAPTER 3

Building the Stone Mountain Circle

Against the bleak and secularized backdrop of our outer landscape
comes, unbidden, a new cast of actors, costumed, beaded, and
bearded, ready to act a play for which as yet we lack the script.
Dancing they come, to an irresistible music, prying loose the gates
of dream, unwilling to live without a myth.

—Dr. Stephen Larsen, *The Shaman's Doorway* (Inner Traditions, 1998)

Jaki and I have been connected with the Center for Symbolic Studies since its beginnings in 1990. Our relationship actually started that May, when Dr. Stephen Larsen attended a cordwood masonry workshop at Earthwood, accompanied by his son, Merlin. Steve was the student doing the Tai Chi exercise at sunrise in our Earthwood Circle.

A LITTLE BACKGROUND

With Steve and his wife, Robin, as codirectors, the Center for Symbolic Studies offers programs and draws renowned guest lecturers in a variety of disciplines. While myth is the core of the program, other related workshops and camps have developed over the years, such as dance, meditative and martial arts, and seasonal festivals keyed to ancient observations. The Center also hosts professional training seminars, led by teachers involved in growing a "new medicine from deep shamanic roots." I mention all this because it gives a pretty good idea of the kind of people you'd expect to be interested in stone circles.

Around 1992, Jaki and I stopped in at the Larsen's home on the 350-acre Stone Mountain Farm near New Paltz, New York, and met Robin for the first time, beginning a friendship that has continued over the years with mutual visits and projects, such as our conducting cordwood masonry workshops at the Center. During the winter of 1997 Robin and Steve came to Earthwood for a couple of days of cross-country skiing, and that was when we first discussed the possibility of building a stone circle together at Stone Mountain Farm. Steve felt that he had the right kind, size, and shape of stones right on the farm and the Larsens had what they considered to be the perfect site for a circle.

LUGNASAD, 1997:
OLD FRIENDS AND STONES

We tossed the idea around for a few months and finally agreed to conduct a stone circle workshop at the Center on the Midsummer Festival weekend of August 1st through 3rd, 1997. August 1, known as Lugnasad by the Celts (and later Christianized as

Lammas), is the Firefeast of Lugh, the sun chari-oteer and musician. In 1997 the actual celebration was scheduled for Saturday night, August 2, mid-way during our stone circle workshop. This turned out to be a very special weekend at Stone Moun-tain, with several events magically intertwined.

Prior to the workshop, Steve did some careful measurements on the site, including alignments for both the summer solstice sunrise and sunset that he took on June 21st. Aided by a friend, Jim Carman, who owns a large log skidder, Steve also brought eight large candidates for standing stones down from an old rockslide on the farm and distributed them around the outside of the circle site.

I contacted my old friend Russ Keenan, who had been so valuable at the construction of the Earthwood Stone Circle, and he agreed to come and help once again.

Jaki, son Darin (then 11), our German Shepherd Holly, and I arrived at the farm just before 5 P.M. Russ had arrived a couple of hours earlier and had already looked at the stones, gathered together on a lovely grassy platform that overlooked excellent views to the west, north, and east. The site was idyl-lic. The wood-lined mowed fields and meadows strewn with purple loosestrife reminded me of southern England or, perhaps, a French river valley. The view south from the circle site was uphill to-ward the main building at the Center. On the left was an occupied teepee, to the right was the danc-ing area and a built-up stage under a large tentlike canopy. Straight south of the stone circle site, on a line with the Center, was the firepit where neo-Celtic revelers would construct and ritually burn the "Wickerman" on Saturday night.

My first impression of the stones was very favor-able. They displayed a variety of wonderful shapes, the stones were in scale with each other, and all but one were strong standing stone candidates. I esti-mated that all eight stones were in the one- to three-ton range, and later calculations proved this to be a good ballpark guess. I sketched the stones as they lay, and measured some of their important features, such as their full height and height when socketed. Two of the stones, it seemed, had the potential to stand on their own bases, without need for socket-ing. These original notes are reproduced in on the next page.

Drs. Stephen and Robin were caught up with other pressing activities, Steve with patients, and Robin with haying the fields, substituting for an injured farmhand. We arranged to meet with the Larsens at 7:30 P.M. for my slideshow on matters megalithic. Meanwhile, Russ, Jaki, Darin, and I re-paired to a nearby pub to renew old acquaintance over pints of Sam Adams Summer Ale.

At the slideshow, I met old friends and new. Rich-ard Ross, who'd built a round cordwood masonry house nearby, and who'd attended celebrations at the Earthwood circle, had come to learn about building with the stones. Rich proved to be one of the most valuable members of the team, ready to jump in on any task that was required. Later, I learned that he means to build his own small stone circle near his round house.

Among new friends were Anne McClellan and John Rosett from New York City, who, together, are known by their collective name of "ElfWiz," proclaimed even on their license plates in block capitals. ElfWiz have or has—I'm not sure of the grammatical protocol here—a close relationship with the Center, and they were up from the city this weekend to celebrate ten years of marriage with a reaffirmation of vows. Their beautiful ceremony turned out to be intricately caught up in the stone circle construction.

My slideshow was a personal history of my in-terest in megaliths over the years and documented the construction of the Earthwood Circle almost ten years earlier. I discussed briefly what I thought we could accomplish over the next couple of days. After lots of good questions and high-spirited con-versation, the Friday night session broke up about 10 P.M. Our core group, consisting of the Roys,

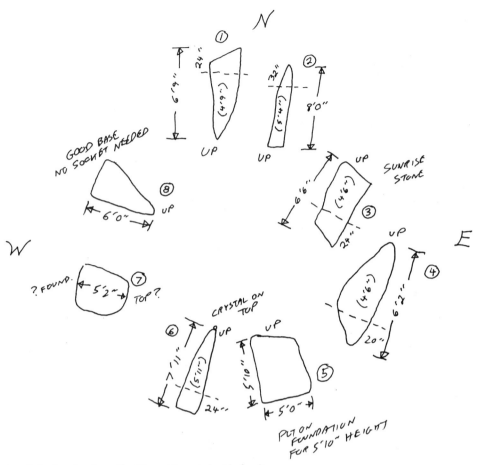

The author's original notes from the Stone Mountain Circle site.

Larsens, Russ, and Rich, carried on for another hour over a sumptuous late-night supper of prayseux at a nearby restaurant.

Preliminary Investigation

At breakfast the next morning, I was concerned about being out on the site by 9 o'clock, the scheduled time of beginning, and I couldn't help noticing that the Larsens did not seem to share my apprehension. At first I explained this (to myself) as the more casual way of doing things at the Center, at least compared with our Earthwood workshops. But, gradually, it became apparent that there weren't really any paying workshop registrants to be concerned about. ElfWiz were caught up in their own

party doings, other slideshow attendees were Farm dwellers and were out and about on Farm and Center business, and mutual friend Richard Ross was a nonpaying guest. I had envisioned a discussion with workshop students, in which we would chart the stones on site, discuss alignments, stone selection, layout, the megalithic yard, the purpose of the circle, et cetera. In the event, Stephen, Russ, Rich, Jaki, and I simply wandered over to the site and discussed the practical standpoint details of building the circle.

Our eyes were drawn first not to the stones, which we'd already seen, but to the tiny backhoe on tracks that had been rented to help erect the circle. The machine had a 16-inch hydraulically articulated bucket, an excellent tool for ditch digging, and a small bulldozer-type blade at the other end. Did it have the power to hoist two-ton stones and set them in their sockets? Steve had chores to do in order to make his pickup truck available, so Russ and I took the time to do some testing. Steve had expressed concern that there might be bedrock close to the surface, which could prevent us from socketing the bases of the stones to the required depth. With Russ operating the little backhoe just outside the circle, we dug a sample hole 32 inches deep, with no sign of bedrock or very large stones. *Encouraging.*

After filling the test hole, we tied the chain to a 3,000-pound stone, and found that the machine struggled mightily to lift it to a standing position. There was no hope of lifting the stones off the ground and lowering them cleanly into their sockets. When Steve returned with the farm's trusty old Ford pickup truck, we advised him of our findings, positive and negative. Russ said that we ought to be able to do the job with the machine, but that it would require a little more thought and manual work.

That We May Pull Together . . .

At my request, we spent a half hour discussing the stones, deciding which ones we wanted where, and finding out what alignments Stephen and Robin were hoping to incorporate into the circle. I numbered the eight stones on site, beginning with the stone lying on the north side of the circle. One of the stones, #7, was 5'2" by 4'6" and rather roundish. With no good base, it had little potential as a standing stone. Another, #4, a nicely shaped menhir not unlike the one that the great French comics character Obelix the Gaul carries around on his back, seemed a little short at 6'2", as its shape demanded that 20 inches be buried in the ground, leaving a potential of just 4'6" standing above grade. This left us with six good stones for the four cardinal points, and the summer solstice sunrise and sunset alignments, which Steve had earlier staked out.

By consensus, we decided that a beautiful 7'10" stone (#6) with a natural quartz crystal right at its very apex would make a splendid North Stone. With luck and careful planning, we figured to get that quartz crystal exactly under Polaris, the North Star, when viewed from the South Stone. This new North Stone would have to be dragged across the circle to replace #1, the pointed stone already lying on the north side. Our first task would be to set #1 as the West Stone, to make room for #6. The first two stones we erected would have to be socketed.

At Earthwood, I had spent weeks becoming familiar with the candidate stones, and planning a circle that would bring into harmony their various shapes and colors. At Stone Mountain Farm, the stones we had to work with had already been delivered to site, and good ones they were. Color was not a factor, as all of the stones were a fairly uniform gray, with other colors mixed in metamorphically. With people assembled to work, planning was a much quicker affair than it had been at Earthwood, where I had no deadline.

At Stone Mountain Farm, we needed to figure the positions of only six standing stones from a pool of eight candidates already gathered neatly at the site, a layout problem with only a fraction of the permutations I'd had to deal with at Earthwood. As Steve and Robin had already done a good job of preplanning in their selection and delivery of stones, and had given some forethought about positioning, our "planning" consisted of final considerations and selections. Steve expressed his and Robin's rather general design ideas, and I suggested how their plan could be refined by integrating special features of the stones into the plan. Jaki, Russ, and Rich realized that the final okay for positioning the stones would be made by Steve, based upon my recommendations, but their own observations were useful. Many eyes are more likely to spot important features that might be overlooked. Consensus worked well this Saturday morning. Energy and spirits were high, and we seemed to pull together well. This might not always happen, however, and it's important that all participants recognize the final authority on decisions of this kind, generally the person or persons whose idea it was to build the circle in the first place, and certainly the person paying for it.

The Magic Stone

During this midmorning discussion, I asked Steve what he and Robin had planned for the circle's center. Did they want a standing stone there, like Boscawen-Un in Cornwall or a firepit like at Earthwood? Neither, Steve replied. He told us that he had found a truly magical stone to go in the center, and would like to show it to us. We piled into the back of the old pickup and Steve took us to another part of the farm, "the Grotto," where the indigenous Shawangunk conglomerate had slipped down a steep slope, now overgrown with mature hardwood trees. The site did not appear to be a quarry, as there were no drilled holes or other obvious quarry marks. It looked more like an elephant's graveyard in stone. The Shawangunk rock, large chunks of which littered the slope, is well known to world-class rock climbers. With no clear bedding planes or obvious regular crystalline formations, it does not cleave well. The 300-million-year-old metamorphic conglomerate is composed of ortho-quartzite and a variety of igneous and sedimentary stones and pebbles, all fused together under great pressure to make the unique—and very hard—"Gunk" stone. Iridescent pink and greenish streaks make the stones even more interesting when viewed close up.

Steve took us to his magic stone, and we soon saw that he had not been exaggerating. A first glance revealed a lichen-covered stone lying on a roughly 40° slope, the lower edge of a much steeper slope. A rough hexagon about five feet across and a foot or so thick, the salient feature was an incredible carving on its exposed face. Someone, more probably ancient than modern, had taken the trouble to carve into the hard flat exposed face a nearly circular depression of perhaps 3½ feet in diameter and an inch or more deep. The perimeter of the circle was a little deeper than the central plain, as if to more easily draw liquid around the edge. At the lowest portion of the circle—lower in terms of depth as well as its position downslope as the stone lay there—was a carved trough about an inch thick and several inches long, connecting the deepened perimeter of the circle to the edge of the stone. It was clear that if balanced correctly, liquid could be made to exit the little pond by this outlet.

I could not see any evidence of chisel marks, suggesting the circle, the perimeter trough, and the outlet were all pounded out of the hard Gunk stone by, perhaps, an even harder stone axe. Alternatively, we considered that iron or even steel chisels may have been used, but that evidence of such has simply vanished through exposure over a very long time—how long, we did not know.

We immediately began discussing the possible original uses for this stone—for it was clearly made for practical as opposed to aesthetic purposes. Bloodletting, either for sacrifice or during normal homestead slaughtering, was the first idea that came to mind. Blood was valuable in the production of "black pudding" in ancient times, and still is in many European countries today. Crushing apples under stones and gathering the apple juice was a possibility; or crushing berries; or even pounding hemlock bark and gathering the tannin, valuable for tanning hides.

Whatever it's original purpose, it is unclear whether the carved stone was ever actually used. To us, the stone appeared to lie in its original position on the slope of the hill. Had it been taken elsewhere and used, it is unlikely that anyone would have bothered to haul its estimated two tons back to the place of origin—and this certainly looked like that place. The stone appeared to have been "plucked" from the hillside by the accumulated effects of frost, gravity, and time. Perhaps it had been used, whatever the purpose, right where it was

Darin Roy inspects the Magic Stone in situ at the Grotto.

found. In any case, it would be used now, without doubt, as an extraordinary feature of the Stone Mountain Circle. Our minds reeled with the possibilities, but the one that won out was to place the stone in the center of the circle, and use the circular depression as a reflecting pool.

Two Menhirs

Steve wanted us to size up some other candidate stones in the same area. One was 11 feet in length and weighed about 4.5 tons. Larger than anything yet drawn to the circle, this stone was a beauty, generally rectilinear in shape and of fairly consistent width and breadth. Steve thought that his friend Jim might be able to drag the monolith to the site with his log skidder.

Several other stones at this site were in scale with the circle as far as weight was concerned, but they were lumpy in shape and we could not see how they would grace the circle's aesthetics in any way. While looking around, however, we came across a menhir that would have graced any major ancient stone circle. At 18 feet in length, 8 feet in breadth (at its thick waist), and 5 feet in thickness, I quickly calculated its weight at 16 tons, a figure that brought vocal allusions to the Tennessee Ernie Ford song of the same name. No normal log skidder would move this stone, and we could only look at it in admiration and despair. I felt like a street urchin gazing upon a beautiful queen. Nonetheless, for fun, we took notes of her measurements. One is entitled to dream.

Gathering Jammers

For the next hour, we searched for smaller stones to serve two purposes back at the site. I hoped to find a couple of larger lumps with at least one good flat side for use as foundation stones for the two standing stones that we'd identified as potential free-standers, but which would benefit from a deep and fairly heavy foundation, like the Bunny Stone at Earthwood. Also, we would need a copious supply of jammers for the four or five stones that would be socketed. There were only a few small stones from a campfire ring at the site, so we'd need a lot more, preferably wedge shaped. Unfortunately, while helping to lift a 200-pounder onto the bed, I hurt my back, which only worsened through the weekend. With a previous history of back trouble, I should have known better.

Our journey took us along farm roads that led to tracks which led to narrow trails and finally to heavily overgrown fields with no worn way at all. We came across a lot of stones, identified a few for possible pickup later with Steve's farm tractor, and loaded a number of fairly large jammers on the bed of the old truck. En route, we visited a couple of huge caves that seemed almost frigid inside on this hot midsummer day. One had a pond of unknown size inside. We threw pebbles in to hear the splash. Steve told us that at least one diver had entered the system of caves and never returned. We also had a look at the turn-of-the-century Rosendale cement works kilns, huge round brick kilns like nuclear cooling towers, some built on extraordinary platforms of stone that made our work this weekend seem insignificant and somewhat feeble by comparison. One structure perhaps seventy-five feet high had stone quoins criss-crossed in the corners, said quoins weighing a ton, two tons, and more. Some of these were in position thirty feet up the wall. We do not have to go back five thousand years in time to find incredible feats of megalithic construction. Nineteenth-century rail, industrial, and bridge en-gineering, in many cases, rivaled the stone-quarrying and moving feats of Stonehenge and Maes Howe.

The journey back to the site was an exciting trip though thick overgrowth that reminded those of us standing in the back—everyone but Steve—of an African safari adventure, chasing, or being chased, by rhinos or wildebeasts. We were having a great time, particularly Darin and his dog Holly.

CONSTRUCTION: DAY ONE

Despite all the fun we were having, I suggested that we'd better start standing some stones or we'd lose the day. We began with . . .

The West Stone

We actually started building about 1 P.M. on Saturday, beginning with the West Stone, #1, which was resting near the north stake. If north is 12 o'clock on the circle, this west stone had to be dragged to and erected at the 9 o'clock position. In my original short study of the stones, I thought we'd be able to expose no more than 4'9" above grade, not too impressive. In the actual event, an opportunity presented itself to gain some height.

Steve dug the socket hole with the little backhoe, just the right location for the stone, but maybe a few inches too deep. I remembered the 200-pounder that we'd loaded on the truck, the one that had done my back no good, and thought it would sit perfectly in the hole that Steve had created. Perhaps we could gain six inches or so of apparent height that way. We dumped the pyramid-shaped stone off the truck and manhandled it to the edge of the hole. We positioned it so that one flip would—hopefully—land it upside down, its flat base up. This worked almost perfectly and just a little persuasion with an eight-pound sledge tightened it up and leveled it off very neatly.

With a chain tied around its end, Steve dragged, prodded, pulled, and cajoled #1 to the edge of the hole we'd prepared. This was one of the smaller stones in weight, right around a ton, but there was no way the little backhoe was going to raise it clear of the ground and place it neatly in the hole, as we'd hoped. To our disappointment, Russ was not going to be able to make use of his "megalithic hitch," but he formed a simple chain loop around the stone's top. Although the hired machine could not lift #1 clear of the ground, it could quite easily tilt it upright onto its heavy end. And we almost always put the heavy end down, making exceptions only for beautiful faces that could not be buried in the ground. Making exceptions for beautiful faces, I might say by way of aside, is a character flaw that has gotten me into awkward positions more than once in life.

We stood old #1 on the foundation stone we had laid in the socket, using the boom and bucket (and guided by manual labor), but its bottom was rounded, not flat, so we quickly jammed around the edges of the standing stone's base to provide some temporary stability. I showed Richard Ross the art of jamming, a skill that he gained quickly and practiced with relish throughout the rest of the project. If I'd shout "Jammers!" with the exclamation point clearly in my voice, Richard immediately rushed to the task, found a stone, and hammered it in place with the sledge. As an experienced cordwood mason, he had a good eye for size and shape.

Rich, Jaki, and Russ adjusted the jammers while I plumbed the stone with a 4-foot level. Steve held the stone safely in place with the bucket. When I was satisfied with the position and alignment from the center, we began to pack with boulder-shaped stones all around the standing stone. About a foot of the stone would be buried below grade—this isn't much—and it was critical to tighten the stone in its socket on all sides. The clay soil prevalent throughout the site was dry from a long drought period, and provided good support when we jammed rocks between the standing stone and the edge of the hole. Final tightening came from ever

smaller wedge-shaped jammers, sometimes hammered in with a sledge, sometimes with an ordinary claw hammer. A two-pound stonemason's hammer would have been handy, but we didn't have one available.

The stone was tight, even before final backfilling, but for safety's sake we hand-backfilled with the excavated clay, and tamped with the sledge before signaling Steve to let up with his bucket. We untied the chain and he backed the machine away. Another few minutes of hand-tamping the clay, with just a little water added for plasticity, returned the excavation to the firmness of undisturbed earth.

The author checks the West Stone for plumb, using a 4-foot level. Richard Ross readies a jammer.

We all stood by and admired our handiwork. Setting #1 in place had gone perfectly, even better than hoped for, as we ended up with 5'10" exposed above grade, a foot more than my original estimate. And it was solid. The gain had come from using just the right stone deep in the socket as a foundation. Good teamwork and excellent jamming gave me the confidence that the West Stone was not going to come tumbling down easily. One down, five to go!

Renewing Vows

We had coordinated with John and Anne (ElfWiz) to stop work during their tenth anniversary ceremony to take place at 2:13 P.M., and the time had come. About sixty guests were in attendance, most festooned in bright costumes. Robin had even decorated the horses and the scene had a decidedly Celtic atmosphere.

Robin, dressed in a long bright red and brown dress, and another lady in bright blue, led ElfWiz (riding two brightly caparisoned horses) up from the fields and through the stone circle site en route to the tent.

Anne and John are led through the construction site on horses.

Anne and John were guided to the nearby tent, where two priestesses conducted a ceremony that combined Native American, Pagan, and Quaker ritual. Evoking the spirits of the North, South, East, and West, ElfWiz reaffirmed their commitment to travel life together as soulmates. They recited Tolkien's poem of the Ent and Entwife, an analogy of mythical male and female trees whose connection is destined and forever intertwined. The ceremony was as beautiful as the day and the surrounding scene. Flowers were strewn liberally as the couple renewed their vows.

With but one stone standing, the Stone Mountain Circle was already an integral part of and a backdrop to a spiritual and joyous celebration: this enduring love between two special people.

As we hadn't eaten since breakfast, we welcomed ElfWiz's invitation to join them for snacks. After the reception, a college friend of Anne's performed original neo-Rennaisance folk music, setting a kind of surrealistic background to the resumption of equipment noise at the circle.

On the way back to work, we passed by the Wickerman construction site, the Man's skeleton being shaped by an enthusiastic crew. The Wickerman, made of sticks and straw, was being prepared for his short life and immolation at the Lugnasad celebration later that night. The stone circle construction added yet another dimension to the whole circus, and people began to take interest in it in exponential proportion to the number of stones standing.

The North Stone

From here on, our work was backdropped with music, drumming, dancing, and other frivolity. Revelers had the good sense to stay well clear of actual construction, but more and more stopped by to look or to have a quick word with members of our crew during the infrequent lulls in the action.

With #1 out of the road, the way was clear to drag the longer and perhaps slightly heavier #6 across the circle and to its new position as the North Stone. I measured and drew the foundation shape on the ground, defining it with four small stones. The small stones were actually placed 6 inches out from the true corners of a perfect socket, so that they remained in the operator's view during excavation. While I did this, the others were busy chaining and fetching the 7'11"-long stone.

This was the stone that featured the quartz crystal right at its apex, and we thought that it would be spectacular to position this crystal under Polaris, as viewed from the future South Stone. Unfortunately, in positioning the stone with the bucket prior to dragging it, the crystal was broken off. I could sense the disappointment from clear across the circle, 16 megalithic yards away, and went to see what the lamenting was about. Steve was already down from the machine, and found the crystal. We all decided that it might be possible to reinstall it in its original position with the aid of some polyester resin. Steve put the crystal in his pocket and we returned to our respective jobs.

North went in as easily as West, but this time we used the fully calculated socket of 24 inches. This left 5'11" exposed above ground, virtually identical to the first stone we'd set, and although North was just as firmly jammed and set as #1, I figured that we could make a slight adjustment the next day, if necessary, to get the point exactly under the North Star.

The first stones had gone in with ease and precision, and they looked beautiful, to me reminiscent of several of the stones at Callanish, which, in a certain light, took on the shapes of ancient cowled figures walking slowly across the the mysterious Lewis landscape.

The East Stone

We'd deliberately commenced with the two easiest stones, to get into the pattern of things and to try to get the crew working together as a team. I felt a strong responsibility to do the job right, and, after all, it had been ten years since I'd built a stone circle.

Although our crew was small, it was proving to be more than capable. Steve Larsen turned out to have excellent judgment in operating the backhoe and became our first-string operator from the start. Russ Keenan, of course, was an experienced mover of large stones, and capable of operating the little backhoe in Steve's occasional absence. His hearing, unfortunately, had deteriorated since we'd built the Earthwood circle, which made communication almost impossible while the machine was on. Still, Russ carried on doing the important little things that he knew had to be done: watering in and tamping the filled sockets, and cleaning up the individual stone site locations with the bladed end of the machine. Richard Ross was a natural and would have been right at home during the Stone Age. With a name like Ross, his ancestors may well have been involved in the construction of Callanish. Jaki Roy, was involved throughout the day in all the various tasks save equipment operation and chain tying, and managed to take a couple of rolls of excellent pictures, some of which appear in this chapter. It occurred to me that, like Anne and John, who were ElfWiz in the collective sense, we, Rob and Jaki, were also thought of by many others as RobandJaki. We work well together.

The East Stone candidate, #2, was heavier and much more awkwardly shaped than the first two we'd erected. To complicate matters, the end of the stone possessing the decidedly superior character was also thicker and therefore much heavier than the opposite end. Like the red Aku-Aku or Easter Island Stone at Earthwood, we'd plant the narrow end deeply into the earth, a full 32 inches in this case.

Choosing the depth of the socket is not an exact science. Depth varies from stone circle to stone circle in Britain, but, on heavy stones with fairly regular shape, a rule of from 25 to 30 percent buried seems to be the norm. Perhaps stones that have fallen down over the centuries had lesser sockets. It would be hard to tell from the physical evidence and it must be remembered that many stones were deliberately torn down by humans.

On smaller stones, of course, I want to get the greatest height above grade without creating a safety hazard, and, often, I socket as little as 20 percent of a stone's total length, particularly if the heavy end is being buried. The West Stone, already discussed, was an exception, with only about 15 percent buried, but it did have a substantial stone foundation to help protect against frost heaving. As a general rule, if the stone has a good heavy broad base, it requires shallower socketing than one with a narrow light-weight foot.

Number 6 had a long narrow base, and it didn't taper symmetrically. The part to be buried would have to be offset to the southern edge of the socket, so that the interesting surface would face the center. Russ suggested that we might be better off breaking 8 inches off the narrow bottom end and socketing the stone 2 feet deep instead. This was a sensible suggestion from the viewpoint of structure, as was to become clear in a few minutes, but this Gunk stone looked to me like it was not going to break where we wanted it to, and there was the very real danger of cracking it farther up its trunk, rendering it useless. Russ allowed that this was indeed a possibility and we decided to go with the 32-inch socket, shaped to take the offset.

I outlined the socket with small stones as usual and showed Steve where the deepest side should be. He cleared most of the clay with the machine and Rich and I finished up the bottom of the hole with a spade, trying to create the same shape as the lower portion of the stone itself.

About this time, an old friend of Steve's, Bill Klopping, arrived serendipitously on the site. Steve and Bill had not seen each other in over twenty years, and, coincidentally, their previous meeting had involved backhoe work. Bill has been involved in heavy equipment construction ever since, and immediately took an active interest in what we were doing. He came at a good time. The East Stone was not nearly as cooperative as his fellows had been, and we needed all the muscle and wits we could assemble. Bill was probably even more experienced with backhoes than either Steve or Russ, and many of his suggestions about gaining the maximum mechanical advantage from the little machine proved valuable. Like Russ, Bill also possessed a seemingly natural "common sense," although I do not believe it is something you must be born with and cannot develop. Ask Russ. He'll tell you that during our teenaged years I had very little (if any) common sense, but I like to think I've developed a fair amount in the thirty-plus years since. Leaving home and having to fend for oneself helps a lot in this regard. Whatever the source of his abilities, Bill was a great help for the rest of the afternoon.

Were it not an obvious male stone, according to Robin, I would have called #6 a—rhymes with *ditch*. A decent backhoe in combination with Russ's megalithic chain hitch would have set the stone right in place, no bother. Trouble was, we had to slide the stone into the socket from the side, and it got hung up on that awkward off-center point in such a way that we could not even rotate it into the back edge of the socket where it belonged. With respect to center, the stone was twisted twenty degrees or more away from the proper alignment. After ten minutes of struggling with the thing, it pivoted on its long tip and fell over toward a point midway between the center and the North Stone. *Groan.* We managed to get it upright again—this 3,000-pound stone taxed the machine to slightly beyond its limit—and we decided to jam it where it was, relax, and take stock of the situation. It didn't take me long to decide that I couldn't abide leaving it as is. After the perfect positioning of the first two stones, this one spoiled the balance of the circle. When Steve came down from the machine and had a look himself, he agreed that we should have another go.

We eased #6 out of its socket and tried again, this time using a fifty-pound boulder on the west side of the socket to prevent the stone from rotating into the position that it seemed to want to occupy. In retrospect, I believe it was the top-heavy and asymmetrical shape of the stone that was so hard to overcome. Lots of debris had fallen into the hole, too, which didn't help, and Rich cleaned the hole while the stone was held safely out of the way.

This time, #6 had no choice. Our strategic boulder in combination with levering the stone with a steel bar while lifting it back into place persuaded the stone to go where we wanted it to go. Even so, the irritating point, 32 inches down, was still taking practically all of the weight. Balanced on this point, #6 could still fall easily north or south (right or left as viewed from the center). At my command—for my back was now in pain and my brawn, limited at the best of times, wasn't much use—Rich laid some powerfully effective jammers under the base with

his sledge, and the stone began to steady. Every available pair of hands, including Darin's, began to fill and tamp the hole, while Steve patiently stayed at the controls. When the stone was fairly steady, but not completely so, we had a final look, plumbed it a little in both directions, and finished our jamming and tamping.

The East Stone had been a struggle, but, to a man—and woman—we were glad we'd taken the time to get it right. Today, it is one of the most intriging stones in the circle and when I visit this stone—for we have become friends—its appearance reminds me of the wicked queen in Disney's classic *Snow White,* when she appeared at the dwarves' cottage in her old crone disguise. The illusion would be complete if the stone's early evening silhouette included a bony hand holding forth the poisoned apple. Robin says that this is another male

Expert jammer Rich Ross tightens the East Stone.

stone, but when I think of the trouble we had standing it–him–her up, I think now that the rhyming word in question is probably *witch*.

The Dance

As soon as we had made the East Stone safe, Steve began to work on my back, including therapeutic massage and a spinal adjustment. While I was lying on the ground in pain, most of it coming from Steve's "massage," a magical event took place. Thirty or so brightly costumed revelers from the Lugnasad festival and the ElfWiz ceremony had joined together into a long meandering line of quasi-humanity. The fantastic figures entered the infant stone circle on the west side, and weaved in and out among the three stones now in position, dancing to the beat of nearby drummers. Some wore brightly colored costumes, others were dressed in black or white. One winsome young lady wore very little, but her body was covered in black mud or paint. A key player in the vows renewal ceremony had Pan-like horns sticking out of his head.

The dancers wove through the stones in snake-like fashion, sometimes bowing to the stones, for, indeed, the three already standing had definite human characteristics, and sometimes blessing them with the touch of a feather or outstretched hand. As I lay there near the center of the circle, I was reminded (between agonies) of the Merry Maidens stone circle I'd visited near Lamorna in Western Cornwall, and how (according to legend recounted by Robert Hunt during the past century) the circle came into existence:

> In the Parish of Burian are the . . . dancing stones commonly called "The Merry Maidens"; and near them are two granite pillars, named the "Pipers." One sabbath evening some of the

The author visits with his new friend, the East Stone.

Lugnasad revelers dance through the building site, giving their blessings.

thoughtless maidens of the neighbouring village, instead of attending vespers, strayed into the fields, and two evil spirits, assuming the guise of pipers, began to play some dance tunes. The young people yielded to the temptation; and forgetting the holy day, commenced dancing. The excitement increased with the exercise, and soon the music and the dance became extremely wild; when lo! a flash of lightning from the clear sky transfixed them all, the tempters and the tempted, and there in stone they stand. (From *Mermaid to Merrymaid: Journey to the Stones,* Ian Cooke, p. 142)

The sky was clear this Midsummer's Day, which, according to the legend, did not preclude a lightning strike. If these revelers turned to stone, great sadness would overcome the length and breadth of the land, but a most fantastic stone circle would be left for the centuries, and my back would cease its suffering. Perhaps Steve and I would become the mystic centerpiece, one stone upon another.

The South Stone

The revelers passed all too quickly, so quickly that Jaki was only able to shoot one or two hurried pictures just as they were departing.

The South Stone proved to be as difficult as the East stone. The problems, however, arose from a completely different set of circumstances, primarily the stone's great weight, and the fact that its modest six-foot height coupled with its wonderful flat base just begged that we erect it without benefit of a socket.

Bill, the new arrival, was into the spirit of things now, just as if he'd been with us since the previous night (so long ago). We were joined by another volunteer who provided some valuable manpower, as mine was now almost completely subservient to back spasms, and the South Stone, the biggest on site, was calculated at three tons. In addition to being the largest, #5 (for those who are scoring at home) was also one of the most interesting. When I'd first seen it the previous evening, the stone's shape reminded me of several of the larger stones at the great Avebury stone circle, particularly the "Devil's Chair."

Steve drove to the farm to get his tractor, which had an attached front-end loader, and the rest of us decided to maneuver the three-tonner into position on rollers.

Russ lifted one end of the stone with the blade of the little excavator, while we inserted wooden rollers measuring four feet long by about three inches in diameter. After more levering on fulcrums, we were able to get larger-diameter rollers into place and the stone easily pushed along the hard dry ground with the machine.

Meanwhile, I measured the base of the stone, transposed the footprint (64 inches wide by 16 inches broad) onto the clay soil, filled the shallow excavation with #1 crushed stone, and tamped the area with a 45-pound concrete block. This—the actualization, not the idea itself—was the dumbest thing I did all weekend. My lower back was feeling temporarily okay thanks to Dr. Larsen's ministrations, but I really should have asked someone else to do the tamping, for I was physically useless the following day. Anyway, I was pleased with the base I'd created.

With Russ operating the equipment and Bill supervising the ground crew, the Avebury Stone (as I thought of it) was in position to tilt up by the time I'd finished the "foundation." From experience, I knew that if we tried to tilt it from its nearly flat position, the base of the stone could very well slide south (away from the center). We drove a 5-foot-long iron bar deep into the ground at the midpoint of the 64-inch base dimension, to act as a foot against the stone's natural predilection to slide.

Then, with the blade, we began to raise the north (top) end of the South Stone.

The excavator's blade could only raise the stone a few inches, but it was enough to jam in rollers and get another purchase. Again, Russ lifted the 6,000-pound stone, and we substituted six-inch rollers for the three-inchers. Unfortunately, a small roller was stuck in place near the middle of the stone's underside, and it was in the way of tilting the stone. The stone threatened to slide south on the errant roller despite our iron bar foot. We decided that the roller had to go, and by repositioning the blade and lifting from the side, we were able to remove the problem piece. Often, moving large stones is reduced to calmly, slowly, and patiently doing the little things necessary to avoid potentially serious problems. You can't be too careful.

With the offending roller out of the way, the stone was quite easy to lift off the ground, just a few inches at a time. By the time Steve arrived back with the tractor, we'd managed to jam the full 16-inch length of a concrete block under the north end of the stone. The whole process reminded me of raising the 7-foot-high, 2-ton South Stone at Earthwood, almost a decade earlier. Coincidentally, the South Stones of both the Earthwood and the Stone Mountain Circles are the heaviest, and both stand on their wonderful flat bases, without sockets.

Once the top of the stone was 16 inches off the ground, the base began to take some serious weight. The iron bar held fast, although it was beginning to bend under the strain. Russ now turned the machine around and used the backhoe bucket for further lifting, while from his position on the south side, Steve used the tractor's front-end loader to resist the stone's tendency to tumble over completely once it reached its apex. Steve and Russ coordinated beautifully, while the men who still had useful backs steadied the stone manually. I stayed clear, but kept a careful eye on the proceedings. Someone has to do it.

The South Stone, #6, the Avebury Stone, the Big 'Un—for, like Strider, he was known by many names—stood nearly vertical, supported by the machinery. Using the 4-foot level (easy to do on this very regular stone) I called for jammers on one side or the other until the monolith stood as plumb as was humanly possibly. Gravity is the universal bonding agent and the stone seemed solid as soon as it was plumb. Nevertheless, we kept it supported mechanically until we'd driven in a couple of dozen small flat jammers around the base to tighten the stone and remove any tendency toward instability. The jamming complete, we carefully released the machinery. I tested the stone manually and found that the inertia of its three tons was powerful indeed. The stone had no more tendency to move under my feeble efforts than the Devil's Chair itself.

Holly takes a break at the South Stone, a miniature Devil's Chair.

Finishing Up for the Day

Our main work done for the day, we allowed ourselves the luxury of admiring the fruit of our efforts. The four primary compass point stones were now in place, an impressive sight. The uninitiated might say that we had only built a square so far, a Scottish "four-poster," but bumper-sticker wisdom reminds us that "Reality is for those without imagination." All of us could see the circle in the square.

Nearby, the Wickerman was being built. The music and the drums continued, building slowly toward the evening's celebration. Concerned with safety (for who could say what craziness might possess the revelers this Midsummer's Night?), we propped the freestanding South Stone with the four-foot rollers and the iron bar, surrounded the stone with logs, and wrapped it with yellow safety ribbon that, somehow, had magically appeared on site. In the meantime, Russ, our designated safety director, smoothed the other three stone sites by backblading with the excavator.

THE WICKERMAN AND THE NORTH STAR

There were other builders on site besides Megalithic Construction Company. Bill Saragossa (who had become the Wickerman *man* over the years) and his companions were putting the finishing touches on the Wickerman, a 24-foot skeleton built primarily from an upside-down tree trunk. Two large branches formed the legs, and the bones of the articulated arms were composed of other large sticks tied crosswise to the fellow's main spinal column.

I wondered about the relationship between the Wickerman and the Midsummer Festival of Lugnasad. Who better to ask than the Doctors Larsen, experts in symbolic study. Robin is particularly strong in Celtic and Arthurian myth and legend.

"The Wickerman is the opposite of the Green Man, his twin, his tanist, the one who dies while the Green Man lives," explained Robin. "He is the chaff, the stubble in the fields, the husk. He offers the chance of casting away that which is outworn."

It was near dusk and the Wickerman was in position to be raised. A couple of dozen brightly festooned helpers were on hand for the task, and I found it impossible to remain detached, busybody that I am. I told myself that I would not lift in any circumstance and figured that I might be able to contribute in some other way.

The potential problem with raising the Wickerman was almost identical to one we had to consider with raising the South Stone. As the Man was lifted by his "head," the legs could slip out. I suggested driving stakes into the ground to foot each leg. Darin ran to get our sledgehammer from the stone circle, and stakes were driven in at each foot. Using long sticks as props both for lifting and for resisting overturn, a dozen lifters soon had the Man standing on his peg legs. He was braced in position with long wooden props and gradually the human support crew removed themselves one by one until the Wickerman was standing safely to Bill's satisfaction.

The next couple of hours were spent in tying straw bales and loose straw all over the Man's frame, fleshing him out for his one and only performance. It was like fattening a turkey before slaughter. The music, provided by Iabas Arte Brasil, Ritual Motion, and the Sister Adsun drummers, became more intoxicating. Wine, beer, and strange aromas all helped to paint the scene.

During the final preparations of the Wickerman, as the clear sky began to darken, I positioned myself to observe the first appearance of the North Star. I climbed over the safety barrier and leaned gently against the north side of the South Stone. I centered myself by reaching out with each hand to find the east and west sides of the roughly five-foot-wide stone. I knew exactly where Polaris should be, based upon years of observation at Earthwood. The star's declination at New Paltz would be only 3 degrees lower than at Earthwood, so it should

be easy to spot. The night appeared to be clear—no clouds darkened the still dimly lit northwest sky—although no star was yet visible. I focussed on the magic spot and waited. And wondered. How far off would we be with our alignment, just 16 megalithic yards or 43.5 feet in length? A small error, a couple of inches, would be critical with such a short axis.

Then I saw it . . . or thought I saw it. As I did not have my glasses with me, unfortunately, I made a little telescope with my hand. You can sharpen the focus on a point by decreasing the "aperture" of your slightly unclenched fist. *Definitely*. It was the Pole Star! (If not, we were really in trouble, for there is no other bright star anywhere near Polaris.) I sighted downward and, there, clear as could be, was the point of the North Stone, shining in the artificial light from the nearby music stage. Up and down I sighted, and could not perceive an error. Ecstatic, I began to call anyone who happened to be in the area to have a look. I pulled total strangers over to the stone, positioned them, and explained the alignment. It never occurred to me that they might think I was a few stones short of a circle, and, in the magical spirit of Lugnasad, none of them did. Everyone seemed to enjoy the apparition as much as I. After a while, I decided to get a more accurate reading than by simply dropping my head.

An hour earlier, I'd met Jim Carman, Steve's friend who'd delivered the stones to the site with his big log skidder. He had come for the Midsummer celebration. Now, Jim, Merlin Larsen, and I put the alignment to the test with the plumb bubble of a 4-foot level and an actual plumb bob on a string. Of the two, the plumb bob worked better, but it was hard to focus simultaneously on both the nearby string and Polaris, many light-years distant. As near as any of us could tell, the alignment was spot on! For the next hour or so, lots of people took turns experiencing the incoming starlight, including Jaki and Darin, Steve and Robin, Russ and Rich, and ElfWiz. This narrow beam of light had left Polaris seven hundred years ago, about the time that Marco Polo left China on his way home. We'd lined up the stones for this beam of light to land on Earth in just the last instant of its galactic journey.

The laid-back expression of wonder in the parlance of the 1960s would have been appropriate in the here and now of this Lugnasad celebration: Far out!

Meanwhile, back at the firepit, the Wickerman was nearing completion. Bill Saragossa outdid himself in creativity this time, and, at his signal, the awaited event commenced. Assistants lit the straw hands at the end of articulated arms, and they swung down in an arc to ignite the legs. Within thirty sec-

The Wickerman, Lugnasad at the Stone Mountain Circle.

onds the Wickerman was in full flame, spectacular in life and death. The heat drove the assemblage back several megalithic yards from the immolated figure. The drums kept up their intoxicating beat and people danced in circles around the fire. Sinewy maidens undulated to the music in exotic movement, as if the fire were not already hot enough. Time ceased to exist, but I know that the sleeping portion of the night was severely curtailed.

CONSTRUCTION: DAY TWO

Most of our stone circle construction crew enjoyed a wonderful sweat at the Larsen's cordwood masonry sauna on Sunday morning. I'd been keen to try this sauna, having been instrumental in its design and construction a couple of years earlier. Happily, it performed perfectly, and we were primed for Robin's excellent breakfast.

We were back at the circle by 11 A.M. to meet Jim Carman and his log skidder. While waiting for Jim, Russ and I washed the base of the South Stone and mixed up a bag of rather wet Sakrete concrete mix and forced it under the stone between its many jammers. We followed with two more bags of Sakrete, mixed much more stiffly this time, and created a 6-inch-wide sloped and smoothened concrete skirt to broaden the base somewhat and shed water away from the stone. Later, when the concrete was stiff, I shoveled a few inches of crushed clay onto the skirt, providing a natural waterproofing cap over the area. Rainwater would now run away from the stone in all directions, greatly diminishing the possibility of frost heaving this unsocketed stone. Clay expands upon contact with water, and the resulting hydrostatic pressure within the clay itself prevents further penetration of moisture.

Modern underground houses are often waterproofed with a refined clay covering, but the technique was known to the ancients. The Maes Howe burial chamber in Orkney was covered with clay after its construction, and the clay kept the chamber fairly dry until the Vikings broke in.

The Sunrise Stone

The next stone to stand was #3, the Sunrise Stone. Steve had taken two sunrise observations from the center of the circle, one on the Summer Solstice, June 21st, and one on August 1st, Lugnasad, the Celtic Midsummer. Wooden stakes marked these alignments at the perimeter of the circle, 8 MY from the center. The two stakes were about thirty-six inches apart; the sunrise hadn't changed all that much in the past six weeks. But Lugnasad, like Imbolc on February 1st, marks the beginning of the two quarters of the year in which rapid change occurs, both in daylight hours and in the azimuths of sunrises and sunsets. Conversely, Samhain (No-vember 1st) and Beltane (May 1st) begin quarters of the year in which change is slow, and daylight hours are fairly stable: long in summer, short in winter.

One end of the Sunrise Stone was squarish, about thirty-seven inches in breadth, and the other end had an off-center point. We decided that if we exposed the broad end and buried the point in a socket, this single stone might be able to mark both sunrise alignments: summer solstice and Lugnasad. If we were careful—and a little lucky—the sun would rise over the left-hand edge of the stone on June 21, and over the right-hand edge on August 1. As #3 was only 6'6" in total length, and nearly two feet would need to be buried to support the greater weight of the squared end, only 4'6" of the stone would stand above ground.

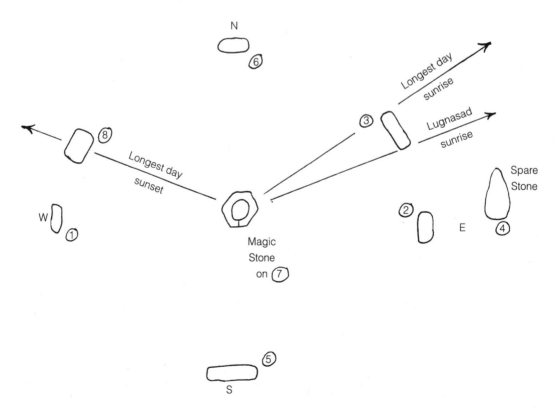

"As built" plan of the Stone Mountain Circle, showing solar alignments.

A Solar Aside

Even before I knew of the Celtic *cross-quarter* holidays of transition, I'd observed them in a personal way. I knew that early November marked the beginning of three months of short daylight hours, while early May began three months of long days and short nights. The farther from the equator, of course, the more pronounced the difference. In the North of Scotland, the year seemed to consist of three months of almost constant light, three months of darkness or very dim light, and two three-month periods of very rapid change.

More recently, I've discovered a very practical application of the Celtic holidays. Earthwood is not connected to commercial power lines. Most of our electrical energy comes from photovoltaic (PV) modules, which charge batteries. The PV array operates most efficiently when the sun's rays are perpendicular to the plane of the panels. The use of an expensive tracking device is not justified at our location in northern New York, so we manually change the angle of the panels four times a year to maximize our output. The best four days of the year to *change* the angle of the panel arrays are Imbolc, Beltane, Lugnasad, and Samhain! But the actual angle to *set* the panels is taken from the noontime sun at the very next solar event: the vernal equinox, summer solstice, autumnal equinox, and winter solstice, respectively. How does it work? The cross-quarter days anticipate the average of the next quarter, be it a quarter of rapid change or a quarter of slow change, while the quarter days (solstices and equinoxes) fall in the middle of these anticipated periods, so we use them—the quarter days—to establish the panel angle. Neat, huh? Our solar panels are right next to our stone circle at Earthwood, and I get a charge out of modern technology standing in juxtaposition with 5,000-year-old technology. So do our batteries.

On Sunday, we had the advantage of Jim's skidder, which gave us more options in maneuvering and lifting the heavy stones. Steve was getting very good at digging just the right size and shape of sockets with the little backhoe, the perfect tool for the job. Throughout the project, we were fortunate that we never had to excavate ground rocks any larger than a bread loaf.

The Sunrise Stone was already close to its final location, so all we had to do was rotate it and position its pointed bottom over the edge of the hole prior to erection. We tilted it up by lifting with the backhoe while resisting forward movement with Jim's skidder. Number 3 slid nicely into its socket, although its square end was not standing perpendicular to the ground. As with the Witch of the East, the Sunrise Stone was hung up on its point, almost twenty-four inches into the ground. Steve repositioned the backhoe to the south side of the stone and drew it toward him while we jammed the clear space that formed in the bottom of the socket. The net effect was that we also raised the stone three or four inches, and we wound up with 4'10" showing above grade, 4 inches more than I'd hoped for, yet the stone was solidly jammed and in no danger of loosening in its socket. Now that it was straight, it looked great and made a fine companion to the East Stone. Only a year's time and clear sunrise skies would tell if our alignments were correct.

Steve and Jim went off to fetch the magic center stone with the skidder, while Russ backfilled the Sunrise Stone and tidied the site.

Of Dolmens and Mushrooms

When Jim and his skidder arrived with the Magic Stone in tow, all attention was diverted to the new arrival, such was her charisma. Throughout the weekend, Steve and I had carried on a running conversation about how the Magic Stone should be featured. We agreed that it had to be in the center, the female object of attention for all of these handsome male standing stones. But Steve felt that it should be supported by three or four base stones, like a dolmen. His idea had aesthetic and historical merit. In North Salem, New York, just forty-five miles to the southeast of Stone Mountain Farm, there stands a prototype, which could very well date from a similar age as our mysterious sacrificial stone. Chapter 16 describes this dolmen in detail, and includes a picture.

I would have loved to have actualized Steve's plan, and, had we an extra couple of days in hand, I might have been less vociferous about what I saw as three very real practical problems with building a true dolmen. (1) We didn't have the necessary pegstones on site, nor had we been able to find any good candidates on our two-hour tour of the farm; (2) it would be extremely difficult to level the Magic Stone on three pegstones, even if we did have them on site; and (3) the centerpiece would not be finished this weekend, for we would run out of time finding, hauling, and socketing the three pegstones.

My suggestion was that we use as a base one of the two nondesignated stones that would be "left over" when the six planned standing stones were complete. The one I had in mind was of similar shape to the Magic Stone, thicker but slightly smaller in diameter. I could envision the Magic Stone supported on this base stone, looking something like a mushroom. From our experience in creating two mushroom-shaped megalithic stone tables at our "Mushwood" Cottage at Chateaugay Lake, I knew that it would be very easy, using jammers and a small hammer, to level the Magic Stone on such a solid base.

Without making a final decision, we decided to position the Magic Stone upon the base stone and see how it looked. Jim, in his element, used the skidder to ease the Magic Stone carefully up onto the base, while Russ stood by with the backhoe bucket to assist as necessary. In no time, the two stones were joined, one over the other, and we shut off all the equipment to discuss the problem in peace. The final decision, of course, was Steve's. He could say that we should wait and, sometime in the future, reposition the Magic Stone dolmen-fashion on legs. I crushed some dry clay into the outlet of the shallow round pond that had been carved into the stone who knows how long ago, and Russ sprayed water into the depression. The stone was already nearly level, although it was still positioned outside of the circle, 10 MY to the southwest. The clay expanded quickly with the water, and the resulting hydrostatic pressure effectively plugged the outlet. Steve studied the effect carefully. Then he said, quietly but with no hint of doubt, "I like it. It *does* look like a mushroom. Let's go with it." I was glad to hear his words, because it meant that we would be able to bring this circle to at least a preliminary stage of completion, and within the allotted time, for all of these people, and the equipment, would soon be heading off in several different directions.

We still had need of the center point to align the Sunset Stone, so we left the Magic Mushroom for a while, and it instantly became an object of great interest to all who happened by.

The Sunset Stone

The Sunset Stone, #8, was one of the loveliest on site, and perfectly shaped to allow us to reveal its full symmetrical beauty. Only 6'1" in length, #8 possessed an almost perfect trapezoidal base, at right angles to its longitudinal axis. In other words, this was the perfect candidate for a freestanding stone, if ever there was one. That the base was both broad *and* wide made it even better than the South Stone, which was standing very happily indeed. Later, accurate measure and calculation revealed the weight of the stone to be around 2,800 pounds, or 1.4 tons.

While the lads were away fetching the Magic Stone, I busied myself in transposing the footprint of the Sunset Stone's base onto the ground, and with help from Jaki and others removing the or-

(above) *Pushing the Magic Stone into place with a skidder.*

(right) *Russ runs water around the runnels of the Magic Stone. (Richard Ross, photo.)*

ganic material, spreading a few buckets of #1 blue-stone onto the little depression, and tamping it into a flat, level, compacted table, just about at original grade.

Old #8, though it was laying right next to its correct position in the circle, did not take it upon itself to stand up on its own during our absence, but we were now well equipped to do the job without too much manual effort. Jim turned the stone so that its leading edge was adjacent to the foundation we'd prepared, and then proceeded to lift it into position with his power winch and a chain, resisting the base with the chassis of the heavy machine itself. Unfortunately, the chain slipped and the stone went over toward the skidder. Fortunately, the chain was still wrapped around the stone. For the second attempt, we decided we'd better use the backhoe to steady the stone and prevent a repeat performance. Jim was able to rotate the stone back

"I like it," said Steve. "It does look like a mushroom."

into approximately the correct position and Russ pulled it upright with the backhoe bucket.

The stone was a few inches away from being perfectly positioned on the foundation we'd prepared, but the machines could not easily slide the monolith into its correct position without digging into the ground and destroying the crushed stone foundation. Again, we missed not having a large backhoe on site.

Years earlier, Russ and I had had a conversation about how the stone heads at Easter Island had been brought from the quarries where they were carved to their sites, sometimes miles away. When asked by adventurer Thor Heyerdahl how the stones were moved, natives were adamant that the stones "walked" to the site. Russ had read an article that speculated how, indeed, the stones might have accomplished this feat (without feet), assisted by plenty of manpower. According to this article, the stone heads were stood up near to the quarry. A long wooden beam was tied horizontally across the middle of the stone, and, with several people at each

end of the beam, the stone could be made to walk. First, the people on the right would raise their end and "walk" the right-hand edge of the stone's base forward, with the mechanical advantage derived from the length of the beam. As the stone came down from this first "step" it would be easy for the crew on the other side to lift up and rock the left edge forward. Once set in motion, the weight of the stone would really do most of the work, like a child's walking toy. A second horizontal beam might be tied perpendicular to the first, so that additional helpers could assure that the stone head did not fall forward or back.

Remembering our conversation of years ago, I decided to tie a 16-foot-long 2 x 8 to the waist of our Sunset Stone to see if we could walk it into position easily and without mishap. Russ chained it in place. The plan worked perfectly! With one or two people at each end of the beam, we encouraged the stone to take its first tiny steps, and in less than five minutes we were able to position it exactly on target. Instead of a safety beam running the other way, the top of the stone was prevented from tipping toward the opposite axis by the backhoe's bucket. I've no doubt that, with practice and a little time, we could have walked this stone to downtown New Paltz.

I'd stretched a rope along Steve's original alignment for the summer sunset, and tacked it down in a couple of spots, as his original stake would be lost during construction. Using the bucket to tilt #8, and plenty of good wedge-shaped jammers, we managed to straighten the stone and get its point right on line with the rope. As with the Sunrise Stone alignment, it would be nearly eleven months before we would find out how close we'd really be to marking the solstice sunset. But if the stone is a little off, it will be an easy matter to walk it into perfect alignment.

The One That Got Away

Steve and Jim took the skidder back to the Grotto to fetch an 11-foot-long menhir that weighed, by estimate, some 4½ tons. While they were away, we skirted the base of the Sunset Stone with concrete as we had with the South Stone: we forced wet concrete between the jammers, and built up a sloped skirt with stiff concrete to shed water away from the stone. Later, Robin drew a dragon in the smooth concrete and included a date, Lugnasad 1997, as an aid to future archaeologists.

Considerable time went by and the skidder did not appear. Russ busied himself cleaning up the site with the blade of the excavator, while Richard and I marked out an 8-foot-diameter circle about the center pipe, the point from which all the 8 MY

This beam, chained to the stone, enabled the team to "walk" the Sunset Stone into place. (Richard Ross, photo.)

radial measurements had been taken. It was a quick and easy task. We simply butted one end of my 4-foot level to Steve's pipe, and placed a hockey-puck-sized stone at the other end of the level, about twenty-four stones in all, roughly every 15° around the circle. Russ, I could tell, was wondering what we were doing, and it must have troubled him even more when I connected all the pucks with fine clay powder, sifted through my hands, until a very clear and tidy circle was evident in the grass.

About the time that we'd truly run out of work, Steve and Jim returned with the skidder, but without the big stone, which was to have been placed as an outlier on the summer sunset alignment. They'd managed to move it about half way to site, mostly along a road, but the uphill journey across a soft

The Magic Stone is an effective reflecting pool.

field proved too much even for Jim's powerful skidder. The beautiful stone lies along the roadside today, a sitting stone for tired pedestrians instead of a spectacular standing stone. Maybe someday . . .

The Magic Center Stone

The Center Stone was perched on a slightly smaller base stone, and Jim decided that he could slide them as a unit to the center of the circle. We tied a chain around the base stone and brought the free end to Jim's powerful winch. The center unit was actually winched into place. Now I could see the light go on in Russ's head as he realized why I'd taken the time to build a pretty 8-foot-diameter circle. He smiled. With the center pipe removed, the only way to place this unit in the center would be to keep it's edge equidistant from the circle that Rich and I had made. Jim pulled the unit slowly to the center, and we told him to stop. The two-tiered mushroom certainly appeared to be in the center, but there was only one way to be sure.

We ran a rope from the center of the East Stone to the center of the West Stone, two of us holding the rope at final table height. Similarly, two others stretched a chalkline from the center of the wide South Stone to the center of the North Stone. The intersection should have been in the middle of the circular depression carved onto the exposed surface of the Magic Stone. It was very close, and we adjusted it a few inches with levers and the skidder's winch. At the same time, we aligned the outlet of the little pool to fall along the North-South axis.

Leveling the stone was a piece of cake. After all, the stone had its own built-in bubble level. We simply filled the round cavity with water and drove jammers between the Magic Stone and its base until the top surface was level in every direction. In point of fact, the depression had not been perfectly carved. Perhaps it had never been finished to the craftsman's satisfaction. The surface is slightly mounded within the circle's perimeter as described by the runnels, so that it is impossible to make a complete round mirror. Too bad. Nevertheless, the stone still delivers its magic, and still makes a wonderful reflecting pool.

We barely had time to clear up the site before it rained. If the abandoned 11-foot monolith *had* made it to the site, we would have had a nasty time trying to erect the stone in the slippery clay. *Just as well,* I thought.

Jennie's Funeral: a Celebration of Life

The rain didn't last. With our major work done for the day, we stood around the pool in wonder, congratulated ourselves on what we'd created, and took in the general magic of the place. Jaki and I took photos. About this time we learned of the death of Jennie, a 28-year-old cart pony with whom Steve and Robin's kids had grown up. Jennie was old and had been sick. That her death occurred during the two-day construction of the circle, and, as a carthorse, at the time of the charioteer Lugh's feast, was an incredibly moving experience for the Larsen family and others on the Farm, and, at Jennie's ensuing burial within the stone circle, the emotion transferred to all in attendance, including the Roy family.

When I learned that the plan was to bury the pony in the stone circle, I was ambivalent. In some ways, spiritually, mystically, emotionally, I could understand why some of the Larsen family felt that it was the right thing to do. As usual, my misgivings were strictly of a practical nature. What would happen when the earth settled months after burying? I let Steve know of my concerns, but he felt that it was a decision best left to those closest to Jennie: Robin, daughter Gwyneth, and Merlin.

The sun came out after the rainstorm—Robin said that there was always a storm when one of the horses died—and we all returned to the stone circle. Merlin, using the little excavator, had dug a grave for Jennie, midway between the magical Center Stone and North Stone. Jennie was lowered into her resting place, and she fit Merlin's excavation perfectly, her head placed to the north. The ceremony at the graveside was beautiful and moving. Everyone who knew Jennie recounted some episode of her life. Wild flowers, grains, and greens were laid in the grave. Steve took a little water out of the Center Stone's pool and sprinkled it into the grave, saying, "So that you are never thirsty." We all threw a shovel of earth into the hole, and departed, except for Merlin and his sister Gwyn, who stayed on together for some time.

The next morning, only a small mound was in evidence where Jennie had been buried. The mound would settle over time, but would probably remain a few inches above grade, good for shedding water away. In retrospect, I realized that the decision to bury Jennie here was the right one. Sometimes I'm too concerned with matters practical.

POSTSCRIPT

It had been an amazing two days. We'd built a wonderful stone circle with six standing stones and a unique centerpiece. We'd witnessed an example of the strength of love thanks to ElfWiz. We'd enjoyed our first-ever feast of Lugnasad, and knew it wouldn't be the last. We witnessed a moving funeral for a much-loved four-legged member of the Stone Mountain Farm family, a send-off of joy, and of a kind not often enough afforded to the two-legged species. Most of all, we'd made new friends and become reacquainted with old ones. There's nothing like a stone circle raising to bring out good people, and the best in them at that.

Two weeks later we returned to Stone Mountain Farm, Jaki and I to conduct a cordwood masonry workshop and Darin to attend a five-day adventure theater camp. I called my old friend Enrique Noguera who lived in nearby Woodstock, and we arranged that he would come to see the stone circle on Monday night. Enrique conducts personal tours, which he calls Megalithic Journeys, to many tantalizing ancient sites in southern New York, including stone slab chambers, dolmens, and ancient inscribed stones. I knew that Enrique would be interested in our work, and particularly in the magic Center Stone. Enrique invited along John Friedman, who had joined Enrique, my son Rohan, and myself on an expedition nine years earlier to document the rising winter sun at the Eye Chamber, described in chapter 16.

Just before sundown, Robin and Steve, the Roys, Anne McClellan, Richard Ross, and several others assembled at the stone circle to get the views of Enrique and John Friedman, whose interest in ancient stones had intensified since Rohan and I had met him in 1988. Enrique remained thoughtful, while John was caught up in the wonder of the

Enrique Noguera aligns himself to the north and contemplates the enigmatic Center Stone.

Center Stone. Immediately, John pronounced that the stone's purpose was sacrificial. He was reminded of the so-called sacrificial stone at the enigmatic Mystery Hill site in North Salem, New Hampshire, and told us that the stone before us was one of approximately fifteen similar stones he has seen, one of which was reliably dated to about 1500 B.C. He also pointed out chisel marks that I had missed, but which came alive in the long rays of the setting sun. (Months later I learned of an alternate theory that might explain the stone's original purpose and use. Bob Goodby, assistant professor of anthropology at the University of New Hampshire, says that the "sacrificial stone" at Mystery Hill looked to him like a garden-variety lye stone, used for leaching lye from ashes to make soap.)

John asked Steve to take him to where the stone was found, to see what he could learn from the stone's provenance. Enrique stayed back to speak with Jaki and me, as his interest in the project was a little different from John's. When I asked him his first impressions of this "sacrificial stone," his reply was intriguing: "I see the stone physiologically." Pressed further, he told us that he saw the stone as a womb with a vagina, a metaphor for the Earth Mother. To Enrique, the stone was here and now, not thirty-five hundred years in the past. His quiet observations were poignant, as I'd come to expect over the years from my old friend.

The others returned from their examination of the place where the stone had been found, and I showed John a picture I'd taken of the stone in situ. None of it made clear sense to John, but he felt strongly that more investigation should be done in the area, particularly at the top of the hill, where, he thought, there might be evidence of an ancient temple.

Since its construction, the Stone Mountain Stone Circle has become a regular part of the outdoor social events at the Farm, with observations at the various important points of the calendar. The circle will grow in years to come, with new stones added for

Midsummer revelry at Stone Mountain Circle.

equinox and winter solstice alignments, and perhaps some of the other important Celtic holidays.

If pride is indeed a sin, then I am guilty, for I do take pride in my part in helping to bring the circle into existence, recognizing completely the wondrous and, at times, almost spiritual teamwork that it took build it. From Steve and Robin's conception and selection of stones, Jim Carman's hauling of them to site, Russ Keenan's and Bill Klopping's expertise and experience in maneuvering the stones, Richard Ross's wonderful aptitude, Jaki's constant support and help, and the positive atmosphere at the Center for Symbolic Studies, the project was, simply, magical.

The next part leaves my stone work aside for a while, and invites you to join me on a megalithic journey to meet some of the world's most fascinating Stone People.

THE STONE PEOPLE

– CHAPTER 4 –

Bill Cohea and Columcille

*There are two kinds of energy at work here, the earth energy of
sacred spaces, and the human energy that people bring with them
to the site. When the two meet, step back and watch for surprises.*

—BILL COHEA

The Earthwood and Stone Mountain Circles are composed of stones of a ton or two. This part of the book tells of people working with stones of up to forty tons, and how they do it. But the Rev. William H. Cohea, Jr. the hero of this chapter, is even more concerned with the *why* than the *how*, perhaps because his own motivation story is so unusual.

A VISION QUEST

Bill Cohea has built a collection of megalithic monuments at the Columcille center in North Bangor, Pennsylvania. The 17-acre site may very well be the largest and most beautiful group of standing stones created by the new Stone People. We'll get to it soon enough, but first it is important to know how it all started, literally, with a dream.

Life begins at forty, goes the adage; in Bill's case, a life-changing event at that age really did foretell a new beginning. It's not that his life had been lacking in variety and interest up until then; a graduate of Northwestern University in Illinois, Bill attended the seminary at Princeton and was ordained a Pres-

byterian minister. He traveled across the country ministering to college students and did a stint as a pastor in Iowa. He founded a layman's academy in New Jersey, started the Chicago Center for Urban Projects, and, later, established an alternative high school. A busy guy.

Iona

In 1967, Bill returned to the mystical island of Iona, just off the west coast of Scotland. He'd been there previously to attend religious conferences, and found himself becoming ever more enchanted with the island on a number of levels: its physical presence, its spiritual history, and its haunting atmosphere. Iona is where St. Columba (Colum Cille) and twelve companions started their monastery in A.D. 563, bringing Christianity from Ireland to Scotland. But, according to legend and fragments of Celtic oral tradition, the island was considered holy long before Columba's time. In the past it has also been named Ioua (after a moon goddess) and Hu (or I or Hy Brassail), for the enchanted island of the Irish otherworld. Although it is only five square miles in size, the island was said to have once had no less than 360 standing stones. Aubrey Burl relates that in A.D. 1560 the Synod of Argyll found it necessary to destroy a stone circle on Iona under whose twelve stones men had been buried alive because people continued to worship at it (Burl, *The Stone Circles of the British Isles,* p. 13).

Dreams

On a warm summer's afternoon in 1967, Bill was hill walking on Iona in the vicinity of Dun I (pronounced *doon-ee*), its highest point. He fell asleep in the soft turf and was visited in a dream by a strange host of characters. In an interview, Bill told me, "I saw a circle of upright stones coming closer and closer to me. And behind them were more and more and more stones! As they got closer, I became

Bill Cohea and Duke at Columcille.

afraid. They were walking, but faceless. They were just stones, but some of them wore ragged clothes. And when they got right up to me, and looked down at me, I was terrified—for a moment. They looked at me and all I could feel was great love from them and I felt great love for them. And there was no fear. It was a new experience, this kind of love that I felt. And I woke up."

The startling dream stayed in the inner reaches of Bill's consciousness. "For years I would look at strange stones, to see if I could see something that looked a little bit like them," he told me. "Then, in 1977, on Iona, I had another dream, a dream that I could get up and walk through. It was my old friends the stones again, and my whole world blew up. I encountered an energy of being in a form that I'd never known before. I was horrified. I asked myself, how do you converse with stones? They're not human. And every question I asked—even if I heard the answer, I can't remember a single one. Now if you're fifty years old, with my background, well, this just blows you away. I woke up, walked to St. Oran's Chapel and thought about the experience for a couple of hours. Then I went back to the hotel where I was staying and ordered a triple scotch!"

Bill returned to Iona twice during 1978, seeking answers to his mysterious dreams. "My real concern in 1978 and 1979 was that I wanted to join them, the stones, in their work, whatever it was. But they did not let me join them, although I asked persistently. 'No,' they said, *'Our work is your work and your work is our work.'* And it's taken me nineteen years to realize what they meant."

The Question Why

When I asked Bill why he built the wonderful "Megalithic Park" at Columcille, he replied, "I was just prompted to do it. Outside of the dream, it didn't make any sense at all. I'd be walking around the living room saying 'I haven't got the money to do this.' And somebody said, 'Since when has lack of money kept you from doing anything?' And I laughed because I hear this voice talking in my head. But it was true. Whatever I've decided to do, I've always figured out a way of doing it. Just like you do, Rob. You know, we must all be crazy. I feel possessed, like that guy in the movie—was it *Close Encounters of the Third Kind*?—who has to make this giant model of that mountain, the Devil's Tower. You must have felt something like this."

I agreed that I, too, had this strong compulsion for standing stones upright. If I see a long stone lying in the woods or by the side of a road, I immediately try to envision it standing. Maybe my—obsession?—is not quite as strong as Bill's, judging by the scale of our respective projects.

Columcille

In 1975, Bill bought a house and twenty acres of wooded land right next to the Kirkridge retreat center in Upper Mount Bethel, Pennsylvania. He had attended this retreat center earlier in his life's spiritual quest, and fell in love with the pristine beauty of the mostly wooded land. So Bill had the place to create any vision that he—or the mysterious walking stones of Iona—might come up with. After his strong dream visions of 1977, the idea began to germinate that he actually had the physical site to create a most extraordinary spiritual center.

A couple of years of difficult questions passed, with answers slow in coming. One day Bill returned from a particularly introspective visit to Iona with a fresh spirit, like Columba arriving on Iona fourteen centuries earlier. But Bill didn't start right in with standing stones. In 1978, with Fred Lindkvist, he started Columcille, Inc. as a not-for-profit religious education research center. Like Columba, Bill was inspired to first build a chapel. In the spring of 1979, in partnership with Fred Lindkvist—who has shared the vision, the work, and the mortgage with Bill for the past twenty years—and the "Friends of Columcille," work commenced on the first structure on the undeveloped land, the St. Columba Chapel. The build-

ing was roofed before Thanksgiving of the same year, a remarkable feat, in my view. Today, the beautiful six-sided chapel of heavy stone masonry exudes a spirituality uncommon even in religious buildings. Soft light enters from a six-sided skylight at the building's peak and from five arched windows of yellow and green bubble glass, which, Bill tells me, came to America on the last clipper ship from Britain. At the center of the chapel is a large round naturally polished black rock, and it is this rock upon which Columcille is grounded to the earth's energy. Bill envisioned Columcille as "a place and space for tired sinners and reluctant saints, a playground of myth and mystery, dedicated to the spirit's work." About three thousand travelers on the path of the spirit visit the center each year, and it has clearly equalled or exceeded Bill's vision. No religion is imposed upon the place. The chapel and, indeed, Columcille itself, derive inspiration from Christian, Celtic, Native American, and other spiritual traditions.

It is important to explain what Columcille is: a place of solitude and spiritual regeneration; a focus of research into myth and mystery; a playground for the soul and spirit—and what it is not: a megalithic Disneyland. I am not a religious person, although, like Bill, I have conducted a lifelong spiritual odyssey, with fluctuating degrees of success. But it is clear to me, each time I visit Columcille, that if anyone has ever successfully drawn the natural earth energy from a site, it is Bill. Standing stones, perhaps, are conduits to the pulse of Gaia. I don't pretend to know the answers, but the question of earth energy seems to come up whenever ancient or modern stone circles are discussed. If you are not receptive to that line of thought, don't worry. Stone circles and their like operate on different levels for different people. And keep in mind that starting an investigation with certainties is unlikely to lead to truth. At Columcille, people walk among the stones and speak reverently in subdued voices. They meditate in the special places hidden along woodland pathways. The way is open to them, as it was for the

St. Columba's Chapel at Columcille.

ancient Irish monks, to "journey to the thin places, where the veil parts and dreams, visions, saints, druids, angels, and other beings become companions and spiritual mentors."

A TOUR THROUGH THE MEGALITHIC PARK

Along with Steve and Robin Larsen of Stone Mountain Farm, Jaki and I visited Columcille on the weekend of our 25th wedding anniversary in late November of 1997. I had met Bill a year earlier and taken a quick tour, and I was keen to share the place with Jaki and my friends, as well as to see it again with a fresh eye. After introductions at Bill's house around 2 P.M., we realized that we'd better get outside soon, as this gray November day would not have many hours of light remaining.

The Stone Circle

Accompanied by Bill, Fred Lindkvist, and Brad Colby, a personable and knowledgable young member of the Columcille staff, we made our way past the pond toward the stone circle, stopping first at Brad's little office/house to get the latest Columcille map. As we approached the circle itself, Steve and Robin could see similarities to the Stone Mountain Circle. Both the size of the circle and the stones themselves were in scale with their own circle, and this one also had a center stone, not a common occurrence anywhere. Bill told us a little about the circle's background and construction.

In 1981 Bill's dream spirits began to manifest themselves on the Columcille landscape. The first standing stones to be erected were the inner eight of the stone circle, and the one at its center. Except for the boulderlike recumbent at the center, the stones of the inner ring are modest—for Columcille—at about four to five feet high. The generous girth of the stones, however, and the fact that Bill tends to socket them pretty well into the ground, suggests stone weights in the two- to four-ton range. Bill was just getting warmed up.

Within two or three years, Bill had completed the outer ring, or Guardian Stones. The later stones were, in general, taller and heavier, and the ring's original sixty-foot diameter only increased by a few feet by the addition of the Guardians. The outer ring is offset from the inner one, giving the effect that the plan of the circle might have been cut out with a pair of pinking shears. Not able to discern a measured or symmetrical configuration, I asked Bill if the Guardians had been placed according to any plan. His answer was vintage Cohea. "They were set where I felt they ought to be set," said the reverend matter-of-factly. So many of the stone circle builders, ancient and modern, locked their rings into a defined pattern or a particular set of cosmic alignments. The Columcille circle was balanced, in form, color, and proportion, but it was the kind of asymmetrical balance that you would expect to find in a Zen stone garden, not the kind that illustrates a geometry textbook. I sense, and Bill has never told me so, that the circle built itself organically.

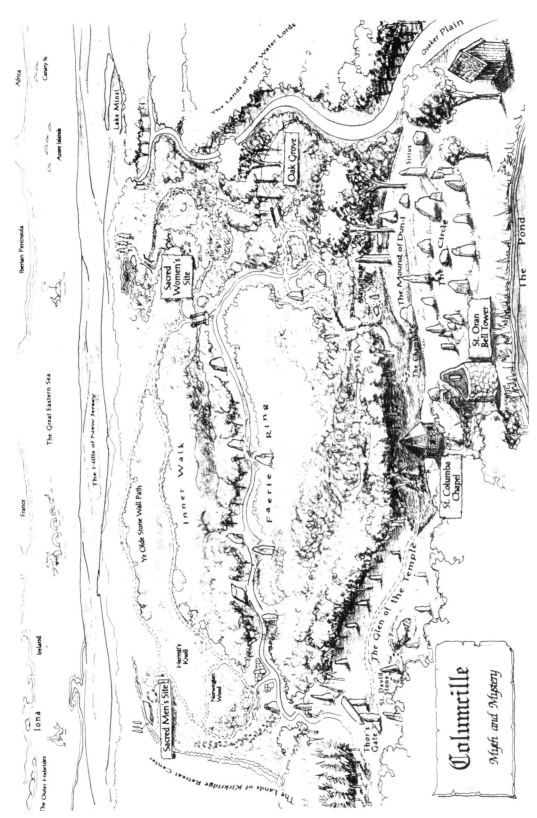

A map of Columcille. (Mark Conners, artist.)

Brad told us that the original eight inner stones represent the eight Celtic cusp holidays: the summer and winter solstices, the vernal and autumnal equinoxes, and the four cross-quarter days of Imbolc, Beltane, Lugnasad, and Samhain. There are no intentional astronomical alignments in the stone circle, although the stones are aligned to magnetic north, south, east, and west.

Backhoes and loaders were the tools of choice for the stone circle construction, and continued as the main equipment for most of the seventy-odd stones that have been set at Columcille. Some of the stones are indigenous to the site, but most came from a nearby shale pit, and were delivered to the site with heavy equipment.

Other menhirs were set during the middle 1980s including the large mounded Sirius Stone and St. Brigit's Stone. Bill was beginning to see the whole property developing almost organically as a megalithic park, not as in the term "amusement park," but more in the British sense, meaning a rural area of fields, paths, woods, and water for people to stroll through and enjoy the pastoral scene.

Mounded Stones

As larger and more impressive (and more expensive) stones were put in place, Bill developed some new techniques. One, which I had not seen before in any modern circle, is to stand the stone at or near original grade, and then to build a mound

The Columcille stone circle is situated next to a pond and adjacent to the St. Oran Bell Tower. (Photo supplied by Bill Cohea.)

Sirius, a mounded menhir, stands guard over the circle at Columcille.

around it with earth. The mound does the same job as a socket; it provides lateral support against the stone's inclination to tip over. The advantage over the socket, however, is that the stone can stand much higher in relationship to eye level. Aesthetically, the mound actually contributes to a pleasing setting. Incidentally, I'd never seen the mounding technique used at any of the ancient sites, either, although the evidence on the land indicates that two of the four so-called station stones at Stonehenge (and possibly the Heel and Slaughter Stones) were originally set in similar fashion, perhaps to derive full advantage of their height for alignment purposes. It is likely that any mounds would erode within a thousand years of a temple's abandonment, and that any stones thus supported

would fall down. Perhaps this is why no examples remain. Still, great minds think alike, it is said (although my neighbor in the Highlands was fond of retorting, "Aye, and fools seldom differ!"), and within a year I saw Ed Heath (chapter 9) use the mound technique on some of his megaliths. And Ed had never heard of Bill or Columcille. Recently, I set a new nine-ton summer sunrise outlier at Earthwood, and used the mounding technique to maximize height.

The other major technique developed around this time was born of necessity. As Bill took on ever larger menhirs, he found that the backhoes that could manage a four-ton stone met their match with ten-tonners. Bill entered the powerful and expensive world of heavy crane technology.

Thor's Gate commands the hilltop at the head of the Glen of the Temple, Columcille.

Thor's Gate

Another kind of megalithic monument had been turning over somewhere in the back forty of Bill's fertile mind, and its name came to him in a dream. Thor's Gate. Thor is most often thought of as a Norse god, but Bill tells me that he also appears in Celtic myth, as a guardian against the north wind. The gate would be a trilithon, like one of the original five trilithons at Stonehenge; maybe not quite as big, but big nonetheless. And it would face the north wind, at the top of a hill.

The traveled way to Thor's Gate passes by St. Columba's Chapel, already described, and we stopped to see it. A visitor was just leaving the chapel as we arrived, one of many who come to Columcille, a place of exceptional peace. Jaki and the Larsens marveled at the quiet spiritual power of the chapel and we stayed for twenty minutes or so. But I must not pause long here in my narrative because the stones are the subject at hand, and the November afternoon grows short.

With any megalithic monument, careful siting is as important as the architecture of the work itself. Although beautiful in shape and proportion, the visual appeal of Thor's Gate is compounded by its position at the top of hill, at the opposite end of an avenue of standing stones that mark the Glen of the Temple at Columcille (see Map of Columcille). The approach to Thor's Gate is along a woodland path from St. Columba Chapel at the Glen's southern end. From the chapel, we could just see the sky-framed trilithon at the crest of the hill and it had the effect of magnetically drawing us along the path. Care must be taken not to miss the almost mystical experience of passing along this avenue of standing stones. Jaki and I were reminded of our stroll along the West Kennet Avenue at Avebury just two months earlier.

As we got closer, Thor's Gate loomed larger and higher in front of us, but its balanced asymmetry kept it kind and friendly, rather than imposing and forboding. I'd visited once before, but this was the first time for Jaki and the Larsens, and I enjoyed the excitement of their first impressions, and Bill's calm yet deep pleasure in sharing Columcille with others, particularly those with their own close interest in matters megalithic.

Bill was hoping to find upright stones long enough that twelve feet could be left above ground, but stones of sufficient length were not available at the quarry. Two beauties of squarish cross-section, three to four feet on a side, *were* available, however, and Bill recognized his posts. The lintel is another beauty, a naturally tapered stone, twenty-two feet long and weighing some thirty tons. The stone looked like many I'd seen standing in Brittany as solitary hand-shaped menhirs.

The large eighteen-ton support stones were set in very deep sockets, five to six feet deep. The holes were partially filled with numerous small loaf-sized stones, and the crane lowered the pointy end down into the hole and held it in place while Bill's crew of assistants started throwing the packing stones in. The backhoe operator even climbed down off his machine and started pitching rocks. He'd never set standing stones before, but really got excited about the task. Later, he even brought his family back to see the completed trilithon.

With two expensive pieces of equipment on site, the lintel was installed as soon as the two uprights were solidly in place. The crane was stretched to its limit, but accomplished the task without a problem. Thanks as much to good fortune as careful planning, the stone sat solidly on the uprights without a wobble. Although possibly unnecessary, a few shimming stones were hammered in between the upright and the lintel to give additional stability.

Sacred Men's and Women's Sites

We continued clockwise around the Faerie Ring, a half mile of maintained woodlands trail punctuated frequently by beautiful standing stones, some socketed, some mounded. We made detours to both the Sacred Men's Site and the Sacred Women's Site, where, during retreats and special gatherings, men and women go to find quiet spots for contemplation, meditation, and fellowship. The men's site features a natural spring-fed pool.

"The interesting thing about this site is that it's so raw and intuitive," said Brad Colby. "Fred discovered the pool, but we made it a little deeper and its form kind of came on its own. The standing Guardian Stone was here, and we set it up. Really, most of the components of the site were already here. When you stand back, you have the sense that all you've done is liberate the spirit that was already around."

My sense was that Brad's comments about the Sacred Men's Site were equally applicable to all of the megalithic wonders of Columcille. Around every turn, it seemed, Bill and his friends had liberated some of the Earth Spirit. Many people believe that standing stones are a conduit to the earth's energy, and authors John Michell, Paul Devereux, and others provide anecdotal evidence (see bibliography).

The women's site, a natural cove in the woods, has a small stone circle where women have left offerings, such as crystals, shells, holly berries, even a small carved replica of the palaeolithic Venus of Willendorf goddess statue, dating to 25,000 B.C. Jaki and I were immediately reminded of similar offerings we'd seen recently at the Boscawen-Un circle in Cornwall. And what was I doing at the Sacred Women's Site? In defense, I must say that one of the ladies went up the path ahead, to assure that we menfolk would not be interrupting any Sacred Women going about their Sacred Business.

After leaving the women's site, Bill told Steve and me how 17 acres of Columcille—all of the part where the megaliths are—is now a preserved site, with a conservation easement from the state of Pennsylvania. Although the tax advantages are slight, the importance of this designation is in the protection it affords the property against development; it guards the trees and the stones. In order to gain the conservation easement, Columcille, Inc. had to identify all further stone-setting and site development at the time of application. This, they have done. Steve was particularly interested in Bill's experience with the conservation easement, as he and Robin obtained a similar kind of designation for Stone Mountain Farm in New York. You might like to inquire in your local jurisdiction to see if a conservation easement might be worth pursuing, which it may well be, especially if you represent a not-for-profit organization.

As we walked along the Faerie Ring, back toward the major complex of megaliths, Bill's dog, Duke, liked to show off by climbing most of the standing stones en route. These were not easy stones to climb, I might add, particularly without rubber grips on one's feet, but Duke relished his work, and couldn't wait for Bill to give the command.

The Mound of Dun I

After a while, we came out of the woods and arrived at another complex of stones at a clearing and knoll called the Mound of Dun I, after the highest point on Iona. This group contains some of the grandest stones in the megalithic park. Our eyes were drawn first to Manannan, a forty-ton menhir standing guard next to the recumbent Coronation Stone, where Bill has performed wedding ceremonies. Steve, master of ancient myth and an amateur magician, asked Bill if the stone hums, referring to the Celtic story of the Irish coronation stone that would hum when the rightful heir sat upon it. Bill said, "Well it depends who sits on it." I was told to take my place on the natural seat, which I did, and I definitely heard a low hum—but it turned out to be Steve, the magician, throwing his voice.

Manannan, named for the ancient sea god of Ireland, stands about twenty feet tall, with about six feet in the ground. I do not know of a taller or more impressive menhir erected during the past three thousand years. Originally, Bill had his eye on a 300-ton stone at the quarry, which he now calls "the one that got away." This huge stone, on a similar scale to the legendary Grand Menhir Brisé in Britanny, was actually dug out of the pit and found to be coated with other stones, so its structural integrity was suspect. And the quarry folk wanted $50,000 to move it, definitely beyond budget. So Bill got the idea of splitting the stone into three 100-ton pieces.

"Lengthwise?" I asked.

"Yes, lengthwise," Bill replied. "Each piece would be maybe ten feet wide, twenty-something feet long, and six feet thick, something like that. But they overpacked it with explosives and they blew it up into 20-ton pieces."

"I can picture twenty-ton stones floating around up in the sky," said Steve Larsen.

Bill chuckled. "Yeah, I've got it all on videotape. But the soundtrack has two classic comments. First you hear, 'Oh, God!' Then, after a perfect pause, you hear, 'Oh, shit!' These two expressions could be my epitaph."

Ed Prynn, who you'll meet in chapter 5, has an uncannily similar story about the origin of his Seven Sisters stone circle.

Manannan was erected using two cranes—with 30-tons and 40-tons capacity respectively—and a backhoe. "Must have been expensive," I observed. "Yeah, but they charge me less and less as they get to know me," said Bill. "Manannan cost $5,000 to set in 1992. Now they charge me maybe $500 a day for the crane, maybe $800 for two days. They like the work so much that the drivers fight to see who gets to come here. Still, we've spent about $250,000 on the megalithic park in eighteen years." I later learned that Bill had cashed in life insurance policies and remortgaged his home to help pay for the crane. Some building funds have come from donations.

A far cry from the $500 I had invested in my stone circle, I thought, but there was a considerable difference of scale. Manannan alone weighed twice as much as all the stones in the Earthwood circle combined!

Another large menhir called Sirius stands about seventy yards or so toward the stone circle. I asked if the two great stones followed a particular alignment. "Not that I know of," said Bill, an answer that was sounding more and more typical of the man as I got to know him. Bill is less concerned with alignments, megalithic yards, and esoteric geometries than he is with finding and preserving the spiritual and aesthetic balance of the site. In this respect, it is as if the landform and the stones tell Bill how to actualize their relationship, instead of the other way around. This contrasts markedly from my analytical approach to stone circle construction, and may explain why Bill and perhaps some of the other stone people in the next few chapters are more receptive to any earth energies that might be present than I am.

Manannan is lowered into its socket by cables fastened to the derrick of a mighty crane. (Photo supplied by Bill Cohea.)

There is no particular right way to build a stone circle. If I could offer any single useful piece of advice, it is: *Stay true to yourself.* If you're a mystic, avoid overanalysis. You won't enjoy it anyway. It would just take you away from your right path. If you're a geometrician or astronomer, build the circle with the features that are important to you, and do it accurately and well. I wish I could pick up the energies and messages at megalithic sites, ancient and modern, like some of my friends. I *can* find the perimeter of a stone circle with dowsing rods, or have so far, at least. More on this in the next chapter.

The last major monument is on the side of a hill facing the Mound of Dun I. It is called the Chamber and it is composed of the twenty-ton blocks of stone floating in the sky sometime between *Oh God* and *Oh shit.* The composition is meant to be symbolic of death and burial. You really can't enter the chamber, but a natural spring emanates from beneath it, very much along the lines of many ancient "sacred wells" in Cornwall and Ireland.

St. Oran Bell Tower

As the afternoon light began to fade, we walked back through the stone circle and arrived at one of the most interesting structures at Columcille, the St. Oran Bell Tower. The tower is a round unroofed building of beautiful rude stones, gathered from all over, even from Iona itself. The project is forever unfinished, and yet it is always complete at its present stage of construction. At the center of the building, a 2-inch pipe is plumbed and held straight with wire fastened here and there to the walls. Masonry work had stopped for the winter, so the site was tidied up and we could all gather comfortably inside the tower, with its inner diameter of, perhaps, sixteen feet. In the spring, Bill and Fred and Brad and others will continue to build it higher. The plumbed pipe will guide a bull's ring, to which will be fastened a mason's line and a plumb bob. The length of the assembly will be the same as the inner radius of the building. Each stone is laid up to the plumb bob stretched from the center pipe, so the inner surface of the tower walls are truly round and plumb. Another similar assembly can be used to check the outer diameter, if desired, so that consistent wall thickness is maintained. Or, knots on a single rope or chain can mark both the inner and outer radiuses. This ancient building technique has been around for millennia, and some builders of round cordwood masonry houses still use it today.

I asked Bill to tell Jaki and the others about the legend of St. Oran, although I suspect that I didn't need to ask. The story personifies one of the basic truths that Bill has discovered in his own spiritual search, and it is one that he enjoys sharing, with a twinkle in his eye.

"Columba was trying to build a chapel, but the walls kept collapsing. He and his friends, including Oran, would build anew, and the wall would fall down yet again. Finally, to appease the spirits, Oran agreed to be buried alive beneath the chapel floor. This seemed to work. The next attempt was successful. The walls stood firm. Three days later, Columba was feeling a little bad about Oran being buried alive under the building, so he ordered the monks to dig the poor guy up. Very much alive, Oran recounted all that he had seen on his journey to the Other World, such as the nature of heaven and of hell, and gave this advice: 'The way you think it is may not be the way it is at all.' Columba promptly reinterred him into the ground." (A photo of St. Oran laid up in the bell tower appears in the color section.)

Bill then rang the bell hanging in a stone arch on the south side of the tower. Turning to Jaki, he smiled and said, "The ringing of this bell reminds us of the wisdom that St. Oran had found, that *the way you think it is may not be the way it is at all*. So keep your paradigms always open. By the time you lock it up, you may have to unlock it again."

Our spirits were high from the wonders we had seen, but they soared higher still with Bill's telling of St. Oran's tale, so key to the kind of spiritual journey that Columcille helps the weary traveler to make.

Bill excused himself to receive some friends who were due to arrive, and Brad conducted the remainder of the tour in the fading light, showing us a long *cairn* (mound) of small stones that had been recently built and the site of a future hermitage for people to stay.

I'd seen and learned so much more on this second visit to Columcille, but my first impression of a year earlier had not changed: The megalithic park was like a gigantic and harmonious landscape sculpture; it did not impose itself upon either the landform or the visitor. We knew that there were many other paths and standing stones and private places to visit on the site, enough that many people seeking quiet contemplation can do so without having their reverie interrupted every few minutes.

Later, I had a chance to sit and chat quietly with Bill in his study, and that's when he told me, with reverence, the story of how he had been inspired in a dream to build Columcille. Of all the ingredients needed to build a megalithic stone circle—site, stones, basic equipment, manpower, knowledge, and motivation—the most important may very well be motivation. Bill's extraordinarily lucid dreams about the stone spirits and, in particular, the powerful nature of the love that he felt during those encounters, have resulted in a motivation of the highest order. A similar sort of almost mystical motivation appears as an important part of some of the other case studies in the next few chapters.

That evening, Bill, along with Brad and the Larsens, helped Jaki and me to celebrate our anniversary at a fine local restaurant. Toasts were made to Jaki and me for our quarter century together, to Bill for his wonderful hospitality, and to the hoped-for success of this book. "End your book with that great lesson from St. Oran," said Bill, "and you can't go wrong!" We knew the mantra. *Remember, the way you think it is may not be the way it is at all.* The evening's festivities ended in raucous laughter.

— CHAPTER 5 —

Ed Prynn and the Seven Sisters

Mystic stuff is all gentle and soft. It's not rough.
It's all easy and soft.

—EDWARD PRYNN

As I begin this chapter, I am suddenly struck by the odd and serendipitous ways that I've met each of the modern stone builders profiled in this book. I learned about Bill Cohea and his stones from a visitor who just happened to see our Earthwood Building School sign at the bottom of the road. It turned out that he was interested in sustainable housing, and I showed him around the place. When he saw the stone circle, he told me that he had seen something similar in Pennsylvania. He'd stumbled across Columcille in much the same way as he'd found Earthwood, and neither place is on the beaten track. I asked the fellow if he could trace a contact for me, and he phoned a few days later with Bill Cohea's phone number.

My introduction to Edward Prynn was even more coincidental. And I hadn't yet read *The Celestine Prophecy,* the best-seller by James Redfield that tells how to cultivate coincidences.

We were staying a few days with Jaki's sister in Truro, Cornwall, at Christmas of 1990. I was passing time at the Truro Book Shop, browsing through their excellent antiquities section, and noticed a lady,

thirty-something, leafing through one of my favorite books, *Megalithomania* by John Michell. Extrovert that I am, I recommended the book highly. We got to talking about things megalithic and I mentioned that I'd built a stone circle myself. "Well, then," said Glynis Kent, "you'll have to come and meet my friend Edward Prynn. He's built a stone circle, too." She gave me Ed's address and phone number at St. Merryn, near the ancient seaport of Padstow, just twenty-five miles from Truro. Two days later, with oldest son Rohan, then fourteen, I motored to St. Merryn in our hired car, while Jaki and wee Darin stayed back in Truro.

At that time, Glynis was helping Ed to finish up a bungalow at a place called Tresallyn Cross, on a minor road a mile from the village. It took a bit of finding in 1990, but now there are neat little roadway signs with arrows on the main roads, saying "The Stones." Ed and his full-scale megalithic museum are on the map!

Edward Prynn is the most outgoing character I've ever had the pleasure of meeting, and a genuine Cornish eccentric, cast from a now-broken mold. Ed welcomed a fellow "stone builder" (as he called me) like a long-lost brother, and, over the years, I have come to think of him as a Brother of the Stones, and a dear friend.

When Rohan and I first met Ed, his astounding collection of megalithic monuments in his garden had already attracted national attention. I didn't know that an illustrated article about Ed's work had recently appeared in a popular book titled *Eccentric Gardens* (Jane Owen, Pavillion Press, 1990) and that his site had been visited by some pretty important Brits, including Prime Minister Margaret Thatcher. Ed had already taken thousands of visitors through his garden before I arrived, and new stone settings were being added each year.

Ed Prynn, mystic, and Brother of the Stones.

What I remember most about our first visit with Ed was his generosity and his genuine eccentricity. He generously gave us several hours of his time, showing us his stones, and guiding us to the Nine Maidens Stone Row and up onto Bodmin Moor to look at—and rock—"the greatest rocking stone in all Cornwall," a 300-tonne behemoth on Louden Hill. To cap it all off, on the way back to St. Merryn, he insisted on buying us supper, refusing any contribution on my part.

Some people who don't know Edward Prynn might accuse him of playing at eccentricity, but this view would be in error. *Eccentric,* when speaking of people, means "one who acts or operates in an unconventional manner." It is derived from the Greek words *ek* (out) and *kentron* (center). Even though Ed enjoys and even cultivates publicity, this in no way trivializes his authentic Cornish character. Publicity is simply Ed's way of meeting people, and this is how he stays in such remarkable touch with the world. You see, Ed is legally blind and can neither read nor write.

THE MAN

To understand Ed's stones—to *really* understand them—you have to know something of the man himself, for the two are inextricably intertwined. As Ed told me in the Cornish dialect, "It's people (that) make the stones, Rob." And he wasn't referring to the physical construction of them. The stones reflect the personality and motivation of their builders. This can be seen clearly at Columcille. Stonehenge, I'm sure, tells us something of what was going on inside the mind of its designer, certainly a different line of thought from that of the designer of, say, Boscawen-Un. My stone circle, I have to admit, tells something about the way I think. It can't be helped; in fact, that's how it should be. Ed's unique rock collection is a function of his deep interest in the stone monuments of the past and his desire to share this interest with the world. And, like Sinatra of the song, he does it *his* way.

Born in 1936, Ed lost most of the vision of his right eye when, at the age of two, he speared it with a "steel," the filelike knife-sharpening tool. When he was ten he had what he describes as "his first mystical experience at an ancient site." While on the nearby St. Breock Downs with some local workmen, Ed wandered off and found the row of standing stones called the Nine Maidens. As he wandered around the stones, he felt that he wasn't alone, that other people were there with him "in a spooky kind of way." This was the birth of Ed's lifelong fascination with stones.

Reading and writing never caught on with Edward Prynn. He'd rather wander down to the many bays and search for spoils from seawrecks than go to school, and by the time he began to attend with anything like regularity, he was far behind his classmates academically and none of his teachers managed to bring him around. An intelligent man with wide interests, Ed later authored two compelling books about his early life, by telling his story into a tape recorder. Transcribed and edited by Jo Park, *A Boy in Hobnailed Boots* (Tabb House, 1981) paints a vibrant picture of what it was like to grow up in rural Cornwall in the 1940s. The sequel, *No Problem* (Tabb House, 1982), takes Ed's story up to 1980, including the accident in the stone quarry when a shard of steel cost him all vision in his "good" left eye, and how he has coped with this new disability. "It's like driving down the motorway in daylight, but it's bucketing with rain and your wipers have stopped. You just get by. It's the easiest way I can describe my sight." Ed speaks of his short bout with "religious mania" and his continued interest in matters spiritual and mystical, including faith healing and fortune-telling. Ed considers himself a Christian, but, like Bill Cohea, he has embarked on a lifetime journey of the spirit, a journey whose way is marked by giant standing stones. Ed's original tapes, which provided the material for his first two books, and the word-for-word transcript of them by Jo Parks, have been lodged with the Insti-

tute of Cornish Studies at Pool. Ms. Park describes the autobiography as "a fascinating story of an extraordinary, ordinary man" and "a record of Cornish village life from the late thirties to the late seventies and of one Cornishman's great courage against odds."

The third part of Ed's personal odyssey is in manuscript form as I write and covers Ed's "Stone Age." It has been transcribed by Glynis, who now lives with Ed at the completed bungalow among the stones at Tresallyn Cross, and I, with many others, await its early publication.

THE STONES

When he was working at the quarry, Ed would tell his mates in the tea hut how, when he retired, he would someday build a stone "henge" of giant monuments. Little did he know that retirement would be imposed upon him at the age of thirty-two, when his world went dark. His disability compensation allowed him to build the shell of his tidy bungalow at Tresallyn Cross, and, later, an inheritance from an uncle allowed him to finish it up. The bungalow sits on a small flat plot of land, which Ed had been *tealing* (tilling) since he was a lad. He finally bought the property, about a quarter acre, for fifty pounds.

The Rocking Stone

In 1982, Ed entered the world of megalithic stone monuments by erecting a replica of a class of stones that is probably more natural than manmade: rocking stones. Cornwall is the home of several rocking stones (also called *loggan* stones), including the massive one that Ed had taken Rohan and I to see. Although the stones are probably natural in origin, many legends are connected with them, and it is probable that the stones have been kept in good working order by the Cornish people since time out of mind. Ed thinks of the rocking stones as ancient druidical ceremonial stones, where archdruids would swear their oaths. It was said that any who tell lies on a rocking stone will be punished now and in the hereafter. Another legend says that anyone with wicked thoughts will not be able to rock the stones. Some of Ed's older friends and neighbors told him to be careful fooling around with the kind of power that these stones can wield, and they weren't talking about getting your fingers pinched cracking a nut. Ed considered himself to be a strong enough Christian to set up his stone without fear.

Like most of the stones in Ed's Seven Sisters Temple at Tresallyn Cross, the Rocking Stone is a giant granite boulder from a quarry at nearby St. Breward. The ten-tonne (eleven ton) stone with its rounded bottom cost Ed £40, almost as much as he'd paid for the property many years earlier, but it was a bargain nonetheless. Ed had installed a flat concrete foundation in front of his bungalow for the stone to rock upon. "We had two big tractors to pull this stone over the top of the concrete," says

The Rocking Stone.

Ed, "and to my amazement, and everybody else's, we never had to do any work on it at all. It rocked at the first attempt. It would even rock in the wind."

The Seven Sisters Stone Circle

The major stone setting at Tresallyn Cross is the magnificent stone circle, but I'll let Ed tell you about it.

"After the Rocking Stone had been in place for a while, a lot of interest was being shown and I was asked what would be next. I had always dreamed of building my own stone circle and with the help of my stone team, my dream came true in June of 1983. My original plan was to build a stone circle with the tallest monoliths in Western Europe. The first stone we blasted measured 36 feet long and weighed 75 tonnes (84 tons); it was beautiful. The idea was to put six feet in the ground and have thirty feet standing, but when the stone was moved at the quarry it broke into seven pieces. I was heartbroken. Brian Nottle, the quarry manager, told me not to worry; he would sculpt me a stone circle out of those seven pieces. As we needed eight, for one in the middle, you will see that the stones I called Monica and Marjorie are twins, one stone split in two.

"I visited the quarry nearly every day for three months to watch Brian at work. Even though I have little sight, I had a vision of how I wanted this circle to look, and it was as if Brian shared this vision with me. It was a great day when the stones were loaded to come home.

"The lorries took the route from St. Breward through Wadebridge, and stopped in the roadside lay-by next to the Nine Maidens Stones on St. Breock Downs, where I'd first met standing stones so many years ago. I had been told by a wise man to collect soil from an old stone site and place it under the center stone of my new stone circle, and that's exactly what I did. When we arrived at my garden, the digger went to work on the holes and one by one the stones were lowered in the ground by a 45-tonne crane, clockwise of course, and backfilled with soil."

"Why clockwise?" I asked.

"I was always told by old people in Cornwall, 'Work clockwise when you build a stone circle.'"

I asked where the old people got this information, as no stone circles had been built for thousands of years.

"Well, Rob, it's there. You know, when you pull a boat in most Cornish Coves—and these are old druidical customs—when you turn it around to go into the water, you always turn it clockwise, the way the sun goes. If you turn it the other way . . ."

"You'll be in trouble," I guessed.

"Bloody right! You won't go to sea. You'll put it back up again. We did it once. Me and me mate Don and me brother pulled the boat down one Sunday. We turned the boat the wrong way, so Don says, 'Well, we won't go to sea.' So I said, 'Fair enough, we won't.' So we pulled it up again. This woman on the beach said, 'Excuse me, you worked so hard to get your boat down, and you've worked harder to get it back. Why didn't you go to sea?' So Don says, 'We put a curse on ourself.'"

I smiled, but Ed, whose life was intertwined with the ancient and the mystic, had told the story matter-of-factly. He resumed the story of the stone circle.

"The stones were shaped and set to catch your eye, and the way we angled 'em was intentional. It's really a stone sculpture, which I call the Seven Sisters, and I named the outer stones after special ladies in my life.

"Marion is is the tallest stone. Marion was a Bluebell dancer and she drove me to and from the quarry. I told Marion I was going to name a stone after her and she had the choice of them. She was the tallest of the ladies and chose the tallest stone. It measured twenty feet and there's fourteen feet standing out of the ground today.

"The Music Maker is a Cornish girl, singer and songwriter. She has given many charity concerts at the stone circle and I feel one day her music will become very famous. Great Auntie Hilda was the youngest of my mother's father's seven sisters and to be born the seventh is very lucky! Monica is a very saintly lady and I felt I had to call a stone after her. Marjorie I called after Mother. The Secret Lady? Now, that would be telling!

"The last one is Jackie. I put Jackie there because before ever the bungalow was completed or any of the stones came, she once said to me, 'Ed, don't ever sell that little old bit of land of yours; I could stay forever.' I decided she could and there stands twelve tonnes of granite named after her.

"The stone in the middle I call the Peacemaker. It's like a phallus. It represents Man and the outer circle represents Woman. That's the fertility rite. And when we do the dancin' and the drummin' and the chantin' here, always it makes everybody sexy as anything on the night, got it? On occasions, we've had it where they've left one another's partners and gone for somebody else. And I've no doubt—and all the early writers say so in their opinion—that stone circles are for dancin'. I can only talk about this one. I can't talk about stuff thousands of years ago, because it would only be my opinion. It's all hit and miss, cause there's never nothin' written down."

I admired the Peacemaker.

"There are nine hundred stone circles in Great Britain," Ed told me, "and not many of 'em have a center stone."

All of the stones were lowered into their sockets with a 45-tonne crane. A chain was tied as near as safety allowed to the top end of the stone, so that the bottom end would find the bottom of the hole. Ed brought in three experienced quarrymen for this job. Each stone was socketed anywhere from 20 to 30 percent in the ground, depending on its length and balance. The jaunty angles on some of the stones makes the whole composition much more ancient, more photogenic. The Rover automobile company has even used the Seven Sisters as a backdrop in their brochures. "I set this particular stone circle up for presentation," says Ed. "I left 'em

The Seven Sisters Temple was built easily in a day, thanks to a 45-tonne crane and a backhoe.

The Seven Sisters.

lie over a bit on purpose. I cheated a couple of thousand years."

We talked about ancient building methods, and Ed spoke of the possibility of making a large wooden boom with baskets at one end for putting stones in place.

"You could make yourself your own crane, just about. It'd have to be a strong basket, and you'd need a helluva big tree, but it coulda been done that way. Another way is that they coulda shifted a lot of the longer ones in the winter, on the ice. Get a lot of ice down and slide 'em. And, Rob, all the writers I heard of in my lifetime, none 'em of talk about animal power. And in prehistoric days, there were a lot of big animals that were capable of pullin' a tonne. That one animal would pull the weight of—what?—say, forty people would. So a few of these animals could have pulled these stones without loaders, just brute force. And, with animals, you wouldn't need so many ropes."

I asked Ed about the cost of the stone circle and how long it took to build.

"I spent several hundred pound to build me stone circle, three hundred for the dynamite and the stones. I had to pay for three low-loaders and the hire of the crane and the JCB [backhoe]. It took about six hours, the whole lot. The biggest job was in the bringin' of 'em and gettin' 'em here. The actual puttin' 'em in, it's just plunkin' 'em in place. I had enough men to be slingin', and I had the JCB for the diggin' of the holes. It took, say, three hours to dig the pits. And you know, Rob, it were a pleasurable afternoon!"

The Judgement Stone

Although this book is primarily about building stone circles, the best-known and, perhaps, most numerous of the megalithic monuments in the British Isles, many of the stone builders get caught up with other stone settings, some related, some not. Edward Prynn has an undying interest in all stones, manmade or not, large or small, ancient or mod-ern. He is the quintessential stone man, born, perhaps, in the wrong millennium. Two months after the completion of the Seven Sisters, Ed was back to work on the rocky north coast of Cornwall, to fetch his Judgement Stone.

"Merlin the great magician talked of judgement stones and he said, 'In the event of something great happening, kings and queens and leaders of the world would meet at a judgement stone.' I knew there were many judgement stones around the world, but I felt I needed one to go with my collection. During a visit to an old wrecker [Cornish term for the men who risked life and limb to salvage cargo from ships wrecked off the treacherous coast of Cornwall], he told me about a ship called the *Petrel* which was carrying a load of I-talian marble bound for church work in Scotland nearly a hundred years ago. When the ship got off Trevose Head a terrible storm come up and Captain Pengelley said, 'I'll save the ship and I'll save the crew,' but the ship was washed into Wine Cove at Treyarnon, which is a dungeon of death. The captain was washed overboard and lost, but all the crew were saved.

"Over the next few months 250 tonnes of marble were hauled up to the top of the cliff with steam engines and slings. On the last day, a piece fell out of the sling. The wrecker told me he was sure it was still at the bottom of the cliff. So with the help of my stone team [Ed's friends and workmates] we tried to find it, and after a few visits at low water we did. On August Bank Holiday, 1983, with the help of me mates, the St. Merryn Coastguards, and a big winch tractor, we hauled the marble slab out of the cove. It's about four feet long, three feet wide, and four inches thick, and weighs about five hundredweight [560 pounds]. And there it is, laid to rest between the Rocking Stone and the stone circle."

Glynis showed me a copy of a TV documentary made in 1983 by TSW at Plymouth called "And Merlin Said" about Ed's Judgement Stone being brought out of the cove, very much in the long Cornish tradition of wreckers.

On a subsequent visit, Jaki and I took our chances and stepped aboard the Judgement Stone. We closed our eyes and awaited revelation. Apparently, the time of our judgement had not yet come. But the story of this stone is one of the most interesting of Ed's many stone stories.

Ed collects stories of other people's stones, as well.

"There was a-l-w-a-y-s been stonebuilders, always. The two most people that worked on stones was vicars and military people. Have you ever heard about a guy called General Conway? He was governor over on Jersey in the Channel Islands. He was such a good governor, and when he was going to retire, the people said, 'Well what would you like? A clock? Name anything that you want.' He said, 'What I want you wouldn't let me have.' They said, 'Tell us what you want.' He said, 'I want your stone circle.' And they said, 'Well, that's a tall order, but you can have it.' And the Royal Navy provided three ships, 150 tonnes like, to that tune, and it was brought back, that stone circle. There's photographs. It's in Henley-on-Thames. You might be allowed to go see it. There was a rich pop star owned it at one stage. A lady told me about this and they printed eight copies of a book about it, and I've got one of those copies."

I asked Ed when all this happened.

"Oh, nineteenth century, like. And, Rob, now they want it back. There's hell up. The Jersey people want their stone back."

The Healing Stone

Again, Ed tells it best:

"In 1984, I was taken ill with a bad back. I tried many cures and different treatments without success. Then I heard of the Men-an-Tol stone in West Penwith near Land's End. When I went there I walked to the stone with the help of two walking sticks and knelt down and rested my body through the hole. Although I was in a lot of pain, I managed to crawl out through the stone and, on my hands and knees, crawled around and went through the

Ed now has his own Men-an-Tol to help keep his back in shape.

stone twice more, as I was told I had to go through three times. There was no instant cure, but after two days my back was a lot better. So I decided that the next stone I would have would be my very own Healing Stone, a replica of the Men-an-Tol. The following week, Brian Nottle, the stone mason who had sculpted all my stones for me, went to West Penwith and took the measurements.

"After I had got my healing stone in place, I wondered how I could make it special. I didn't have to worry. Within a fortnight, some eighteen faith healers found their separate ways to 'un. As I watched them rub their hands over the stone, they all left their little bit of magic behind. This is how I believe I was cured at the Men-an-Tol. Through time, many healers have visited that stone and touched it and left something special behind."

The Islanders

Only Edward Prynn could have pulled this one off. His friends told him he was wasting his time.

"When the Falkland Islands campaign was on I knew there would be a lot of people who wouldn't be coming home again. I felt there should be a

Men-an-Tol

On September 25, 1997, Edward's 61st birthday, Jaki, Darin, and I visited the original Men-an-Tol stone with Ed. We stopped at Lanyon Quoit, one of the finest dolmens in Cornwall and rushed to the Men-an-Tol lay-by, hoping to catch the sunset. I jogged in along the moorsland path, nearly a mile from the car, and arrived, huffing and puffing, just in time to take some excellent sunset pictures. Jaki, Ed, and Darin came along a few minutes later. At the time, I was suffering from a long bout with *lateral epicondylitis* or "tennis elbow," which I'd actually contracted by playing four sets of tennis after several years absence from the sport. I couldn't hammer, mix mortar, or run a chainsaw, all important to my work—never mind play tennis. I'd tried everything to effect a cure: physiotherapy, ultrasound, homeopathic remedies, psychic healing, cortisone, total rest, you name it. And, before leaving for Britain in September, I'd scheduled an operation for November.

"You got to pass through the stone nine times, Rob, and always come around clockwise," said Ed. The ante had tripled since Ed's visit thirteen years earlier! So I crawled, nine times through, always turning right. This actually takes a while. It was getting late, so the others started back toward the car while I filled my prescription. Upon completion of this remedy or penance (or tomfoolery), I caught up with the others on the moorsland track. I certainly admired Ed's coming out here in 1984 with his walking sticks and bad back.

Was my elbow cured? Well, when I returned from our megalithic journey in October, it felt 100 percent. Of course, I'd done nothing more strenuous during our month's travels than lift a pint of ale to my lips, but I guess I built up some strength in that way. Hoping to avoid the operation, my doctor put me on another series of physiotherapy visits, but, unfortunately, I reinjured the elbow during the exercises and I was back to square one. I saw another specialist who told me that, after a year, only the operation could fix me. I write these words now just three weeks after the operation. The surgeon found a tendon torn right through, and it's hard to imagine how any magic stone, even one with four thousand years' experience, could have knitted it back together.

But today, while researching the Men-an-Tol, I read in *Journey to the Stones* by Ian Cooke that for curing purposes, the supplicant needs to *crame* or crawl on all fours nine times through the hole "going against the sun's course" (*widdershins* or counter-clockwise!) to cure lumbago, sciatica, and others "cricks" and pains of the back. Had I gone the wrong way? No. A psychosomatic approach to a malady requires a sympathetic frame of mind. In this regard, Ed has the advantage over me.

After our visit to the Men-an-Tol stone, we rushed down to the Acorn Theatre in Penzance to hear a lecture by antiquities expert and author Craig Weatherhill, who passed on an interesting piece of information that I had not seen in any books on Cornwall or stone circles. It has now been determined that the three obvious stones at Men-an-tol are on the arc of a circle, and sure enough, other stones (some buried) and stone holes have been found on the same curvature, describing a medium-sized circle like so many others in Cornwall.

memorial stone brought back from the Islands for those who had fallen and I asked our government if they could bring a piece of stone back when the ships came home. The answer was no. So I had a letter written to Sir Rex Hunt, the governor of the Islands, and it was delivered by hand by a man from Saltash who was returning there to work. A few months later, Sir Rex wrote and told me he liked the idea and that one day I would get the stone. In 1985, while the contractors were building the new runway, they blasted two one-tonne pieces of rugged quartzite. They packed 'em into two wooden crates filled with sawdust and put 'em in a container on the *M. V. Lesterbrook* for the journey home. They sent two in case one got broken, but they was packed so good that I got 'em both intact. The contractors brought 'em all the way from the docks for me."

In fact, the contractor's resident area director at the time, Mr. Bill Bloomfield, was an amused on-looker at Tresallyn Cross as the stones were uncrated and lowered into place by crane. Nylon straps were used instead of chains to prevent any damage to the long slender monoliths. Dozens of Ed's friends and supporters came to witness the festivities, and the whole day became quite a media event. Reporter Robert Jobson, of Cornwall's *Western Morning News,* wrote on August 20, 1985: "Cornish eccentric Eddie Prynn savoured the greatest moment of his stone collecting career yesterday when two crates completed an 8,000-mile journey from the Falkland Islands." Ed basked in the emotion and excitement of the day and held court for his friends

Ed supervises the careful removal of the stone from its sawdust-packed crate.

The author scans for incoming spirits atop the Angel's Runway. The Marriage Stone lies beyond the dolmen.

and the media. As quoted by Jobson, Ed said: "This is the ultimate for me. I am so happy. I feel honoured and privileged that so many people should have gone to so much trouble to make it possible. These rocks will be a memorial. One is dedicated to those who did not get back from the Falklands, the other to those who did."

The tall and lovely stones are set at rakish angles to each other, and stand close to Ed's front door.

The Marriage Stone

This triangular holed stone is a replica of the Tolven Stone, a tall granite monolith pierced with a 16-inch-diameter round polished aperture. The ancient stone is in the yard of Tolven farm, a mile north of Gweek at the head of the Helford River. J.T. Blight, the Victorian antiquarian, described a healing ritual at the Tolven Stone that consisted of passing a sick child nine times through the aperture, alternately from side to side, and then laying it on a particular low grassy mound to sleep with a sixpence under its head (Cooke, *Journey to the Stones*, p. 48).

Permission is required to visit the original, but at any reasonable time you can visit Ed's excellent replica—again, carved by master stone-shaper Brian Nottle after a visit to the original. In fact, Ed will even throw in a wedding ceremony. Says the gaffer:

"This stone was to change my life in 1986 in more ways than one. It was a custom in days gone by to marry for one year and one day at the summer solstice, at a holed stone. Once the stone was in place, I had a certificate drawn up and soon a couple came to make wedding arrangements. This couple had been married for fifty years already and

Lanyon Quoit, an ancient dolmen near the Men-an-Tol stone, West Penrith, Cornwall.

wanted me to perform a ceremony for them. I told a journalist about the wedding and she told me I would need a white hooded druid ceremonial robe! I told her I was a quarryman at heart, not a vicar, but she told me that someone has to perform the ceremony. So quickly a robe was made for me and the great day arrived. I must admit I was very nervous; I had to hold on to the stone to stop me shaking. There was nothing written down; I just used me own words. The couple knelt down either side of the stone and at the end of the ceremony, they kissed through the hole.

"The weddings I perform are not legal, as I've altered the rules and the couple marry just for fun. The secret is you don't commit yourself too heavy to start with. Just get married for a week. If its going good, have another week, and then a month.

"When I started wearing the robe, I thought 'This isn't me,' but as time went by I found the robe, the druids, and the stones do go together, even in the 20th century at a new stone henge, so I appointed myself as Archdruid of Cornwall."

Why not?

The Angel's Runway

By now, it must be clear that the story of Ed's stones is very much the story of his life. The Angel's Runway, Ed's largest monument, was built in 1987. It is a modern-day replica of an ancient dolmen, not any one in particular, although Cornwall has several of the best to be found in Great Britain, including Lanyon Quoit near the Men-an-Tol stone.

A dolmen is a mysterious type of neolithic monument. In general, it consists of three or four massive upright support stones (although other numbers are not unknown) that support a giant flat capstone. Some dolmens may have been the core of megalithic burial chambers. Time, perhaps aided by animals and humankind, eventually eroded away the earth covering. Others may have been built as is, without earth sheltering, for purposes unknown. There is a well-known—and disputed—dolmen in North Salem, New York, which was certainly never intended to be earth-sheltered. (See chapter 16.) While a detailed discussion of dolmens is beyond the scope of this book, the one in Ed's front yard is certainly worthy of description, and it has planted the seed with me to construct a dolmen at Earthwood sometime soon.

As with any of Ed's stone settings, his dolmen has its own fascinating story, and, as usual, there is no one to tell it better than himself.

At the quarry, a 45-tonne crane loads the capstone onto a low-loader.

The low-loader transports The Angel's Runway through the streets and lanes of Cornwall.

Volunteers tamp the sand around the deeply-socketed legs, while the crane holds the stone in place.

The capstone is eased onto the giant stone legs.

"When I built this monument in 1987, I decided to call it the Angel's Runway as a lot of people had told me about their encounters with angels and I thought the angels needed somewhere to land. But it wasn't until 1990 that I had my own encounter with an angel. He came to my bedside and told me that life would never be the same for me again. At the time my heart beat so fast and then so slow and I thought I was going to die. But the next morning I found my dear Uncle Bert had died outside the back door at Mother and Father's house, where he lived with 'em. He'd had a heart attack and died just as I come 'round the corner, and I was the one who found him. Life for me never was quite the same again, because he left me enough money to finish off the inside of my empty bungalow. So, at the age of fifty-four, I had my first home to call my own."

"Mystic stuff," as Ed calls it, is fun and interesting, but I also wanted to know about the construction of the dolmen. In answer to my questions, Ed gave me the hard facts. The front leg, the tall one, weighs fourteen tonnes (15.7 tons) and the two shorter back posts weigh about four tonnes (4.5 tons) each. The capstone measures 15 feet long by 10½ feet wide and weighs 18.5 tonnes (20 tons). "The capstone is a *grass rock* found just below the plow line, whereas the legs were quarried fifty feet down in the pit. The ancient peoples would have used grass rocks to build their stone monuments."

I asked Ed how deep the uprights were socketed.

"Well, Rob, I had to play it on the safe side. I'd never built a dolmen in me life and I haven't spoken to anyone else who ever did. I knew people would say, 'Hey, that's dangerous.' So I went to the ultimate down into the ground—half of 'em is in the ground, Rob—so that nobody could ever say they was going to topple over, and no one ever did. In fact, when I put that front stone in, there was a hundred people came that night and everybody chuckin' pound coins in there—most prob'ly a hundred pound in pound coins in the bottom of that stone.

"I asked this farmer, Rob, if I should backfill with the same rubbly stuff that came out of the earth. He said, 'No, we'll backfill with the yellow sand from the beach.' I said, 'sand! It'll fall over!' He said, 'Ed, when it's dry, the sand go down; when it's wet, the sand go down. If it don't work out, I'll put that monument right on me own expense.' And we backfilled with sand, and when I put that capstone on, I thought the legs would be buckling all over the place."

"Did you tamp the sand?" I asked.

"Yeah, but not a lot, because you couldn't get it tight. It's got to go down by water. I watered it down and it took me a ton and a half to refill the pits. But those uprights were solid, Rob; the capstone didn't move 'em, because there was so much down into the ground. We set the capstone with a 45-tonne crane and a sling. There was no cock-ups or nothin'. It was magic."

Glynis and the Hall of Fame

"A fortune-teller had told me, twice, that a young woman was coming into my life," says Ed. "I just had to be patient and wait."

The woman turned out to be Glynis Kent, and she came into Ed's life in 1989, just two years before I'd met her in Truro. At first, Glynis became part of Ed's "stone team," but this soon blossomed into what is known today as "significant other." The first project they worked on together was the Hall of Fame, hundreds of inscribed pieces of slate hanging, first, in the garage, and now on the exterior walls of Ed's bungalow. Glynis routs the names onto the inch-thick slate slabs with an electric drill and bit. Then the recessed names are painted according to a color code. Originally, silver was reserved for the important historical figures of Cornwall. But his "stone team" (loosely defined as anyone who has even vaguely helped, encouraged or been a part of Ed's temple or home construction in any way) gets special treatment, "red for those who had helped a little, blue for those who had helped more, and yellow (or gold) for those who had helped beyond the bounds of duty." Promotion and relegation is possible. Once, Ed fell out with his bank manager, who was then demoted to red.

Recently, the Hall of Fame has become more eclectic and international in scope. Jaki and I are up there now, on the home's front wall, between two bedrooms. Our inscription is: "Rob and Jaki Roy, 20th Century American Stone Builders." The names are in white and our designation is in green, and I am hesitant to ask Ed and Glyn about the new color code system. On the positive side, we share the panel with Sir Charlie Chaplin, Tchaikovsky, Michael Faraday, Harry Houdini, Dr. Billy

Graham, Captain James Cook, and Dr. David Living-ston, as well as lesser-known luminaries such as Harry Ramsden, "owner of world's largest fish and chip shop." In front of "our" panel stands an old stone cross that was given to Ed. "There was always a cross at the crossroads near here," says the archdruid. "Legend says a saint from Ireland and a saint from Britanny met there and put up a cross. For many years people came annually for a spiritual gathering. In 1836 a man came up the road on his horse, which was frightened by the cross. He fell off the horse and was killed, and the cross was removed to Tresallyn Farm. One day, up at the quarry, Marion [one of the "Seven Sisters"] saw a cross that was thrown away on the scrap heap, and she gave it to me as a present for the Temple. And that's it."

The Womb of Mother Earth

Glynis was a full partner in the construction of the most ambitious megalithic project at the Seven Sisters Temple, a project that took six months to complete. So I'll let her tell the story:

"A *fogou* is a magic underground chamber. There are six fogous in Cornwall that the ancient people built, so Edward thought, 'Why not build one by 20-century people that can be used?' We started by going to the beach and drew a circle in the wet sand sixteen feet in diameter to represent the floor of the chamber. We then marked the circle out into blocks one foot wide and two feet, three feet, and four feet in length. Then we knew how much stone we had to order.

"On May 12, 1992, the first piece of turf was dug out. It took two weeks with a digger and a rock drill to excavate the pit sixteen feet across and twelve feet deep. We thought we'd taken on more than we could expect the stone team to complete, but over the next few weeks, with the help of friends, holidaymakers, and people from all walks of life from all over the country, we got the two-foot-thick granite walls built. Working from 8 A.M. to 10 P.M. six days a week, we were ready for the great day—

Ed's bungalow is covered with slates inscribed by Glynis, their "Hall of Fame." The cross was salvaged from a stone scrap heap.

A seven-tonne roofing slab is raised from the flatbed and placed over the fogou. Ed guides it into place.

July 27th—when MacSalvor's 30-tonne crane arrived to lower the four massive capstones over the chamber. The largest weighs seven tonne (eight ton.)

"The next day work started on the passage which leads into the fogou. First we dug a shaft down to the chamber's entrance, and then we built the passage walls and the steps, working our way up from the floor of the chamber. It was the end of October before the fogou was finished and the garden was back to normal.

"The chamber is called 'The Womb of Mother Earth,' as some of these chambers around the world are used for rebirthing ceremonies. The lucky touchstone of white marble which stands at the center is a piece from the *Petrel,* rescued from the sea thirty years ago. It stands erect at the center as a phallic symbol. A slate on the wall says:

Deep down in the Womb of Mother Earth,
Men and Women come to the chamber for
   rebirth,
To touch the stone that will make them lucky.
Stripped off, they're guaranteed better luck
   Drekly.

"The phallus at the center is called the Drekly Stone. Drekly, in the Cornish dialect, means maybe today, tomorrow, whenever, as in, 'I'll be returnin' that plow I borried drekly.'

"People ask Edward what else these chambers were used for and he answers that no one really knows. All we know is what we and our friends use the chamber for. Edward has done a few rebirthing ceremonies, but he loves to have a bit of mystic fun, too. He has a band called Eddie Hardrock and the

Mystics, and, for instruments, they use oil drums, dustbin lids, bells, pipes, anything that makes a noise. Anyone can join in, and after five or ten minutes the rhythm and beat of the drums can be quite amazing."

STONE CIRCLE ENERGY

We visited Ed and Glynis once a few years ago when Darin was small. Ed was showing us how to "dowse for energy" with a forked hardwood stick, one of several he kept handy in his garage. Jaki was recording the lesson on videotape.

"Now, Rob," says Ed, "I'll show you how to find the edge of the circle with the sticks."

"Well, okay," I said, "But it's only fair to tell you that I don't really believe in this sort of thing."

Ed brushed me off. "It doesn't make a difference," he declared. Then he showed me how to hold the Y-shaped divining stick between my fingers in such a way that, as he put it, "the stick comes alive." You can apply tension to the stick's handles so that it easily snaps either downward (for a *positive dowser*) or upward (for a *negative dowser*.) I quickly found the right way to hold the stick, and Ed had me close my eyes and approach the circle from without, holding the stick in tension in front of me, parallel to the ground. "Make the stick come alive, Rob," he said. I felt somewhat foolish, but played along. There was a wide enough gap between two of the stones that I knew I wouldn't walk into an immovable chunk of rough granite. I walked slowly toward the circle, and, after a while, the point of the stick—the base of the Y—suddenly moved downward. I did not feel a pull, but felt the tension that I was applying to the stick suddenly release as the stick decided which way it would go. I was lucky, I guess, to be a positive dowser. Had I been a negative dowser, like Glynis, the stick might have thumped me in me pronounced Gaelic nose. I opened my eyes and the stick was pointing right to the outer circumference of the circle. I laughed long

and hard, while Jaki recorded the event. Edward wasn't laughing. He'd seen this too many times before. He was the scientist here and I was the skeptic.

When I'd stopped laughing—*this is so ridiculous, I thought*—I came at the circle from another direction. Again, same result. Down goes the stick at the very perimeter. Again, I'm laughing uncontrollably at this absurdity, Ed's not laughing, and Jaki's still recording.

"Now, Rob," says my instructor. "How old is your stone circle?"

I told him that it was about six years old.

"You must have had hundreds of visitors to your circle by now," he says, to which I agreed.

"They all been touchin' your stones and puttin' their energy into 'em. Now, you take this stick home with you, and you'll find the energy at your own stone circle, Rob. It's there now. You'll find it."

I thanked him for the special gift—I still have that divining rod—and told him I would try it as soon as I got back home.

Ed teaches Jaki and Darin to dowse for the aura or energy given off by the megaliths. In this unchoreographed snap—honest!—all three dowsing sticks simultaneously leap upward as the group passes under the perimeter of the Angel's Runway.

Before we left Ed on that particular visit, he tested Darin, then about five, to find his "aura." Ed reported that Darin had a very strong aura about him and Ed was able to pick it up from twenty-five or thirty feet away.

We went home, and I tried out the divining stick at the Earthwood circle, still skeptical, still thinking that this is really silly. But every time I tried it, the stick came down at the perimeter of the circle. I don't pretend to understand it. But I still approach these types of situations with what I consider to be a healthy skepticism. This is the only stance, I think, that enables us to learn something. At Guy Fawkes Night bonfires, I'll sometimes tell this story (often after a beer or two) and someone will ask me to fetch the divining stick and demonstrate, something I've always done, but always with the fear that I would make a fool of myself. So far, I have not. But still I have the fear, every time.

One sunny summer's day, I was visited by a couple from Montreal, who had come to see the house, being interested in cordwood masonry. As soon as they saw the stone circle, however, their interest shifted immediately. I told them the story of dowsing for energy, and they listened with intense interest.

"Do you still have the stick?" asked the lady. I said that I did, and she asked if they could borrow it. I fetched the stick from the house and left them alone with the instrument, to conduct their own experiments.

About twenty minutes later I returned to the circle and asked them if they had had any success. They said that, yes, they had. But they had not been curious about finding the edge of the circle. I suppose any damned fool could see where that was. Instead, they had independently used the stick to measure the energy of each standing stone in the circle, all twelve of them.

"And what were your findings?" I asked.

"We agreed that two stones gave off considerably more energy than any of the others," said the lady, the spokesperson of the couple.

I asked her which two stones they had identified, and she indicated the Bunny Stone and the Cat's Back Stone right next to it, which we also call the East Stone. Instantly, a chill went down my back, as I realized that they had agreed upon the two stones—the *only* two stones—that the children play upon, the only two that they *can* play on. Led by Darin, he of the strong aura, the small kids climb the Cat's Back and leap onto the back of the Bunny Stone. Had they put their strong children's energy into these stones? I had not mentioned the children to the Montreal couple before their tests, but they were very interested to hear of it afterward and smiled knowingly as if it all made sense. Which it certainly didn't to me.

Later, I worked out the math, using standard probability theory. Assuming that we know which of two stones (out of twelve) the children play on, what are the chances that the lady, acting alone, will accidentally identify those two? Remember that she is not actually looking for the children's stones, but for stones of high energy. Well, I'll save you the tedious math. Her chances are 2 in 132, (or 1 in 66 or 1.515 percent), and that assumes that she is looking for exactly two stones, which was never a given. As she might have identified one or three or five stones, her chances of finding just the two children's stones are actually much less. Now, the gentleman has the same chance of identifying the children's stones (assuming he is going to choose two at random out of twelve), so his chances are also 1.515 percent. However, the probability of *both* coming up with the two children's stones is compounded. Again, the math would take a little explaining but the bottom line now is that their chances of each identifying the two children's stones is 1 in 4,356 (.0002295, or .02295 percent): pretty slim. Admittedly, it is true that one of them might influence the other with their results, but they told me that they arrived at their conclusions independently. They had no reason to be anything other than honest, as there is nothing extraordinarily impressive about

The Bunny Stone and the Cat's Back, at Earthwood, identified as having exceptional "energy."

their identifying the same two stones for highest energy, which is, again, a 2 in 132 likelihood. It's the conjunction with the third circumstance—the child's play factor—that makes the numbers extraordinary.

I told Ed about this apparently higher energy in the two stones where the children climb and play and he nodded knowingly.

"Yeah, I've always been convinced of that," said Ed. "There's hundreds would go along with that." He didn't need to know anything about probability theory.

Ed says that many mystics and dowsers monitored his new stone circle very carefully after its construction. "The power came to my stone circle over a period of three years and now it is fully charged, as if it had stood here for thousands of years," he said.

I also told Bill Cohea all of this, and asked his view of the energy in a stone circle. Bill's view is that the energy is there the moment you build the

circle, at least he feels that this has been the case at Columcille.

I'll finish this little diversion into the—what should I call it? mystical? pseudoscientific? transrational?—by telling you something that happened during the construction of the Stone Mountain Stone Circle, something that I did not want to tell you in chapter 3, thinking it would be more appropriate now.

On the first day of construction we only put up the four directional stones. That night was the Wickerman bonfire, and we tested the north-south alignment against Polaris. About midnight, ElfWiz (Anne and John, you may recall, who were renewing their marriage vows) gravitated to the center of the circle, which was really as much of a square or a diamond at this point, as only four stones were standing. There, according to Anne, they "picked up a phenomenal energy, strongly magnetic and deeply spiritual." They could feel these vibes focused on them at the center. Their experience, which I have no reason to doubt, could be seen as support for Bill Cohea's contention that the stone circles have energy as soon as they are created, as opposed to Ed's view that the energy comes over time, and from hundreds of people visiting, touching, and embracing the stones. I considered yet another possibility. Perhaps the stones were like an "energy battery" and the high energy on the Stone Mountain site that day—from the builders, the snake dancers, and the blessings of children—were enough to give the stones a cosmic jump start.

And this from a skeptic!

THE SEVEN SISTERS
TEMPLE TODAY

Ed uses his stone circle, and is very interested in how others use theirs. When he visited Earthwood with Glynis in October of 1995, we built a bonfire at our stone circle in their honor and several of our friends came over to meet the archdruid and druidess. Ed and his friends dance around their stone circle and achieve a kind of high that way. "I am turned on by rocks," Ed says. "If you run around the circle five times, you feel as high as anything, like you'd had a pint of beer." A scientist might suggest following the same running pattern without the stone circle and comparing the results.

People come to the Healing Stone to cure their backs. They come to the Rocking Stone for good luck. Some have won a lot of money on the lottery after rocking the stone. (Quite a few others haven't.) Couples young and old come for Ed's special nuptials at the Wedding Stone, even if it isn't strictly legal. Some have come to remember lost loved ones who died in the Falklands War. Some just come out of curiosity and to meet the eccentric. Ed doesn't care. All are welcome.

There is no admission fee to see the Seven Sisters. As a temple, Ed doesn't feel that people should have to pay to see it, and he is obviously disappointed about the visitor's situation at Stonehenge. (He has never visited the famous monument, although I have tried to talk him into going for years, by way of a special access permit.) In a pamphlet that they leave for visitors when they're not at home, Ed and Glynis say, "If you were expecting an ancient site, please don't leave before you have looked around. Remember, Stonehenge was 'new' once and how wonderful it must have been to be amongst the first visitors on the day it was finished and during the years that followed."

I think that Ed uses the circle most as his eye on life. Although he does get around for a man with very limited vision, he does not really need to go out and see the world. Eventually, the world comes to see Ed. I can't think of a better use for a stone circle.

— CHAPTER 6 —

Ivan McBeth and the Swan Circle

It is vitally important for a human-made structure to honour and make connections with the physical environment. This creates stability, acceptance by the spirit of the land, and very real anchorlines which allow energy flow between the structure and surrounding landscape.

—Ivan McBeth, *The Swan Circle*

September 21, 1995, was a bright warm perfect English day at the great Avebury stone circle in Wiltshire. Jaki, Darin, and I made our way inside the henge, across the trunk road that skewers the circle, and into the sheep field that contains the five remaining stones of the south inner ring. Immediately we saw that some special event was underway. Dozens of brightly festooned individuals were gathered at the south ring, chanting, drumming, dancing, and generally celebrating . . . the equinox! At home, I never fail to be aware of the four important solar days of the year, the equinoctial and solsticial beginnings of the four seasons. But halfway through a four-week vacation in Britain, with no appointments to keep or plane to catch, I'd lost track of the date.

The most intriguing celebrants were a young man and woman with didgereedoos, the Australian Aboriginal droning instruments, playing a timeless refrain to one of the major standing stones. The long-haired lady with the top hat particularly caught my attention, as the lower half of her body seemed to be covered with nothing but painted swirls.

Closer examination, however, indicated that the canvas for this compelling artwork was, in fact, flesh-colored tights. The closer approach also heightened the intoxicating effect of the unfamiliar music. We began to pick up the unusual harmonies between the musicians, and we could feel the resonance with the stone itself. Both players held the horn of their instruments right up to the fifteen-ton monolith. That the playing was part of a set piece, possibly an original composition, became readily apparent when the refrain stopped suddenly, but with a musically logical conclusion.

I gave the pair a few minutes to come back from wherever they'd been, and asked the man about the event we'd just witnessed. In truth, I was still a bit in awe of the earth goddess with the provocative stems, and thought, perhaps, that the gentleman might be a little less put off by a middle-aged American. I asked him—Richard—if they had been putting energy into the stone, deriving energy from it, or what?

"It's a bit of both, really," said Richard, and we were soon into a conversation about stone circles

A young couple at Avebury exchange energies with an ancient megalith, autumnal equinox, 1995.

and their energies. I told him my story of dowsing for energy at our own stone circle. Richard said, "You build stone circles? You should talk to Ivan McBeth. He builds stone circles." I asked if McBeth was here today.

"He was here a few minutes ago—there he is, big bloke with the top hat."

IVAN MCBETH

Ivan McBeth, accompanied by a brightly dressed lady, was making his way out of the sheep field. He was barefoot, and wearing a red jacket and a large green top hat, seemingly the headgear of choice at the gathering. Leaving Jaki and Darin behind, I rushed to intercept Ivan just before he exited the

west gate, which led to the Stones restaurant. Bearded and big indeed, Ivan goes about 6'3" and weighs (I learned later) 23 stone. (American readers: This is 322 pounds, but to measure Ivan McBeth in anything other than stones just isn't right.) I apologized for my intrusion, and explained that I was a fellow stone builder and interested in his work. Our conversation was short, but cordial. Ivan gave me a phone number to contact him upon our return to the area from Scotland and, fortunately, drew me a map to find his Swan Circle at Pilton. As we expected to see each other again in a couple of weeks, and he was heading for lunch with his friends, we kept the conversation short and sweet.

Upon our return to Southwest England in October, I tried to contact Ivan, without success. Sometimes, despite his size, he can be hard to find. But we did find his wonderful stone circle at Worthy Farm in Pilton, near Glastonbury.

Two years later, I was able to track Ivan down through the phone number he had given me for Star Child of Glastonbury, a company specializing in healing herbs, incense, candles, and other New Age products. Ivan, it turned out, had built a lovely stone circle for the owner of the company. I called Ivan and told him that I was coming back to Britain on a megalithic research journey, and asked if we could meet again. We could, and we did.

Ivan McBeth is certainly one of the world's foremost experts on designing and building stone circles, and yet I might not have met him, or even heard about him, had we not accidentally visited Avebury on the autumnal equinox, or if I had not extended myself to Richard, the standing stone musician. This tale of serendipity is a perfect example of what James Redfield calls "cultivating coincidences." By 1995, I had read Redfield's *The Celestine Prophecy,* which simply reinforced my natural predilection toward extroversion. I speak to strangers, always have, and it has made a huge and positive impact on my life.

When I caught up with Ivan again, he was staying in Bristol with his lovely girlfriend (and druid priestess), Julie Britton. I arrived at Julie's on the day before the autumnal equinox, and found out that a grove ceremony would be happening the following night at the little stone circle in Julie's back garden. Twenty or so druids were expected for the event, known as Alban Elued in the druid circle of the year, and I would be welcome as a guest. Of course I was delighted to have the opportunity to see a stone circle in proper use!

I spent most of Saturday and Sunday in Ivan's company, and Julie often joined in our conversations, although, as priestess of her grove, she also had a lot of preparation to do for the Alban Elued ceremony. During our first long session, over herb tea in the sitting room, Ivan and I compared notes, stories, and photographs of the various stone circles with which we'd been involved. It was like a couple of cricketers discussing batting (or baseball players talking about hitting).

A grizzly bear of a man, Ivan might be intimidating were it not for his smiling eyes and gentle manner. Very early in a conversation with Ivan, it is apparent that he is far more concerned with the invisible than the visible, the magical as opposed to the physical, the transrational instead of the rational. Nowhere is this demonstrated better than in the design and construction of the Swan Circle at Pilton.

In a way, it was good to see the Swan Circle two years earlier without its creator on hand to explain it. My family and I could explore the ring for what it is, just as we might explore an ancient circle whose designer has been long forgotten. Intuitively, we sensed that some magical geometry was meant to

Ivan McBeth.

be incorporated into the design, but we could not solve it in a short visit. The "circle" was clearly not circular, as pacing confirmed; I guessed an ellipse. We still had our dowsing sticks from Castlerigg, and my diary reminds me that "I went two for two in finding the edge of the circle."

But now I had the man in front of me, and his 52-page self-published book, *The Swan Circle* (1992) as reference. What if we had such a book written by the main architects of Stonehenge? I wondered. How much work and silly speculation could have been avoided? As it is, we'll never know most of the answers about Stonehenge, a situation with which Ivan is very comfortable.

Stonehenge

Our conversation began with Stonehenge and the druids, an appropriate place to start given the preparations that were going on. I asked Ivan and Julie about the connection, if any, between druids and stone circles.

"We have to go by what feels right," said Julie. "We know when the druids were first written about, but we don't know how long they were around before that. There's this assumption that they arrived when the Romans reported them."

"There have been people who have been dedicated and committed to finding out the secrets of

The giant Druid's Oak stands sentinal near the Swan Circle.

the universe since the beginning of time," observed Ivan. "And, in order to do that, you need to increase your personal power, to increase your life force, so you can understand more. So these people would have gone to the places in nature that afforded them the best chances of doing this, which we call sacred spaces or power spots or whatever. Waterfalls, wells, sacred mountains, tree groves, and, eventually, stone circles. And there's something about stone which is incredibly powerful. These people created the stone circles, but, at that time, maybe they weren't called *druids,* which is just a label. They were the shamans of the day."

"So the line goes straight back," said Julie, matter-of-factly. "Dead easy."

I brought up the common complaint among archaeologists that the druids had nothing to do with the building of Stonehenge and that they find the annual druidical presence there on important solar dates to be a joke.

"Those are people with territorial problems," replied Ivan.

Ivan respects Stonehenge on every level: spiritual, intellectual, astronomical, inspirational, physical, aesthetic. He goes there by special access—sometimes alone, sometimes with Julie or others—to visit in an intimate way. He goes there to celebrate, to get answers to life questions, to tune into what he thinks of as the transrational reality.

"Whenever I go to Stonehenge," Ivan explained, "I feel that I have an appointment. I know that I wouldn't get anywhere near it unless it was inviting me in. And when I go there, there's always a message and it comes when I relax. I greet it. I honor it. And, somehow, I just open myself up, if you like, and go into a sort of dream state."

The Question Why

This seemed like a good cue to ask Ivan the question that I always ask a stone builder. *Why do you build stone circles?*

"It's my passion," stated Ivan without hesitation. "I was brought up on Dartmoor. I didn't have a particularly happy childhood, but I found a lot of peace and solace inside these structures. It's always stayed with me and I want to be able to recreate this sort of environment for myself and for others. As I got older, I became even more passionate about them. I believe that there are lots of different ways that people transform themselves into peace and joy, and building stone circles is *my* way. Someone else might feel the same way about playing a trumpet."

A passion for the stones. Exposure to the wild stones on the moor at an early age. Getting messages from the stones. It all sounded familiar. *Is there a pattern starting to develop here?* I wondered.

THE CIRCLE IN JULIE'S BACK GARDEN

The sun was now shining over the hedge into the back garden and I was keen to see the small ceremonial circle and photograph it while the light was good.

The circle is small, perhaps fifteen feet in diameter and composed of twelve small standing stones of varied color, texture, and character, each standing between one and two feet tall. Ivan told me that he had gathered most of them from the surrounding countryside, while one or two had come from as far afield as Cornwall. The individual stones are roughly aligned with the directions from which they came. The circle is in scale with the small garden and is enclosed by wonderful greenery: ground plants, flowers, shrubs, and a grape arbor. It is charming, friendly, and—as I was to find out the next evening—functional. It shows that a stone circle need not be built of multiton monoliths to be powerful and beautiful. *Build a small circle.* Perhaps it will be a prelude to something on a grander scale, perhaps it will be perfect as it is and will satisfy your wants and needs.

Ivan discusses building a stone circle for Joan, right, while Julie and her son Lorien look on.

Certain druidical features are incorporated into this circle, such as a processional way through the grape arbor and into the circle from the west gate, defined by a pair of stones. A small firepit in the center is the focus of ceremonies, and next to the firepit is a 13th stone, the Anchorstone, which grounds the circle to the Earth.

While I was learning about and photographing the small circle, a brightly festooned druidess of about sixty years arrived on the scene. For the next half hour, I witnessed a meeting that might have taken place in ancient times. Joan, the rainbow-draped visitor, had come to speak to Ivan about building a stone circle in her garden. She described the features she wanted to incorporate. "It needs a water feature, but I don't want the responsibility of a pool I've got to maintain; I just want a circulating pump and some rocks for the water to come down."

"That's fine," said Ivan, smiling, conceding that perhaps a modern circle might include modern technology.

The two discussed stones, their availability, and how to transport them. Joan said she would like Ivan to come and look at her site. They discussed cost. "I can tell you how much I cost," said Ivan. "I work on donations."

This extraordinary exchange—stone circle builder and contractor—got me wondering if there would have been experts in ancient times, designers and construction crews that might have gone from place to place on commission to build circles. I mentioned this to Ivan later. "Sure," he replied, "But I believe that the people then were very much different from how we are now. It's dangerous to project our way of perceiving the world to how it was four or five thousand years ago."

Chanting a Stone into Place

Before we went back inside, Ivan indicated the largest stone in the circle, and told me how he and a group of druids had applied chanting to the process of installing it in its socket.

"You chanted the stone into its hole?"

Ivan laughed. "Yeah, sort of. The stone weighs a couple of hundredweight, and I wanted to put it in myself. Everyone made a horseshoe around me. Along with the stone, I was in that place of power that you see at the opening of a horseshoe magnet. Everyone started chanting, including me, chanting the druid mantra *Awen,* until the resonance was just right and I felt their power concentrated on me and added to mine. When it felt just right, I picked the stone up and dropped it in the hole, simple as that. I have no recollection of feeling any extraordinary weight or difficulty."

The story reminded me of how rowing teams use a stroke to bring harmony to racing sculls, improving their efficiency. Or how, in many parts of the world, music is used on large projects to unify the workforce. I could well imagine that drums or chanting were used at the pyramids or Stonehenge or at Easter Island to expedite the work.

BUILDING
THE SWAN CIRCLE

After Joan left, I got a chance to learn more about Ivan's involvement with the Swan Circle, his major work to date (although "commissions" are coming his way thick and fast as I write). I learned even more upon reading his fine booklet. Much of the information in this section comes from *The Swan Circle,* with Ivan's kind permission.

During the 1989 and 1990 Glastonbury Music Festivals, held at Worthy Farm, Pilton, in Somerset, Ivan worked in partnership with Iain Sika Rose to create sacred space. In 1990, they actually built a small "pixie" circle, "delicate and beautiful." The circle only existed for a year, because it had been built on land designated "agricultural use" and it was decided that the stones might be hazardous to the cattle. But during its short existence the circle was visited by hundreds of festival folks. In *The Swan Circle,* Ivan tells of the pixie circle's short but stellar existence:

On Midsummer's Day it became the venue for many colourful ceremonies and celebrations, including an overnight vigil, the sunrise ceremony itself, naming ceremonies for children, and a marriage. Mistakes were made in its creation, and were noted for the next time. Our successes were honoured, and insights remembered. Its beauty and success as a functioning sacred space became the inspiration for the Swan Circle. Some of its main stones were incorporated into the Swan, including the white quartz central stone which carries the memories and feelings of the first stone circle at Pilton into the new.

The Spiritual Reality

Ivan approached the design of the Swan Circle with a belief that stone circles have three main purposes:

a) to provide a network to regulate energy flows between the heavenly bodies and the earth, a sort of self-regulating acupuncture system;

b) to create a network of sacred space over the earth, within which humans can do things that actively affect the life of the planet as a whole;

c) to be guardians of the secrets through which humanity can realise its divinity.

Pretty heady stuff, and a tall order, indeed. But Ivan approaches the task with the tenacity of Hercules, tempered with the patience of Job. He does not allow himself to be distracted by the physical at the cost of the transrational. Neither does he ignore practical problems.

Measuring instruments and pocket calculators are the tools to work with the physical levels [of stone circle construction]; feelings and the ability to open oneself to the Dream, or the Silence, are the tools to investigate the invisible realms.

Before the reader dismisses such talk as New Age claptrap, consider that Ivan's way of looking at and talking about the problems of stone circle construction and the methods of solution, are probably much closer to the way the ancients approached stone circle construction than a more "rational" approach that might be taken, for example, by a contractor asked to erect a few monoliths in front of the Standing Stones Motel. Which approach will come closer to the multileveled realities of a Callanish or a Brodgar or an Avebury? Continues the bard:

Our obsession with physical evidence necessitates us, collectively, to deny this living network [of earth temples and stone circles] which is "plugged in" to the sacred. Over the relatively short period of time that this increasing lack of consciousness has held sway, many of the most beautiful and powerful temples have been badly damaged or destroyed. This is reflected in the current state of health of the land, and our lives.

Ivan uses the term sacred space to describe "nodal points" in the Earth's energy or vortex, corresponding to specific physical places on the landscape.

Yet before the dominance of the "rational" left-brained world view, the land and all elements of life were sacred and meaningful. There are still peoples in diverse areas of the planet whose everyday lives are lived in accordance with age-old traditions which re-create "the world" in every action. The Australian Aborigines, for instance, live simultaneously in the everyday world and the "Dreamtime." They are aware of how both aspects are essential for Life in their interweaving flow; regularly, when they hear "the Calling," they leave their homes and go "Walkabout."

Their bodies physically walk, yet their awareness is primarily "in the Dream." They make a pilgrimage through the living sacred land of their Ancestors, recognising their journey as a return and reconnection to their spiritual roots. At special places on the way, they will make ceremony to honour and re-enact the mythological creation story pertaining to that specific space, "knowing" that what they do helps to keep their sacred World alive and healthy. Sadly, under conditions of "spiritual genocide," and the destruction of their sacred "Dreamscape," their unique, gentle way of life is severely threatened.

Siting the Circle

On small lots, or a backyard, there may not be much choice about where to position a stone circle, but the owner of Worthy Farm, Michael Eavis, gave Ivan a large canvas to paint, a large triangular field called the King's Meadow. Ivan naturally gravitates toward "safe and calming influences" and his first thought was a flat site area "between the Druid Oak and the totem pole."

While walking around the field with Michael, searching for the right place, the pair arrived at a long flat area in the middle of the field, what is called in parts of America a *bench*.

He felt sure it was right; I withheld judgement until I had spent more time in the field. This second potential site had very different energy; it was higher up the field, giving it a view of the valley. It had a feeling of wildness that the other space didn't have, as if anything could happen there. It seemed more in contact with the elements, and was far more exposed. . . . I asked the opinion of many people whom I trusted for their sensitivity and knowledge of earth energies, and finally decided on the second site. Only when the stones entered their holes in the earth, did I realise how right it actually was.

Designing the Circle: The Coming of the Swan

Ivan's first design was a true circle of stones, standing about three or four feet out of the ground, and all surrounding a taller central stone. He had in mind to position the stones by hand, using rollers and pulleys. He ordered nineteen stones from a quarry thirty-five miles away, stones of the size he'd envisioned. Unfortunately, the stones were badly damaged during transport and Ivan and Michael were not at all impressed with them. Disappointment did not last long, however, for the very next day they heard of some much larger and much more beautiful stones available from the Torr Works, very close to Pilton.

They were indeed wonderful, and I chose nineteen of the most suitable. They were delivered to the farm individually, and so suffered minimal damage. I had to keep a mental picture of the shapes and sizes of the stones I had chosen while supervising the unloading, as I only had one shot at dropping them off at the right places. It was important for the different "characters" to be placed close to their final positions. Moving a ten-tonne chunk of rock is very difficult.

The new stones changed everything. Ivan had to abandon any hope of moving these much larger stones into position by hand—he had only four weeks to complete the task, not four months. A second major change was that the much larger stones simply wouldn't fit in the flat site available. The largest stone circle that could be built there would be twenty-two yards in diameter.

I was sitting by my fire on the night of the May full moon, pondering the dilemma. I absent-mindedly watched the moon rise—such beauty as the field became bathed in silver! I remembered that many so-called stone circles are in fact egg shaped, and realized that an egg would solve the problem. An egg! The stones would move out in two directions, and stay within the boundaries defined by the contours of the field. Within a couple of minutes of this realization came the final piece of the puzzle. The Swan!

This was not an entirely illogical association. Ivan has always been a keen student of the night sky and one of his favorite constellations is Cygnus, the Swan, which he describes as one of the few star pictures that one can actually see, "without twisting one's imagination into horrific knots." You can find Cygnus flying through the Milky Way in the Northern Hemisphere sky, according to Ivan, "on the cusp of Capricorn and Aquarius."

Nor did the idea of incorporating the Swan enter Ivan's considerable consciousness for the first time on the silvery rays of May's full moon. A year earlier, he had wondered how his favorite star group would work with stone ring geometry. He transposed his paper design onto a wooden pallet that was covered with plaster. At the star positions, he placed pieces of slate. After the plaster dried, he painted the swan stars white and the beak star yellow. Earth was sprinkled over the entire model, leaving the painted slates extending upward, and grass seed was planted.

The major stars would not fit into a circle, but perfectly into an egg. It looked and felt very beautiful; I spent hours gazing at it, giving it the occasional haircut!

On that special evening, as the full moon rose higher, a peaceful harmony filled me, and I knew exactly what I had to do. What had started out in an innocent, bumbling and slightly apologetic way had changed into a dream of grace, beauty and power. This was to fill me with an unusual confidence and trust that lasted until the work was complete.

The layout of the Swan Circle, including how Ivan incorporated the Cygnus star chart into the egg shape, is shown on page 135. In chapter 11, I show how to use simple geometry to lay out "eggs" and other shapes.

The Midsummer Sunrise Alignment

A key feature of Ivan's stone ring design was that the long axis of the egg would align with the rising sun on the longest day. And the most trying, the most frustrating, and probably the most stressful part of the entire operation was not knowing where the summer solstice sunrise point would actually appear on the horizon. Ivan speaks woefully of "the agonies I had to go through" because of this missing information. Had the center of the ring been known a year earlier, and an observation taken, it would have been an easy matter to align the axis of the ring, but Michael had not indicated his desire for a permanent circle until October 1991, and Ivan did not get the final go-ahead until April. By mid-May, time had become a critical concern. Contractually speaking, the circle had to be available for use as part of the Glastonbury Music Festival on June 26, but from Ivan's point of view, it had to be fully "on line" by sunrise on June 21st. In the event, Ivan marked out the alignment and positions of the stones three times. On May 17, based upon pure guesswork, he made a "practice run" in order to become familiar with the equipment available. After spending time at the site of the original "pixie" circle, where he still remembered the sunrise position, he did a second layout in the King's Meadow on May 20, based upon a transposition of the pixie alignment. He still wasn't happy, but the stones could not wait any longer and they began to arrive while Ivan made his second alignment. *Panic stations*. Ivan choreographed the placement of the stones near their estimated positions.

The first holes were dug, by hand, on May 21, based upon Ivan's second estimate. But Ivan continued to work on the sunrise problem, worried that he might be digging holes only to have to fill them back up again. On the morning of May 25, Ivan's confidence was given a cosmic boost.

Serena (my girlfriend) and I were sitting by the fire, talking about the ways in which the stones would come alive and awaken. Suddenly a very strange pulsing, whistling sound filled the air and both of us froze. It was quite frightening as its volume and pressure increased. Then, seven swans, flying in V-shape, flew across the King's Meadow over the stone circle, from west to east. From that point on a great weight was lifted and I knew that what I was doing was confirmed by the sacred and fully supported. Although there were still crisis points and worries after this, there was an underlying certainty that all was in order and on schedule.

Fortune shined when Ivan got his hands on a book about the nearby ancient and sacred hill site of Glastonbury Tor, with a diagram of the positions of the sun throughout the year. He had observed a sunrise at the King's Meadow (Swan Circle) site on May 18, and had marked the position. Then, with the help of the Tor diagram, he worked out the angular distance traveled by the sun between May 18 and the solstice on June 21: 8°. As both the Tor site and the King's Meadow site enjoy fairly level horizons, Ivan deduced that he could subtract 8° from the sunrise azimuth that he had observed on May 18. (The sun rises farther north each day until the summer solstice; the true north azimuth is defined as 0°. The midsummer sunrise azimuth at Stonehenge, on exactly the same latitude as Pilton, is approximately 49° east of north.) The third layout, then, was based on science—but Ivan was still nervous, and Michael's giving him a hard time of it didn't help Ivan's peace of mind.

I don't know if he (Michael) was being serious or just winding me up for the sake of it! He would repeatedly state, confidently, categorically, that he knew the land like the palm of his hand and my estimates for the sunrise were way out. "Maybe, if you are lucky," he said mischievously, "the sun will rise over the stone to the left of the Sunstone, and all will not be lost, but the position you have guessed is miles off!"

Sod's Law [Murphy's in America] was an integral part of the process. Having committed myself irrevocably to the final alignment, the sun, of course, refused to show itself for nearly three weeks. With Michael's comments, and the well meant advice of various "experts," I had to restrain a strong urge to realign the egg to their predictions. Finally on June 13 there was a perfect sunrise. Michael arrived as the sun rose, and we watched it together. It appeared a bit to the right of my estimated position. Allowing for the slight movement remaining before Midsummer, my alignment was exact! I danced a wild jig in celebration, the weeks of constant uncertainty dropping away like an outgrown, heavy skin into the dew-spangled grass.

The Physical Reality

Preferring human involvement whenever possible, Ivan, along with volunteers, dug all of the sockets by hand. Hand digging also allows a more precise hole, tuned in to the stone's shape, and makes for a tidier site upon completion. "Both Michael and Keith, the digger driver, couldn't understand why we wanted to expend so much energy, when it 'would be so easy' to use the machines—I don't think they ever totally understood."

Unfortunately, four of the holes dating from the time of the second alignment had to be filled in and dug again when Ivan made his final alignment on May 26. Work proceeded at a feverish pace, and there was no regular team of helpers, just Ivan and whoever was foolish enough to happen along. The stones were too large to move by hand; Ivan decided that it would take a crane to build the swan. Ivan booked the crane for May 31, and the sockets for stones 1 though 18 were completed by the 30th.

On the morning of the 31st, Ivan and friends performed a simple ceremony at each stone socket, leaving gifts and offerings that they deemed appropriate. At each hole they left: "Chalice Well water, a home-made medicine made from St. John's Wort, a crystal from the central chamber of Crystal Mountain, a scattering of crystals from Southwest Ireland, a pinch of tobacco, and assorted flowers and herbs."

Ivan had put out word that he wanted stones or crystals from places all over Britain and around the world to put under the stones, and he wound up with offerings from South Africa, Ireland, Peru, Land's End in Cornwall, Dartmoor, the plains of Armageddon in Syria, and virtually every island group in Scotland including the Hebrides, Orkney, Shetland, and Skye. Sacred places were well represented: Stonehenge, Iona, Mt. Snowden, Tintagel (reputed location of Merlin's Cave in Cornwall), crystals from Knossos where Jason slew the Minotaur, water from the Ganges River, ash from a lion temple in India. Some of the more exotic stuff included a Herkimer diamond, a meteorite, crystals of uranium, and a fragment of the Berlin Wall. Archaeologists in the year 4000 are going to have a cosmic puzzle on their hands. Much of the ceremony was completed before the crane's arrival, although additional offerings were made by Ivan and his friends just before each stone was dropped into place. In addition to linking the circle to the energies of these otherworldly power points, Ivan was hoping to create a mood of peace and harmony.

At noon the crane arrived with a very negative and bad-tempered driver. I knew it was going to be an onslaught and hoped I could hold it all together. The driver refused point blank even to attempt to lift some stones. Reluctantly he readied the crane and, unfortunately, we started with the Wobble Stone. It was a pig and refused to respond to our attempts to move it. It was our first try, and, lacking experience, impossibility hung thickly in the air. Eventually the stone lay in its hole at a horrible angle and I knew that it would be very difficult to maneuver later. At least it was in the hole and that was a start.

The next stones were large and received some knocks from the digger that churned my stomach and set my teeth on edge. Yet, as we gained experience, the stones went in easier and easier. Soon, after five stones and numerous squirts of Rescue Remedy later, I knew the crisis point had been passed. [Note: Rescue Remedy is a homeopathic healing infusion made from flower essences. According to Ivan, it "rectifies shock."] As we moved around the circle, the crane operator became more and more enthusiastic, jumping out of the machine, offering tips and asking questions. What was at the beginning a torture became a deep and satisfying creative process.

In conversation, Ivan shared additional details of the crane operator story.

"His heart wasn't in it to start, but after a while he says, 'What's all that stuff you're putting in the holes?' I explained that it's really important that each stone goes in in a beautiful way. Like, the North Stone is aligned to Iona and Callanish, so we've got little bits of crystal from each of those places. He says, 'I live west of here. Can I put something into that hole over there?'"

"So you had him," I said.

Ivan laughed. "Yeah, I got him. So then he really became a part of the whole thing, and with each stone it got better."

And who can say that the offerings had not done their job?

Packing the Stones

What the crane did that day was to lower the stones into their holes, using 5-tonne-test nylon slings to prevent damage to the stones. The next operation involved straightening the stones with a mechanical digger, known throughout Britain by a trade name—JCB—and in America as a backhoe.

The next ten days were spent adjusting the stones with the JCB, and ramming clay down around the sides with scaffolding poles—exhausting work—to set them firmly in place. The crane came again on June 8, and we reset both the Wobble Stone and the East Stone.

What a sight the stones had been, plonked in their holes before being straightened and set firmly in place: megalithic drunkards, lolling around at all angles; stone beings seeming to defy gravity in a frozen chaotic dance. It was slightly unnerving to see them like this, but once they stood tall and proud, feelings of embarrassment and humour changed to awe and reverence. By this time it was obvious we needed more stones, and four new ones were found, all between twelve and fourteen feet long, with exciting personalities.

A stone is lifted with the loader so that a nylon sling can be installed.

The nylon sling enables the stone to be lifted by a crane and placed in its socket.

More stones were needed because Ivan felt that the gaps between stones 6 and 4, as well as between 4 and 7, were too great. Also, two stones were deemed to be too small, and had to be replaced. Stone number 21, the Sunstone, was not positioned until after the final sunrise observation on June 13.

A Tour of the Circle

Space does not permit a complete discussion of all of the stones, their meanings, their alignments, and their individual stories, but it is worth highlighting some of the special features.

We'll tour the Swan Circle in roughly the order in which it was built, which also corresponds to Ivan's numbering system. The best stones, the lowest numbers, were earmarked first.

"The four directions are honored by large stones with appropriate personalities," Ivan told me. "I always use Polaris to get my primary alignments. I stood where the Center Stone would go and a friend moved a distance away from me to the north with a really long pole. Even then, I couldn't touch the star, but at least I could see it by eye, and position the pole. My friend marked the bottom of the pole

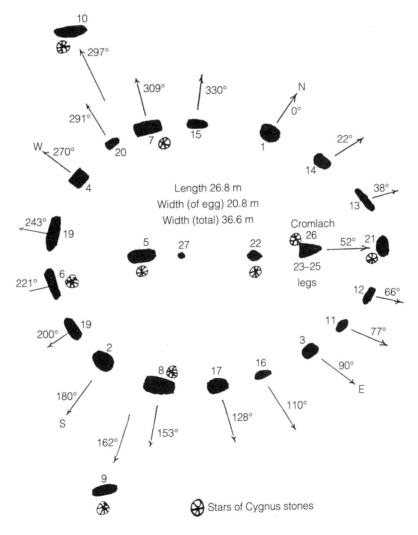

Plan of the Swan Circle, by Ivan McBeth.

on the ground, and I just carried the line through. All the rest gets done from that with simple plane geometry." The directional stones are marked 1, 2, 3, and 4 on the chart.

Stones 5 through 10 are the stones that signify the stars of the constellation Cygnus the Swan. Number 5 is a holed stone within the ring called the Moonstone. The hole is about six inches in diameter and is a foot off the ground. Originally a quarry hole used for setting a dynamite charge, this aperture somehow survived the explosion intact, and it is used to site the midsummer sunset over stone 10, the Wingtip Stone. Later, stones 22 (Starstone) and 21 (Sunstone), and the *cromlech* (or dolmen) were installed, completing the Cygnus star group.

Stones 1 through 10 were a given; they were required by the special astronomical and aesthetic features of the design. After their positions were marked with wooden stakes, Ivan "had to sit for a long time inside the circle, contemplating the empty spaces, and wondering how to fill them." His considerations included the number of stones available, sunrise and sunset alignments, alignments to prominent features on the landscape, and—very important, in my view—balancing the sizes and shapes of the stones. After all this meditation, the remainder of the ring evolved as follows:

Number 11, the Needlestone, aligns with Small Down Knoll, a hump on the horizon.

Numbers 12 through 19 filled spaces, as seemed appropriate. They have names like Diamondstone, Birdie, Wolfstone, Wobblestone, and Teardrop, often descriptive of the monoliths they identify.

Number 20, called Sunset or Deathstone, has a part to play in the midsummer sunset alignment, and 21, Sunstone, of course, had to wait until the sunrise alignment was confirmed.

Number 22, Starstone, is part of the Cygnus stellar configuration, as is the cromlech, composed of a large anvil-shaped capstone supported by three legs that had been individual stones in the original pixie circle. It should be noted that the word *cromlech,* in

Britain and Ireland, most often refers to a dolmen-type monument: three or more stones supporting a heavy capstone. In France, however, cromlech refers to a stone ring, often with closely packed stones, called orthostats, almost touching each other, such as the stone oval at the west end of the huge Menec stone alignment at Carnac. Both uses derive from the Welsh—and, therefore, earlier Celtic—*crom* (bent or bowed) and *llec* (a flat stone.)

From the beginning of the project I knew that a cromlech would be part of the circle. I wanted to build a covered space for the "Little People" who are shy and are reluctant to appear unless there is a place prepared for them, covered and discreet. The cromlech represents a hidden star in the neck of the swan, in astronomical terms an X-variable, a "black hole." There is a certain mystery about a cromlech, especially one that is too small to enter, and it seemed perfect to represent the black hole.

The final stone, number 27, grounds the entire monument to the earth, but its story is best told in the context of the final week's feverish activity, leading up to the critical moment of the circle's birth.

Starting the Circle Spinning

With less than a week to go to the midsummer's sunrise, another celestial event came into play for Ivan and his Swan Circle project.

Spending the night of June 15th, a full moon eclipse, at Stonehenge was important to me; I was able to recharge my batteries, contact deep energies of the earth for guidance and have a welcome break from Worthy Farm. I took with me a gift from the Swan Circle—a strange geode found while digging—and left it there. I found a small stone from the center of Stonehenge—a very small one, honest!—and later buried it under the Sunstone, together with an ammonite found on Glastonbury Tor. As the Swan Circle lies on the Stonehenge-Glastonbury ley, the leaving of gifts is a wonderful way to increase connection between the sites.

Ley is a term coined by Alfred Watkins in the 1920s and refers to straight lines that seem to connect ancient monuments across the countryside, despite any obstacle. Many believe that these "old straight tracks" follow paths of earth energy (Alfred Watkins, *The Old Straight Track,* 1970).

Upon Ivan's return to the King's Meadow, life suddenly began to move at very high speed. On June 19, in just two-and-a-half hours, the last five major stones were positioned, leaving only the 180-pound white quartz center stone to go. On June 20, Ivan and his volunteer crew worked feverishly to repair the damage to the meadow incurred during construction. Keith, the digger driver, removed the excess piles of clay and old turves from the hole digging, and used the material to make two earthworks, one each side of the circle. He also brought in many square meters of fresh turf from the edge of the field, for use in repairing some of the most severe scars. Using "reject" stones, the crew also created a megalithic gateway at the entrance to the field. Later, a remarkable caricature of Michael Eavis, the guardian of the land, was incorporated into this gateway.

The team completed their physical work and downed tools at 5 P.M. on Midsummer's Eve, exactly on target. But Ivan could not relax. He was both the maestro and the midwife; all looked to him to take the circle into its final birthing phase.

There seemed to be so many individual things I had to do, people to coordinate with, archetypal energies to honour and surf on, all the while

Michael Eavis, the owner of Worthy Farm, is captured in stone and greets visitors to the Swan Circle.

trying to monitor my own sacred space and keep centered. I had to be aware of the choreography of the moment; everything had to be done in its right order, while a million things and the energies of a lot of beautiful people seemed to pull my attention. In these situations, where there are no logical solutions, my terror seems to provide me with the energy to dance through. Literally.

The final act in the process of creation was underway: the giveaway of the Swan Circle into the world to fulfill its own destiny.

The conclusion to the Swan Circle story must be told in Ivan's eloquent words alone, without interruption.

I spent Midsummer's Eve in vigil with a large number of people either inside or close to the circle. Waves of silence, music, stillness and activity swept through us and passed on. A large flickering fire slightly up the hill illuminated the magical scene, silhouetting the stones. Wild deep-earth dancers were celebrating their birthday; they moved to the rhythm of the flames, pulsed with the drums, and beckoned people to join them.

At about 3 A.M., the Centre or Ancestor stone found its final resting place at the centre of the circle. It is the focal point through which the Four Directions and the major alignments pass; I recognised it as the "key" to activate the physical energies of the Swan. Very nervous about what I was doing, I relaxed and the magic took over. As I started to dig by the light of a single flare, a ring of silent, supporting people gathered. I invited whoever felt moved to dig out a spadeful of earth. Still in silence, a stream of people, moving in the slow motion of Dreamtime, took turns preparing the hole to accept the key.

Again I felt that some extra energy was needed, and looked around. Absolutely on cue, nine-year-old Tara, Serena's daughter, separated herself from the ring of people and crouched down beside me.

"Rocky the racooon and Bungle the bear are here to help," she said. These were Tara's furry hand puppets. The stone was positioned on the rim of the hole. Rocky, Bungle and I pushed with all our might. With a hiss, the key entered the earth, just six inches of white crystal left above the ground. In a short while, it was secure.

Music and chanting continued until 3:45 A.M., when the Glastonbury Order of Druids made a torchlight procession up the field, into the stones. With a mixture of humour, outrageousness and reverence, they conducted an ages-old ritual, blessing and inaugurating a new sacred space.

Then I sang a song, written for the occasion, although I didn't realise it at the time. With the words flew my dreams, my love for the Circle, and my temporary guardianship of the temple during its creation; back, back to the place from whence it all came. As my energies left the scene, so to speak, I invited in those Beings who, in partnership with the People and Stars, will inhabit the stones. They will help bring the Swan Circle "on line" as an important, functioning sacred space in the earth network.

An hour after sunrise, it was all over for me. The Silence had descended, and I spent the rest of the day on my back, sleeping the sleep of the—dead? The next day, I spent hours inside the circle with Serena, as a visitor. One of my deepest dreams had come true and I gave thanks for all who made it possible, Michael Eavis in particular. It was wonderful just to sit, watching the world go by, with no responsibilities. The circle was alive, had come of age,

The Swan Circle has become an integral part of the Glastonbury Music Festival.

and was perfectly capable of looking after itself. The bird was flying.

And, after such a dream, one has to wake up, and move on.

Pause with me now and reflect on the events we have just witnessed. Is it much of a stretch to imagine similar scenes, repeated hundreds of times throughout the British Isles, during the third millennium B.C.? Not for me, it isn't.

ALBAN ELUED

Sunday, September 21, 1997, was the autumnal equinox, one of the eightfold days of celebration in the ancient Celtic year, and exactly two years since I'd met Ivan at Avebury. I had some business in South

Wales that morning, and Ivan accepted my invitation to join me for the ride. En route to my appointment, we stopped at Harold's Stones at Trelleck in Gwent, an ancient grouping of three large standing stones, and a monument that neither of us had ever visited. Ivan spent time with each of the three menhirs, and sensed that two of the stones were very much alert and energized, while the third was dormant.

"You could come back later tonight, or next week, and find a different configuration," Ivan told me. "Stones sleep, just like we do."

On the way back from Wales, we picked up some liquid paraffin for the garden torches that would illuminate the druid grove ceremony that evening. Ivan appointed me official fire-setter, as he helped Julie with last-minute preparations. By late afternoon, the various druids and guests started to arrive.

The Transrational Musings of Ivan McBeth

I thank the Stone People featured in part 2 for reintroducing me to what I have long thought of as "the other reality," which some call the "spiritual" realm. I prefer "nonmaterial" to "spiritual," which loses some of its meaning when used by people who really mean "religious." But of all the terms that I have heard, my favorite is Ivan's own: *transrational*. And no one is more eloquent in relating experience in the other reality than McBeth himself. Listen.

★ ★ ★

My feeling is that the people who built the stone circles were very different from the people of today. We're trying to pick up the smallest threads of a knowledge that was totally integrated, and a lot of it isn't on what I would call the rational level. And when I say that, I'm not coming from some Erich Von Däniken spaceman point of view. I'm coming from the point of view of actually training myself to work with what I call conscious dreaming or working with the transrational. That's what magic is all about.

★ ★ ★

I was doing an interview with a guy from Channel Four. We were walking toward the Swan Circle. I was telling him about the magic of stone circles, and he just kind of nods and says, "Yeah, sure, but . . ." Suddenly I notice I'm walking alone inside the circle. He's stopped. I went back and said, "Is there anything wrong?" And he says, "I can't—I can't—I can't walk any farther." I said, "What do you mean?" He says, "I literally can't walk past this point." And he was actually getting quite worried. So I said, "Okay, don't worry about it. Just go with it and ask if you can walk through." So I left him and after a while he came up and he says, "I've never experienced anything like that before. It wouldn't let me through. So I asked it if I could walk, like you said." And then he was there. So, you know, you talk about dowsing for energy around the circle. I just walk up to my circle with my eyes closed and—bang!—it's really strong. I know exactly where the ring of force is.

★ ★ ★

Stones don't like to be moved. If people just want to move stones, and the stones don't want to move, *they will not move.* Everything must be right. It must be ordained, as it were, and then they'll move. That's something I've learned again and again, and so I wait.

For instance, you know there's a big hype about the millennium here, so this big firm commissioned me to make a mega stone circle honoring the Thames Valley. And it would have been amazing, so I went there, did loads of research. I've still got the model I made for them. And I made slide shows, and all kinds of things, and they did a brochure saying that Ivan McBeth will be making a big stone circle that will be the centerpiece for the millennium—and the whole thing went up the spout. I haven't heard anything after sending all this stuff off. So, you see, if the stones don't want to move, they won't. It needs people who care, who are really in touch with the needs of the land, and whose commitment is 100 percent. Then the stones will move. If people just say, "Hey, I want to do that—there," well, it won't happen. It's a very magical process.

★ ★ ★

I believe Stonehenge was completed. Whenever I go there, I see it completed, and I don't know whether I'm picking up the actuality or the intent of the builders, but, for me, it's complete. But all of this, you see, all of these kinds of questions—they're important, but they're also not. To me, life is an incredible mystery, and there are certain things that cannot—and *must* not—be answered in a rational sense. And stone circles are one of those things.

★ ★ ★

There is one stone circle still standing on Iona, and I discovered it one day when I got completely lost while trying to get from one part of the island to another. It was actually quite frightening, somehow, to get lost. Then I found my way and I climbed down this rocky hillside, and suddenly here's this stone circle. It is the most delicious and peaceful and beautiful stone circle. The stones are more boulderish than thin. It surrounds a mound, probably a cairn, and it's within about four hundred yards of the Abbey. It's delicious and it really made my heart sing.

Julie called the meeting to order about seven, in her sitting room, and a clear order of events followed. First, on a variation of the "talking stick" method of group conversation, a special "talking sword," called Ashling, was passed around the circle, each person speaking in turn. The first pass consisted mostly of introductions, for there were two other guests at the grove besides me: Joe from Baltimore, who was doing a photographic book about druidism; and Angelique, a raven-haired self-proclaimed witch from South Africa. The druids themselves, eighteen in number and thirtyish of age, were represented equally by the sexes.

The talking stick (or sword) is a great way for a score of people to participate in the same conversation. As the sword went around, the main topic became whether or not Joe should be allowed to photograph the ceremony. Most felt comfortable with Joe's request, having seen a portfolio of his sensitive black-and-white photography, and having heard his assurances that no flash would be used; but three members did not feel comfortable with any form of photography. When Joe received the sword to speak for the third time, he said, "I will resolve the question. A circle, like a chain, is only as strong as its weakest link. Everyone must be comfortable with the outcome, or it will not work. I have decided not to take pictures." End of discussion. I wondered how the situation would have been resolved had Joe not made this decision.

Next, Julie led us all on an imaginative meditation. Her soothing words helped us to relax our

Ivan McBeth visits with Harold's Stones, Trelleck, Gwent, Wales.

bodies, from our toes to our heads, putting us in a receptive frame of mind. She took us to a stone circle in a field and asked us to leave the circle carrying only our harvest with us. Alban Elued is the time of harvest, the time of preparation for the long dark quarter of the year that would begin six weeks later at Samhain. We then processed to a trilithon of stones, a gate to the next phase of the year and of our lives. Symbolically, each person was asked to think of what they should carry with them, and what chaff could be left behind. It was a time of deep introspection, and after Julie brought us out of the dream state, she asked us each to tell of our experience. All of the people had calming and thoughtful experiences to report; some had insights or inspirational messages to help them along life's path. I was impressed by the vividness of their dream states. They were obviously used to this kind of perception. Unfortunately, I had not gained the same sort of insight. When Julie asked me what I had experienced, I said that I had measured the trilithon, and found the uprights to be four megalithic yards high. "The capstone," I said, "weighs about ten tons, by estimate." Julie and Ivan laughed. They knew me pretty well by this time.

After a snack inside, the druids put on their robes and other special celebratory clothing. Ivan donned his green top hat. Various people—the chieftain, the herald, the bard, a maiden, Julie as priestess, the honorary sword bearer, etc.—had speaking parts in the ceremony, which was done as a short play. The nonspeaking participants, including the guests, went on ahead to the stone circle, where my fire was doing a yeoman's job of lighting and warming the grove. Then the main players processed into the circle, candles cupped in hand, by way of a green arch through the grape arbor and into the circle by the west gateway. They took their positions in the circle.

Again, the theme of the ceremony was creating anew. "In the mist of evening, in the falling of the year as night lengthens, wait upon the spirit. Look

clearly and you will see, listen attentively and you will hear." Peace was given to the four cardinal directions, each invocation delivered by a different robed figure. The ceremony was, indeed, peaceful, the language formal, and it built on the theme of preparation for the coming winter. It was certainly a new and fascinating experience for me. For the second time this weekend, and at the same location—inside the little circle in Julie's back garden—I had the sense that I was witnessing a scene from the distant past. This was, perhaps, how stone circles had been used long ago: as places of spiritual renewal.

After the formal cermony, we sat down and the proceedings moved to the Eisteddfod phase, (pro-ounounced *I-steth-vud*), during which each person in the circle shares some gift with the others, a song, perhaps, or poem. Ivan played his flute. While he was still standing, Julie declared her love for Ivan in a moving affirmation that just about floored the big guy. All he could say was, "Wow."

The druid grove was an experience I will not soon forget. There was nothing marginal going on, in my view, that should have caused anyone any apprehension about photography: no nakedness, no lewd dancing, no casting of spells. Nothing was spoken or implied that I would think of as religious dogma. Druidism (at least in the grove I observed) seemed a basic, natural, almost primeval religion. I saw a group of people with a genuine love for nature and an appreciation for the interconnectedness of all things.

I stayed up talking with Joe, the photographer, until 2 A.M., and finished reading *The Swan Circle* before falling asleep. Ivan concludes his booklet with a section called "And Now?" It is a look into his future, and, written in 1992, it seems to be a fairly accurate piece of prophecy.

Whatever happens, I know I'm "on call," and will recognise where I have to go, and what to do, when the time is right. There is a particularly perverse streak in me that directs me to-

The Alban Elued ceremony is over. The druids have departed, to reconvene at Samhain. Only Anubis, the Egyptian jackal-headed god who led the dead to judgment, stays behind, watching over the dying embers.

wards those things that I have the most fear about, and I have found that these situations hold the greatest learning and transformational potential for me. I have already mentioned that my "terror" is one of the greatest energy sources I have to move things; methinks I ain't seen nothin' yet!

THE PROBLEMS OF MODERN-DAY MEGALITHIC CONTRACTING

If the previous paragraph sounds like a classified ad for a mystical stone circle builder, well, maybe that's not far off the mark. After building the Swan Circle, Ivan was retained by Longleat Estates, halfway between Glastonbury and Stonehenge, to design and build a huge stone circle, a full-scale replica of Stonehenge. Who better to supervise such a project

than the gentle giant stonebuilder? Ivan worked over two years on the project, approaching it from every conceivable angle; a planning permission application included feasibility studies, traffic and parking analyses, and environmental impact reports. He showed me a copy of his 220-page study, which, sad to say, was snookered at the eleventh hour by the Historical Landscapes Commission, which spoke strongly against the granting of planning permission. The volume of meticulous work that Ivan had done at Longleat was incredible. At least he was well paid for his effort, and learned a tremendous amount about stone circles in the process.

In a conversation with Ivan and Peter Carpenter, secretary of the British Earth Shelter Association, I expressed the irony that one couldn't get planning permission to build a Stonehenge, which, outside of London, is Britain's most visited tourist mecca.

"No, well you don't want too many of them," said Peter. "They would lose their rarity value."

I opined that it would be great to be able to see Stonehenge as it would have been upon completion; that it would contribute to, rather than detract from, the ancient site itself. I told Ivan and Peter about the full-scale replica of Stonehenge in Washington State. (The reader can learn about it in chapter 15.)

Peter wondered if the Swan Circle had gone through the tedious planning application process.

"No," said Ivan, poker-faced. "Officially the stones are scratching posts for cows, and nonjurisdictional."

THE STAR CHILD CIRCLE

Less lucrative, but more successful in coming to fruition, was Ivan's involvement with the Star Child circle, in a large private back garden near Glastonbury. Although it is not "open to the public," Ivan felt that the owner would be pleased to show it to a stone circle aficionado, such as myself, and gave me directions. Jaki, Darin, and I visited the circle in

Plan of the Star Child Circle.

late September, and surprised a lady guest who was reading at the circle's center. She informed us that the owner was sleeping, so we didn't knock on his door. But we did enjoy a look around the circle, and understood right away what Ivan meant when he said, "The energies, the feelings, the atmospheres between the Star Child and the Swan are so different. Star Child is completely peaceful and harmonious. The Swan is wild, because it's on the side of a hill, and it's—well, it's flying."

At left is a plan of the Star Child circle based upon a drawing by Ivan. Twelve large chunky stones, each four to five feet tall, make a true circle. Six intersecting pathways connect six of the stones and transform the inner space into a six-pointed star. Access is provided to all sides of twelve raised beds. An inner hexagonal space provides a place for relaxation or ceremony. The circle is formal and geometrical, but the beautiful rough stones lend a sense of age and of magic to the site.

Built on the edge of a farm field, the unique feature of the circle is that it provides growing space for healing herbs sold at the Star Child shop in Glastonbury. As the agricultural use of the land was not affected, planning permission was not required. And the herbs seem to do very well in their magical and harmonious stone circle setting.

IN THE WORKS

As I write, Ivan has projects in various planning and construction stages, including a circle on an island off the Irish coast and one called the Dragon Circle in Surrey, where he is fulfilling his lifetime dream of building a full-size ring by hand, without the use of machinery to erect the stones. The story of this circle is the topic of chapter 10 but first you must meet some other Stone People, who are also a very important part of that chapter.

This part of the book is about people, wonderful people like Ivan who have helped me along on my

The Star Child Circle is in a private back garden in Somerset.

megalithic journey. Once these people began to become aware of each other, storylines began to intertwine, and a nice linear plot expanded into many dimensions. The least I can do to assist the reader is to try to keep things progressing chronologically.

Need a stone circle? Build it yourself. That's what this book is all about. But if you absolutely have to have professional help, contact Ivan (see sources), particularly if your dream is founded in *the other reality*.

— CHAPTER 7 —

Cliff Osenton,
Modern Stone Mover

I surprise myself. It's not what I go in with when I do an experiment. It's what I come out with.

—Cliff Osenton

Our megalithic journey took Jaki, Darin, and me north on Sunday, September 28, 1997. We visited Ivan's Star Child Circle en route, as well as a little garden-variety mini-Stonehenge, just five miles from the real thing, illustrated in chapter 12. We were making for Banbury near Oxford, home of megalithic heavy-lift engineer Cliff Osenton.

Of all the serendipitous meetings and introductions that made our journey so magical, none were any more fortuitous than the events that led us to Cliff. Almost a year earlier, Jaki and I had spent a weekend with friends at their house in Ontario. The house was a round cordwood house like ours, and had beautiful standing stones in the garden. Bunny and Bear Fraser, our hosts, told us that there was to be an interview on CBC radio with some guy in England who claims to know how the megaliths were built. At the prescribed time, a strong female CBC voice introduced the piece.

"Did the ancients build Stonehenge just to drive us all crazy with construction theories? Some think it took thousands of laborers. Some think that the builders had help of the E.T. variety. Cliff Osenton

is a medical engineer with a new theory to throw on the pile. He believes that Stonehenge was so easy to build that a handful of people with a few simple tools could do it. We reached Mr. Osenton at his home in Banbury, England."

The six-minute interview hooked me completely. I had to make contact with this man with the grand ideas.

Adding to the serendipity, one of the Fraser's dinner guests that evening had worked for CBC radio, and volunteered to get Cliff's phone number for us. Months later, while planning our journey, I called Cliff to try to arrange a visit. Imagine my surprise when he offered to put us up in his home and take a day off work to actually move heavy stones with us at a local quarry!

A LECTURE IN THE KITCHEN

Slowed down by our various stops, we arrived at Banbury about 8 P.M. and met Cliff's wife, Maggie, at the door. We were led into the small kitchen, and Cliff arrived home just a few minutes later. He

England's justifiably legendary Stonehenge, viewed along its axis of symmetry, which is also the alignment for both the summer solstice sunrise and winter solstice sunset.

Kegadiou, in western Brittany, France. This ancient menhir, shaped with stone mauls, stands 29 feet tall.

ANCIENT CIRCLES AND MEGALITHS

Callanish I, the major stone circle at the heart of a megalithic wonderland on the Isle of Lewis, Scotland.

The Maes Howe burial chamber in Orkney, Scotland. At the winter solstice, the chamber is lit by the last rays of the setting sun.

The Ring of Brodgar on Orkney Mainland, Scotland, one of the largest and most evocative of ancient stone circles. [photo credit: Charles Tait]

The center of the Earthwood stone circle is along an arc whose center is just in front of the raised-bed gardens. All of the round cordwood masonry buildings have their centers on this arc. The photo was taken partway up a wind-turbine tower.

EARTHWOOD
New York, U.S.A.

The winter solstice sunset alignment at Earthwood's stone circle.

Sunrise. The gnomon at the center of the Earthwood circle is an 8-foot-high column of ice that formed below the drip line of a roof, then was sledded over, tilted into position, and held in place by "snowcrete."

Beltane festivities.
[photo credit:
Richard Ross]

STONE MOUNTAIN FARM
New York, U.S.A.

The Wickerman is immolated at Lugnasad festivities.

The meridian alignment. The north stone
is reflected in a shallow pool at the
circle's center.

The massive Thor's Gate trilithon draws visitors to the Faerie Glen.

COLUMCILLE,
Pennsylvania, U.S.A.

Columcille, "where the veil between worlds is thin." St. Oran's Bell Tower and the stone circle are just beyond the pond. [photo credit: Bill Cohea]

St. Oran, laid up in the walls of the Bell Tower, said, "The way you think it is may not be the way it is at all."

"The Farmer's Tomb."

ED HEATH
Quebec, Canada

"Samurai." [photo credit: Don Davidson]

Ed's pond grouping suggests a Japanese garden, while the standing stone in the water is reminiscent of an ancient menhir in coastal Brittany, two miles southeast of Pont l'Abbé, which stands as much as 12 feet above the tide.

The Megalithic Railway.

A hawk's view of the camp and stones May, 1998. [photo credit: John]

A stone circle is born, with Ivan McBeth (right) as midwife. A 4.22 tonne stone is levered away from its fellows. It will be rolled into position and become Center, the focal point of a new stone circle.

THE DRAGON CIRCLE
Surrey, England

The Dragon Circle in autumn, 1998, about halfway to completion. [photo credit: Ivan McBeth]

Standing Stones Perennial Farm, Vermont, U.S.A.

Reva's Ring, Connecticut, U.S.A.

The Ragged Band, a musical group, practice at the Australian Standing Stones, Glen Innes, New South Wales, Australia. [photo credit: John "Trigger" Tregurtha]

OTHER MODERN CIRCLES

The Druid's Temple, Yorkshire, England.

A stone circle of indeterminate size on the beach at Dominical, Costa Rica.

TEMPORARY STONE CIRCLES

A small stone circle with an avenue at Wapta Falls at Yoho National Park, British Columbia. The center stone is white crystal quartz.

A stone circle on the Wilcox Pass, Jasper National Park, Alberta, Canada. Later the stones were returned to their original positions.

looked very much as I expected from our phone calls, a tall, good-looking, and bespectacled man in his late 40s. His smile was wide, genuine, and infectious. Within five minutes, we knew that we had just encountered the living personification of the word *enthusiasm*. We'd no sooner removed our outerwear when Cliff launched into a detailed discourse about his latest megalithic discovery: how to lay out stone circles without a rope.

"Maybe the ancients had ropes, maybe they didn't," said Cliff. "But anyone can lay out a perfectly accurate stone circle without a rope." He described a method that he had been testing at a beach with Maggie and Iain, their five-year-old son. The method involved a sequence of pacing, and I tried it later on sandy beaches on the islands of Lewis and Iona, with great success. The technique is described in a sidebar in chapter 11.

Cliff waxed eloquent for forty-five minutes about laying out circles, while Maggie prepared supper and the rest of us listened enraptured. We interrupted periodically with questions for clarification. Some of our questions were necessitated by our occasional failure to decipher Cliff's strong Midlands accent and/or use of words, and others were precipitated by an almost dyslexic word pattern that crept in somewhere in the transition between thinking and speaking, a condition that I was later to learn had made life a little bit difficult for Cliff. The odd pattern is particularly manifest when Cliff is trying to explain a technical matter, and it shows up in both written and verbal communication. The extraordinary exception is when he appears before a video camera, at which time his thought process and spoken words *sometimes* crystallize into a lucid explanation of the matter at hand. The speech pattern problem is compounded because Cliff is so intelligent and knows what he wants to say so well that he often charges right in at page 40, forgetting that his audience may not have the required technical background to follow his ideas. With all of this heightened by his infectious enthusiasm, Cliff's

listeners are in for a wild but thoroughly entertaining ride.

(Months later, when I'd gotten to know Cliff better, he told me that his strong accent gives a vital clue to his work: "Instead of trying to solve megaliths using modern civil engineering, which will always fail to match original evidence, I went back to my rural roots. I made my first stone axe at age eight, and worked on farms using the old horse-drawn technology. By pitching my megalithic work at this level, I made breakthroughs in a lift system that matched ground evidence. Modern technology will never interface with the Stone Age. To find the answers, you have to walk off the map of modernism, and embark on a journey to the distant past.")

Maggie prepared an excellent supper and the conversation lasted until we went to bed at midnight. We learned that Cliff was having trouble finding recognition among the "experts" in megalithic studies, and I suspected that communication difficulties may have had something to do with the problem, plus the fact that some archaeologists and historians can be rather set in their ways.

"The problem is that the 'experts' are academics, not practical people," Cliff told me. "So they interface and cross-ref each other's theories until they have built a matrix of quasiphysics without a trace of normal scientific test data. Imagine Darin, in his science homework, submitting the objective, then skipping methods, apparatus, and results, and going straight to the conclusion. He wouldn't get away with it, but archaeologists have been getting away with it for a hundred years, and now expect physical science to match untested impractical theories."

THE QUARRY DOLMEN

After breakfast the next morning, we loaded Cliff's van with his "megalithic kit," consisting of a large supply of 3 x 9 timbers in 12-inch, 24-inch, and 36-inch lengths; a number of hardwood wedges; several levers of pressure-treated wooden poles varying in length from eight to twelve feet; and a single oak fulcrum, a refugee from a village bonfire. We also loaded on a good supply of sandwiches, biscuits, fruit, and tea, provided by Maggie. We followed Cliff to Condicote Stone Quarry in our rented car, as nearly every cubic inch of space in the van was occupied with equipment and foodstuffs. I noticed that we drove right past the famous Rollright Stone Circle, and hoped we would be able to stop to see it on the way back.

The young quarry owner was very supportive of Cliff's experiments in megalithic engineering. He had a variety of heavy equipment on site that could move forty tons or more but he delighted in seeing Cliff maneuver stones in the five- to ten-ton range by hand. We were joined by Cliff's friend, Frank, a

Cliff Osenton's megalithic kit.

stocky cheerful Englishman of Italian descent, with a strong interest in the megalithic period.

Our project for the day was to reconfigure a dolmen that Cliff had built during a previous visit.

"It's a Welsh dolmen right now," Cliff explained. "Let's change all the legs around and make it an Irish dolmen."

This was okay with us. We just wanted to move heavy stones. As built, the dolmen consisted of a capstone measuring about 2 feet by 5½ feet by 7 feet—5 tonnes or 5.5 tons—supported by three stone legs, each about 3 feet tall and weighing about 500 pounds. Cliff explained that we would install two legs under the capstone where only one supported it now and replace the other two with a taller upright, giving the monument an Irish flavor.

Cliff explained that the way to understand the building of Stonehenge was to first work out how the dolmens were built, at an even earlier period. Dolmens older than the earliest stone circles were built in Ireland and France with capstones of up to 100 tons, much larger than the largest 45-ton upright at Stonehenge, and many times the weight of the largest lintels of the Horseshoe Trilithons.

Compound Lifting

The key to Cliff's success is a technique called *compound lifting*. I had heard Cliff describe this technique to the CBC radio interviewer, and he'd tried to explain it to me the evening before we'd arrived at the quarry. But my cranial lightbulb failed to click on until I actually saw Cliff at work that morning.

A little basic mechanical physics is needed to explain the method. *Single-stage lifting* is simply lifting an object—called a *load*—with a lever. The lever pivots on a point called a *fulcrum,* which is most frequently located between the load and the opposite end of the lever bar. Illustration *a* on the next page shows a common playground teeter-totter or seesaw. If two 80-pound children are at opposite ends of a 16-foot plank and the pivot point (fulcrum) is located in the center, the system is in bal-

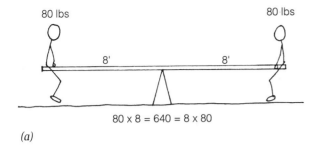

80 lbs   80 lbs

$$80 \times 8 = 640 = 8 \times 80$$

(a)

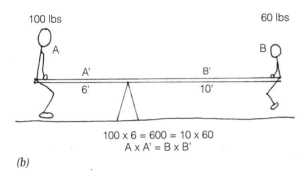

100 lbs   60 lbs

$$100 \times 6 = 600 = 10 \times 60$$
$$A \times A' = B \times B'$$

(b)

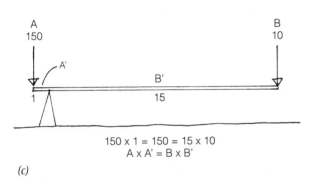

A   B
150   10

$$150 \times 1 = 150 = 15 \times 10$$
$$A \times A' = B \times B'$$

(c)

ance. The children can create the teeter-totter motion with just the slightest imbalance alternately applied first by the leg muscles of one child, than by the other. If Child A weighs 100 pounds and Child B weighs 60 pounds, some kind of adjustment needs to be made to achieve the desirable balance. Child B, we say, needs a *mechanical advantage* (m.a.) to offset A's extra weight. If the fulcrum remains at the center of the plank, B will be left

high and dry. Not much fun. And if A gets off the plank, B will tumble down with a crash.

Let's put the kids on the opposite ends of a 16-foot plank, for easy figuring. In illustration *b,* we'll provide B with the needed m.a. If the fulcrum is placed at a point 6 feet from one end—10 feet from the other—and we position Child A at the end of the shorter lever arm and Child B on the end of the longer arm, we can achieve the desired balance and the kids can have their fun. Expressed mathematically, load A times distance A' must equal load B times distance B' for the system to be in balance. As 100 times 6 equals 60 times 10, all is well at the playground.

Now we must leave the children behind before someone gets hurt and work with pure numbers. These numbers could represent stones and levers—add as many zeros as you like—and, in the field, there is still plenty of scope for danger if the load happens to be a multiton stone and we are not careful. But Cliff *is* careful. He takes pride in that, and always has a contingency plan—usually a safety block—in case he works too close to the limits of his system.

In illustration *c,* we increase the m.a. to 15 to 1. A 150-pound load can be lifted with only a 10-pound downward pressure (called a *reactionary load* if you're technically inclined). Or a 150-pound weight can be balanced with a 10-pound weight. Again, A times A' (150 × 1) equals B times B' (10 × 15) and all is well. Cliff feels that a 15 to 1 mechanical advantage is approaching the limits for hardwood levers lifting heavy stones with single stage-lift techniques, although he may sometimes exceed this limit slightly . . . and carefully, as this is when levers will crush or break. Now let's increase our numbers by a factor of 20. A 200-pound person, someone a little heavier than Cliff, can lift a stone of almost 3,000 pounds by hanging on the end of such a lever. *A ton and a half.* Pretty good. But we can do a whole lot better, with compound lifting.

After we'd unloaded the van, Cliff demonstrated the principle of compound lifting by performing a

Pushing up from this position requires arm strength. About half of Cliff's body weight must be lifted, the other half supported by his feet.

In this position, Cliff·creates a longitudunal axis through his body. His left arm lifts the left side of his body with ease, as most of the weight is supported by his right elbow and his feet. The trade-off is that his body is not lifted as high. The trick to lifting megaliths is to minimize effort.

few push-ups. With his feet supported on a knee-high stone and his hands on blocks of wood, Cliff did push-ups in the normal way: both arms pushing his full body weight up and down, up and down. After doing a few—and that would be a few more than I could do—he said, laughing and puffing, "Now this is starting to get really difficult. But if I rest my right elbow on this block, like this, I can easily push my body up with my left arm, because now I'm working my leverage close to my center of gravity. Before, I was pushing up about five stone with each arm. Now, my own body weight is offsetting the load. Effectively, I'm probably lifting about a stone and a half [21 pounds] now."

In the clay that covered the stone floor of the quarry Cliff drew a picture something like the illustration below.

"The little circle is the center of gravity of the stone. Any line drawn through the circle is called an axis. There are an infinite number of axes that you can draw through that center: longitudinal, transverse, diagonal. The secret is to work close to these axis lines without crossing them."

The puzzled expression on my face spoke more eloquently than words. Cliff had seen similar expressions many times before.

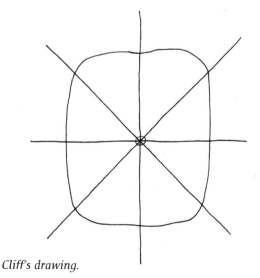

Cliff's drawing.

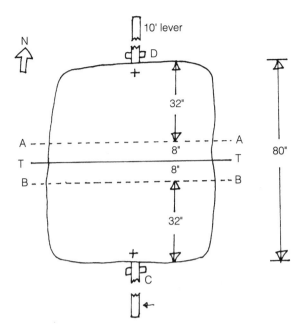

My visual aid to Cliff's drawing.

"Come on," he said. "It's easier just to show you."

And that's what he did, for the next five or six exhilarating hours. Now, with the benefit of our experience, I can lend some explanation to the diagram, but I'll have to add a few visual aids (see above). Let's suppose that a giant balances an 80-inch-long 5-ton stone, similar to the one we worked on that day, on a well-supported on-edge 2 x 8 that lies under—for example—the transverse axis (TT). Clearly it would take very little effort to move the stone up or down once it was positioned like this. A child at either end of the stone, at the points marked with a cross, could rock the stone back and forth like one of the rocking stones found all over Cornwall, including in Ed Prynn's garden.

And what is gained by this exercise? Not much. One end of the stone goes up, one end goes down. Big deal. But, if we rock one end up and put an on-edge 2 x 9 under the stone along line AA, which is, let's say, eight inches from the transverse axis, now we *can* make some progress. The stone is now tilted

with its high end to the north (in our diagram) and the low end to the south. Roughly two tons of the stone's weight is north of the on-edge plank at AA and three tons of the stone are to the south of it. The two tons to the north of AA counteracts two of the remaining three tons south of the line. Effectively, Cliff (or anyone else armed with a lever propped at fulcrum C) has only a ton to lift, the other four tons canceling each other out. If he sets his lever for a 15 to 1 m.a., he need only apply 133.3 pounds on the end of the lever to lift the ton he's responsible for. But he really is lifting the better part of a five-ton stone, even though part of it is actually moving down. Now, with the south end of the stone in the up position, an assistant can slip in an on-edge 2 x *10* at BB, and the stone is let down on the new support. Moving his lever to fulcrum point D, Cliff can again lift the stone higher, prop it again and continue.

Cliff's lifting power can now be expressed as 15 (the lever advantage) times 5 (he's only lifting one ton, not five), a compound lift advantage of 75 to 1. If support line AA is moved even closer to the transverse axis, say four inches instead of eight inches, then 2.25 tons is north of AA, and 2.75 tons is south of AA. In this case, Cliff is only lifting the difference: half a ton. The stone's own weight offsets the rest. Now the compound lift advantage is 15 times 10, or 150 to 1. Now Cliff needs only 66.7 pounds of downward thrust to move the five-ton stone. But Cliff would tell you that he is now working dangerously close to the axis. He prefers to keep on the safe side of a mechanical advantage of 7 on the stone, which limits him, practically speaking, to a compound m.a. of about 100. (15 x 7 = 105).

In the real world, we wouldn't support a 5-ton stone on 2-inch-thick lumber in this way. Even if the wood was strong enough on compression—an iffy question—the construction described would be unstable without fastening crosspieces to both ends of the planks. The example's only purpose is to graphically illustrate the mechanical advantage

The dolmen was originally built in a day by two men using a single fence post for a lever. (Cliff Osenton, photo.)

of compound lifting. The photographs of our work that day, although interesting in their own right, do not clearly show what's happening.

A Question of Order

To reconfigure the dolmen meant to lift it up in the air and change around the three standing stones around underneath it, simply an exercise in heavy lifting. In the event, we actually replaced one of the stones with a slightly longer one. Cliff explained the process in his own inimitable way:

"We extend the stack underneath and transfer the load from this rock onto the stack, but to move the axis from here to the rear stone *in* slightly, so that when we lift the far side we can ease the load. And the whole point of this game is you have to work one step in front. So we're setting up here, not for *this* side, but for the *far* side. That's how we

work our technique, constantly one step in front of ourselves. If I just go to the next step, we'll jam out and we'll have to come back and go through it again. We're constantly working near the balance looking underneath, and assessing where the axis is."

That's what he said. I have it on tape. American readers might be reminded of the legendary baseball manager Casey Stengel, whose manner of speech was similar, and became known as Stengelese. In fairness to Cliff, the quote above makes a little more sense when you are actually standing there and seeing which stack and which stone he is referring to. I will attempt an explanatory translation:

Before work began that day, the dolmen looked like one above, with the entire weight of the capstone—the *load*—on the three stone legs. To reconfigure the dolmen, we needed to take the load off of one or more of the legs and transfer it some-

152

where else. We built a stack of wood—sometimes called a *crib* or *pigstack*—out of Cliff's 3 x 9 blocks. Cliff prepared the ground surface very carefully where the stack was to be built so that the first two base pieces, each about three feet long, were flat, were in the same plane with each other, and did not wobble. This was easy to accomplish in the quarry, as we were working on fairly flat stone bedrock covered with just a little grit, sand, and clay. Nonetheless, Cliff would spread additional grit under the base pieces until he was happy with the quality of his foundation. Jaki asked him about his meticulous surface preparation.

"It's just to get a flat bed. If the timbers wobble, all that movement will be absorbed, so we won't be able to get any wedges in. So I scrape the ground a bit, use this sand to pack it down and get a very solid bed for the wood so that when the load comes on, it's solid every time. Otherwise, the stacking will start doing this—" (he wobbled a nearby stack with his pickaxe) "— and we waste a lot of effort. You soon learn the optimum level of ground prep. If you don't do it, nothing will work for you. If you're too fussy, there's no advantage. You've just gotta feel it through."

Working between two stone uprights, Cliff built a stack of wood that was very solid indeed. The first course consisted of two 36-inch 3 x 9s set parallel to each other and about six inches apart. The second course was made of two 24-inch 3 x 9s perpendicular to the first course. In this way, a 24-inch-by-36-inch table was built up. After eight courses, the top of the table was two feet off the ground. On the opposite side of the dolmen, Cliff built a similar stack of wood. Next, he placed a hardwood fulcrum on the first stack and asked Frank to take some of the capstone's 5.5-ton load onto one of the 10-foot levers.

Wedges, we learned, are extremely valuable tools in the megalithic kit. Cliff keeps at least a dozen good hardwood wedges at the ready. With some of the load off of the stone uprights, Cliff could push a wedge between the right-hand adjacent leg of the dolmen and the other leg on the far side, creating an axis line between these two points. However, as Cliff says, the game is to look a move ahead. It's like a pool or snooker player with an easy shot in front of him. He'll devote most of his thought and energy into how to position the cueball for his *next* shot. Again, here are Cliff's enthusiastic if enigmatic words, recorded on site, followed by my translation:

"As long as there's a positive load here, you can work just one side of the axis and the leader man tells you how close you're running. But you never cross the axis line. If you lift here [indicating], you've wedged it here. Next time, lift up, wedge in there and back here so you're setting up for the next time."

My best sense of this is that the closer that you can place two wedges to a chosen axis line, the easier it will be lift the stone. The stone's own offsetting weight is working for you. Cliff says that he likes to work to an m.a. of about 7, which he considers to be the practical and safe limit. This means that the stone's own offsetting weight helps to lower the lift to one-seventh of the stone's weight. The "leader man," usually Cliff himself, is the bloke who has to estimate the axis line and set the wedges (or a single wedge opposite an already existing support point, such as the point of a stone). If the leader man is wrong in his estimate of the axis of balance, the stone might fall over. For this reason, Cliff always prepares a "safety," a stack of timbers placed so that stone can only fall two or three inches. This causes a temporary jamming situation that is easily rectified by levering off of the safety stack and repositioning the wedges on the correct side of the axis.

"So you've always got to be making a judgment as to where the center of gravity is," I said, "in order to establish a desired axis." By this time, I knew that there are an infinite number of potential axis lines all passing through the center of gravity.

"Always, yeah," said Cliff. "And we're always looking for clues: how much something crunches, something like that. And we're always assessing it.

Cliff explains his methodology to Jaki.

Multiple choice quiz: Frank needs (A) an extra four stone of body weight, (B) a longer bar, or (C) a lower fulcrum set closer to the load. Answer: C.

It's a fictitious place, the center of gravity. And the big problem is, the nearer you get to it, the less you can see, and then you're relying on your lever man, because he can get a better line on it than you can."

After a pause, he added these ominous words, almost like an important afterthought that could not be left unsaid: "The lifting sequence is a mental game of skill. One mistake can mean death when working in this close. This makes megalithic work very exciting."

Walking the Stones

Soon we'd transferred the load—the weight of the stone—to two stacks of wood, with one of the legs acting as a safety. Or, we might transfer the load to a single stack and a stone leg, the other stack acting as a safety. The standing stones were relatively easy to move. They had flat bottoms and we could "walk" them along the hard quarry surface quite easily with two flattened metal pipes and a couple of fulcrums. This method is quite impressive to watch, but simple to do. The flattened end of the pipe is easily jammed under the edge of the stone, and a two-inch-thick fulcrum is placed as close to the stone as possible. Then, with a circular rowing motion—think of the fulcrum as an oarlock—Jaki and I found that we could walk the 500-pound stones along the quarry floor, and had great fun doing it, laughing all the time at how easily we could move them. The only thing that slows progress is moving the fulcrums. Below, scaffolding pipes with flattened ends are used to maneuver one of the 500-pound legs.

It is true that the ancients did not have access to metal pipes or bars, but the same method works with a wooden bar on soft ground. It just takes a little longer to create a fulcrum support with good

Cliff and Frank maneuver one of the dolmen legs with flattened pipes.

The Megalithic Construction Company guys gather on the capstone of the reconfigured dolmen. (Jaki Roy, photo— someone has to do it.)

load-bearing ability. On the hard quarry surface, the pipes just made the work go that much faster.

The ease with which we raised the dolmen's capstone several inches into the sky and reconfigured the three legs to change the character of the monument was exhilarating for all of us. Jaki and I wanted to get on the next plane home and build a dolmen at Earthwood. Frank was simply amazed and enjoyed the day completely. "I'd like to have this in my garden," he said, sitting under the massive capstone. Cliff concluded, "There. You're looking at the first Irish dolmen to be built in about five thousand years."

The whole process, including unloading and loading Cliff's kit, tea and conversation breaks, and stopping for Maggie's excellent packed lunch, took about five hours. But we weren't finished at the quarry. Jaki and I could see now how easy it is to lift a 5-tonne block into the air. Two people can do

it with only one of them doing the actual lifting. But one thing was troubling us: Cliff had already got this capstone up onto its three legs during a previous visit. Now it was easy to move *(once you know how)*. Our question to Cliff was, "How do you get the thing off the ground in the first place?"

Cliff was happy to show us. About a hundred yards away and laying on the hard ground was a long stone weighing about six tonnes (6.5 tons.) We brought a few blocks, wedges, and a couple of levers over to the recumbent stone and the master went to work. First he walked around the stone a couple of times, looking for clues, "to see which way the stone wants to go." He'd learned this valuable tip from a retired auto recovery driver. Next, he wedged under the two ends of the stone to take advantage of the longitudinal axis. Then, as with the capstone, Cliff kept transferring the load to a new axis. He worked his way around the stone, slowly lifting it into the air. When he had the stone comfortably resting on wedges set very close to the transverse axis, he set up a lever for Darin to operate and our eleven-year-old—all eighty pounds of him—lifted the end of the stone with ease.

"There's an eleven-year-old one-handing a six-tonne stone!" said Cliff proudly. I was pretty proud myself.

"That's amazing," said Jaki. "In a way, this is even more impressive than lifting the capstone."

Cliff smiled. "It's simply a case of assessing where the balance point is and working so that you've always got this positive load on the lever. From this position, I can do anything I like with it: Load it onto a sledge, transport it, stand it up, roll it over . . . I can do anything I like!"

"How would you transport it?" I asked.

TRANSPORTING STONES

"I'd just have it straight up in the air, build a long sledge underneath it, load it onto some big rollers, and then I'd go straight across that quarry."

"Not just rollers under the stone?" I asked.

"No, no, no," said Cliff emphatically. "It'd be too unstable. You can't use big rollers like that. The whole lot is going to nose up. This stone is about twelve feet long, three feet wide. I'd put it on rollers about twelve feet long and two feet in diameter. If you look at any big low-loader or any vessel carrying a heavy mass, the actual truck or vessel is probably half or two thirds of the cargo. So we won't be timid about using another two or three tons of timber to move this stone. With reasonably prepared ground in the dry season, we could roll it with just a few people or some oxen, straight across the countryside."

Darin lifts a six-tonne (6.7-ton) stone.

Observations of a Megalithic Engineer

It's like you're riding on a really big horse, that's also really fast. You don't kick it. You just squeeze your leg muscles a fraction, and the horse changes up a gear. It's like you're sitting up on a stepladder moving thirty-five miles an hour across the countryside. *That's* the feeling I get with these stones; high excitement and danger, not brute force.

★ ★ ★

An archaeologist is not trained in the site handling of fifty-ton stones, but heavy-lift engineers are. We're not looking at a big mystery here; 90 percent of most megalithic structures are above ground. From studying that, we can assess the build quality, the tolerances, the techniques. You don't shy away, even from the biggest, most difficult site. If you can solve that one comfortably, you'll find that it will come down to one technique, and that technique is going to build the next difficult awkward site. By going through every single detail of every single process, the technique is going to tighten up every time. You don't *estimate* something. You actually *test* it, every detail, every aspect of surveying. For example, if I lay out a circle which is *too* accurate, I've failed. I want the same percentage errors as they had, and then I know I've matched their technique.

★ ★ ★

I'm working to an international standard of auditing. To test a theory, you have to ask, "What evidence do you have to support that statement? Show me every detail of every process to match that tolerance and procedure." There can't be any gaps. There's no saying, "Well, that's a mystery; we shall never know." We are talking about heavy engineering, and heavy engineering comes after a process.

★ ★ ★

A train crash is not designed. The train has gone straight over the embankment. In the past, you couldn't use a mobile crane of 100-tonne capacity, so you had to use rail stock to extract the train. And that's heavy engineering at its best. You've got no cranes. You've got winches, you've got timbers and levers, that's all you've got. Whatever condition you find those trains in, you recover 'em.

"Twelve feet long and two feet in diameter?" I asked, shocked by the recommended roller size. "They'd weigh a ton, these rollers."

"Yeah, but you can roll them on ahead . . ."

"Which would take an awfully wide trackway so that you can get the rollers past the stone and back to the front."

"Yeah. The big underestimation with all these sites, like Stonehenge, is the ground preparation for the access route. You just pick your route very carefully and from moving these stones, you soon learn the optimum level of ground prep and roller prep."

"And what about going uphill with rollers like that?"

"We've found the limit to be about a one in seven grade," said Cliff. "We put wedges under the rollers to stop them sliding back. You've got a highly forward load, and the problem with oxen is that you've got to double up your force; not for the extra power up the hill, but because you're going to have to work them straight through without a break. You'll have to do the hill in one go. Once these masses start, they must not stop, they must keep going. Redhorn Hill . . ." (halfway along the commonly accepted ancient route from Marlborough Downs to Stonehenge) ". . . is virtually impossible with oxen, because there's a dogleg in the hill. You can't bend the line of oxen around a corner because the forward force pulls across the sharp curve. It tries to straighten itself. So if you had a very long team of oxen, like the South African Boers with their huge oxcarts, you'd have to realign all the oxen, break the load at the dogleg, realign the oxen again at the other end, and take up the load again. You can't run 'em on a curve. You'd use a beam made of a tree trunk as a torsion bar under the sledge to help make the corner."

"And what do you make the sledges out of?" asked Jaki.

"Tree trunks," said Cliff. "You often see pictures of stones lashed to sledges, but there's no need to lash them down. There's twenty tons of stone on wood. The friction is incredible. The stones won't go anywhere!"

"But the two main beams, the ones that run lengthwise, they have to be fastened to each other, don't they?" I asked.

"Yeah, you cross-strap them, front and back, quite heavily in fact. You've got to notch some joints. You spread the displacement of the load over the whole sledge that way."

Cliff smiled pensively. "The whole thing should be what I call an axe job, literally arriving on site with a gang—this would be the ultimate—a gang of blokes and no tools. You make your own tools, you extract your own timber, and you build a fifty-ton dolmen. Every detail—if you want rope, you've got to make it—every detail, you make the lot."

None of us knew then that Cliff would get a chance to test many of his ideas, successfully, a year later in Wales, when he participated in moving bluestones along the ancient Stonehenge track from the Preseli hills. (A picture of a megalithic stone boat that Cliff made by axe in Wales, appears in chapter 13.) Nor did I know that I would be working with Cliff again the following spring on a new stone circle by Ivan McBeth, an experience described in chapter 10.

We packed up the remainder of Cliff's tools and kit, said cheerio to Frank, and had a quick tour of the large shed in the quarry where powerful saws shaped large blocks into building stones. It had been an exhilarating morning and afternoon, and it was time to head back toward Banbury.

THE ROLLRIGHT STONES

Our route took us right by the famous and ancient Rollright Stones, just 2½ miles NNW of the town of Chipping Norton, Oxfordshire. The light was beginning to fail in the early evening as we arrived. While not conducive to good picture taking, the lighting presented the Rollrights at almost their spooky best, just a little mist lacking for the perfect atmosphere. This ancient (c. 3000 B.C.) circle is 103 feet in diameter and consists of about eighty closely spaced stones, which William Stukeley described as being "corroded like wormeaten wood, by the harsh jaws of time" (quoted by Burl, *A Guide to the Stone Circles*, p. 72).

There are many legends associated with the Rollrights, involving witches, druids, Danish kings, fairies, even sacrifices. The village of Long Compton, a little over a mile away, had been a stronghold of witches, according to Burl. A notice posted on a shed near the circle told us that there was to be a Halloween play in a few weeks' time, so perhaps little had changed. People were still gathering at Samhain, the cross-quarter time.

"I often sit here for an hour when working nearby," Cliff said pensively as we walked around the circle. "It's good to see the circle's mood change with the season, and after a full moon, the witches leave signs from their nocturnal activities. Here you forget time, as the connection with the past is still alive. This is not some modern druid fantasy, but something deeper that I have touched a few times."

He seemed lost in time and I ventured to ask him to give an example. His normal smiling enthusiasm took on a more serious aspect and his voice lowered just a little.

"In 1983, right here, on a sunny day, I tested for energy using sensitive medical equipment. I didn't detect any energy. Next, I tried dowsing and walking in a set sequence around the main circle. After a while, I felt dizzy, and the sky began to darken. I touched the stones and they felt as if they were moving, but they weren't. It was me, not the stones, the kind of effect you might feel on a bar stool after a while. It became darker and colder. Suddenly, I was overcome by fear and ran out of the circle, just as a violent thunder and hail storm broke. I sheltered in those trees, there, which were bending near to breaking. Afterward, I left this place convinced that I had touched on something very powerful. Years later, Maggie noticed that there was going to be a midsummer play in the circle, and asked if I'd like to go. I told her that there was no point, as everyone would get soaked just after it starts . . . and they did."

"Spooky," I said. "Do you mind if I tell this story in my book? I mean, as an engineer, you . . ."

"I don't mind. It happened. This is very real. Living locally, I know of many such tales that have never been recorded. I just think it's best to respect all religious sites, as if you are being watched. It certainly feels that way sometimes."

Certainly, of all the circles I'd visited, the Rollright Stones had the strangest "atmosphere." At first,

The Devil's Arrows

The evocative Devil's Arrows alignment of three great standing stones lies just outside Boroughbridge, Yorkshire, near the A1 road. An historical marker told us that the monument dates from about 2700 B.C. A fourth stone, I learned, was destroyed in the 16th century. The heights of the three that remain are 22'6", 21', and 18', all impressively tall. But the features that set the Arrows apart from other standing stones are their curious symmetrical shapes and their long and distinctive parallel grooves along the edges of each stone. The grooves were once thought to be manmade, but are now considered to be the result of natural weathering on the geologically unusual gritstone. We could not tell from personal inspection.

Homer Sykes, writing in *Mysterious Britain* (see bibliography), tells us the local tale of these prehistoric stones: "Legend has it that the Devil, angered one day, took up his crossbow and fired three bolts from his position on Howe Hill at the village of Aldborough where there was a Christian settlement later to become the Roman town of Isurium. He missed the target and the arrows fell where they have remained to this day."

The author examines the 21-foot Arrow.

The mysterious Rollright Stones, near Chipping Norton, Oxfordshire.

I attributed this indefinable something to the peculiar knarled shapes of the stones, but later, I wondered if (as many claim) some ethereal residue of times past can remain at a site in a transrational way, and at this site in particular. A number of books in my megalithic collection told of dowser and writer Paul Devereux, who, with other researchers connected with the so-called Dragon Project, had conducted extensive studies at the Rollrights using sensitive equipment such as magnetometers, geiger counters, and electronic scanners. Among other strange and interesting results, Devereux's team recorded the unexplained presence of ultrasound, particularly at sunrise and at "certain times of the year" (Michell, *New View Over Atlantis,* p. 208).

Earlier in the day, Frank had told us that the Rollright Stones had been purchased about a week before our visit for £60,000, "an absolute steal,"

and that he'd heard that the new owners would be charging 30 pence a head to visit the stones, "with no exceptions." As druids and other neopagans consider these sites to be sacred to their religious observations, Frank considered this dictum to be a bit harsh. This was also an unfortunate turn of events from Frank's point of view, because he had been out of the country at the time the place had been put up for sale, or he would have tried to buy it.

"It's becoming a business," Frank told us "and it's a shame. But I believe that the circle itself is going to remain accessible. They're not going to do a mini-Stonehenge. You won't have to stand back ten yards to view the stones. They just want to earn their money back. Even at what it sold for, it was cheap. There are thousands of years of history there. All you can buy today for sixty thousand is a little semi-detached house. You can't buy history."

We visited "after hours" and didn't have to pay. Or maybe the new owners had not yet set themselves up to collect their 30Ps. I figured that they would have to show 200,000 people through the site to recover their capital expenditure, never mind the costs of site improvements, wages, and taxes.

We arrived back at the Osenton home after dark and were treated to a wonderful vegetarian bolognaise, thanks again to Maggie. I spoke with Cliff for an hour or so in his study after supper while Jaki and Darin visited with Maggie and five-year-old Iain Osenton downstairs. Cliff was still enthusiastic, but very tired, and he had to get back to work early the next morning, so we did not stay up late. We discussed a range of subjects from ancient societies and religion to heavy-lift engineering. Knowing that I was researching a book for the general public, Cliff made a point to say: "Please warn everybody not to try my lift techniques on their own. It can easily be fatal, with just one mistake."

Good-byes were heartfelt the next morning. Although we had known this family for just thirty-six hours, we had hit it off so well with them that it was like parting from old friends.

In Yorkshire, en route to our next stop, we visited the wonderful 19th-century Druid's Temple (see chapter 15), and the ancient Devil's Arrows in Yorkshire. Our megalithic journey was in high gear now, and even greater wonders awaited us at Callanish, far to the north in Scotland's Outer Hebrides.

— CHAPTER 8 —

Margaret and Ron Curtis, Callanish

The study of standing stones and stars is called
archaeoastronomy. It's not astroarchaeology, by the way,
which implies that you're digging up the stars.

—MARGARET CURTIS, LECTURE IN SURREY,

MAY 25, 1998

As with so many of the other stone builders, we met Margaret Curtis accidentally, many years ago, when she was still Margaret Ponting. She and her first husband, Gerald, had already established themselves as two of the leading experts on the great complex of megalithic sites on the Isle of Lewis in Scotland's Outer Hebrides. Together, they had authored *The Standing Stones of Callanish* (G. and M. Ponting Publications, 1977), the definitive guide for several years. In the light of new information, the Pontings wrote and published *New Light on the Stones of Callanish* in 1984, which remains today as the more informative of the two guides currently available.

CHANCE MEETING

Jaki, Rohan, and I were visiting Lewis while on vacation in 1985, and, for me at least, the island's major attraction was Callanish. We ran out of film on the day we arrived at Callanish I, the stone circle made famous in so many tourist brochures of Scotland. Driving through the nearby village, we noticed a very low-key "shop" for tourists in the form

of a sun room attached to a cottage. Postcards, books, and pamphlets about Callanish were offered for sale, and various stones and artifacts were on display. We were greeted by a dark-haired lady in her 40s, speaking with an English accent. She did not sell film, but offered me a roll of her own at replacement cost. I began to piece together the clues and ventured to ask if she was Margaret Ponting, and she admitted that she was. We spoke for a very short time. I bought some postcards, and thanked Margaret for her kindness.

From reading, I knew that Margaret had discovered the missing stone 33A at Callanish I, the last stone in the eastern row of stones that radiates from the central circle. As I'd already "met" the lady, I decided to give her a call twelve years later in advance of our megalithic journey, thinking that her experience in the discovery of the stone and its socket, and the part she played leading to its re-erection, would be an interesting link with modern stone builders. She did not remember me, of course, but transatlantic calls usually lend the caller enough credence to get to the point of the call. I learned that Margaret was now married to Ron Curtis, and that she had discovered a "new" ancient stone circle on Lewis, exciting news indeed! Margaret and Ron said they would be willing to meet with us in early October of 1997.

On October 2, 1997, Jaki, Darin, and I departed from Ullapool on the large car ferry *Isle of Lewis,* and enjoyed a quiet crossing of the Minch to Stornoway, the main town and port on Lewis. We were met by Ron Curtis, seventy-two, a small wiry man with a full white beard. His purple parka and matching knitted woolen hat told me that he was a man prepared for the changeable and often piercing Lewis weather, even in early October. Had his color scheme been red instead of purple, he would have passed as St. Nicholas toward the end of a six-month starvation diet.

Druim Dubh

Ron had taken the bus in from Callanish, eighteen miles to the west of Stornoway, and joined us in our hired car. The wild moorland begins fairly suddenly upon leaving town, and just three miles southwest of Stornoway, on the A859 road to Harris, Ron asked us to pull over at the abandoned Half Way Garage on the left—"half way" to what is unclear; Nowhere, perhaps.

Still seated in the car, Ron asked us, "Can you see the stone circle that Margaret discovered? It's visible from here."

We studied the landscape in every direction. I thought I saw some stones protruding from grassy tufts on a small hill on the opposite (northwest) side of the A859, and made a guess that this was the

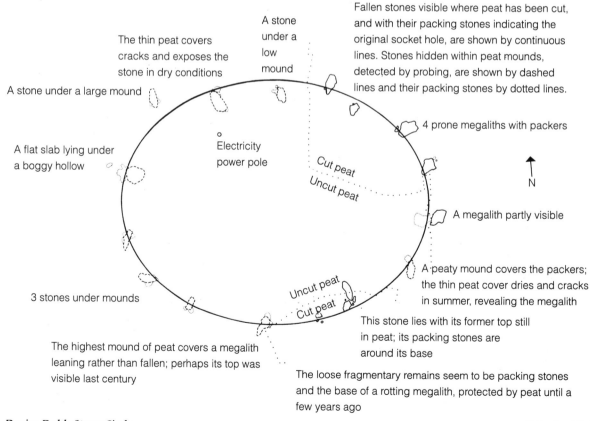

A stone under a low mound

The thin peat covers cracks and exposes the stone in dry conditions

A stone under a large mound

Fallen stones visible where peat has been cut, and with their packing stones indicating the original socket hole, are shown by continuous lines. Stones hidden within peat mounds, detected by probing, are shown by dashed lines and their packing stones by dotted lines.

4 prone megaliths with packers

A flat slab lying under a boggy hollow

Electricity power pole

Cut peat

Uncut peat

N

A megalith partly visible

A peaty mound covers the packers; the thin peat cover dries and cracks in summer, revealing the megalith

3 stones under mounds

Uncut peat

Cut peat

This stone lies with its former top still in peat; its packing stones are around its base

The highest mound of peat covers a megalith leaning rather than fallen; perhaps its top was visible last century

The loose fragmentary remains seem to be packing stones and the base of a rotting megalith, protected by peat until a few years ago

Druim Dubh Stone Circle.

© Curtis 1994

circle. That I had guessed correctly wasn't much of a feat, as Ron had taken us to a place where we knew a circle would be visible. Countless people had passed this spot for centuries, and, in recent years at least, no one had noticed the evidence of a stone circle—until Margaret spotted them from a bus on May 1st, 1992. Some of the stones were exposed where peat had been cut, but that they were anything but random boulders on the moor was not finally proven until the Curtises completed a detailed survey of the site with probing rods. The bare bones of their meticulous survey is reproduced opposite, drawn by Ron and Margaret and reproduced with permission from their self-published *Mini Guide to Druim Dubh Stone Circle.*

Druim Dubh (prounounced *"drum doo"* and meaning Black Ridge) Stone Circle was built on a flat-topped hillock of solid rock. Nearly two feet of peat covers most of the site, and their drawing shows where locals have taken the peat away in recent times. The original packing stones are very much in evidence at several of the sixteen stones that Ron and Margaret found above and below the turf, suggesting that the stones once stood upright. Because of the lack of soils—the Lewisian gneiss bedrock is very close to the surface here—it is possible that some if not all of the stones were not socketed in the usual way. Rather, the one- to two-ton stones may have been stood up on bedrock, supported by stones and packed earth. Over the years, natural erosion may have caused the stones to tumble, although the Curtises suggest that they might have been intentionally felled in antiquity. In either case, it does not seem that the stones were plundered for building material, as at so many sites in England.

The ring's shape is very close to an ellipse. In their mini-guide, the couple say, "Its long axis lies east/west, which may indicate a connection with equinoctial sunrise or sunset, in March and September." No outlier stones or other obvious alignments have been found. Thanks to the couple's survey work, the Druim Dubh ring was scheduled as

an Ancient Monument in 1993. It sits on land presently owned by the Stornoway Trust.

Achmore

Our route to Callanish took us by another stone circle—the island is fairly littered with them—called Achmore, and Ron, knowing now the nature of my book, rightly guessed that this one would be of interest. Like Druim Dubh, Achmore was discovered in this century, largely the result of peat cutting for fuel. By 1981, the site was seen as possibly prehistoric in origin and it was reported to archaeologists already on the island for the major dig at Callanish I. Margaret (then Ponting) did most of the investigation at Achmore in 1981, opening trenches around each of the eleven visible stones. Probing with a metal rod revealed several more stones beneath the peat, and three months after the completion of site work, the Pontings published their report *Achmore Stone Circle* (revised, updated, and republished in August, 1998 by Margaret and Ron). Their work revealed a true circle with a diameter of about 41 meters (134.5 feet), making it the 14th largest in Scotland. At 41.46 meters, the diameter would be exactly fifty of Alexander Thom's megalithic yards. Ancient layout error, regional differences in megalithic yardsticks, or the difficulty of surveying fallen stones today could all account for the slight difference, if, indeed, the use of megalithic yards was intended. Margaret and Ron make no such claim.

Only two of a probable twenty-two original stones remain standing at Achmore today, and one of those is damaged. What is interesting is that the fallen stones are accompanied by their packers in almost every case. (One broken stump is in situ.) Careful excavation could very well reveal the exact shape of the original sockets, so that a very precise reconstruction of the circle might be possible using the original monoliths and packers. No such work is scheduled, but, if ever it is, I don't know of anyone more skilled and qualified for the job than Ron and Margaret Curtis. As will be seen, they have already proven themselves up to the task with a much larger stone at the fascinating megalithic site called Cleiter—or Callanish VIII—just ten miles to the west.

Achmore was our first introduction to Sleeping Beauty, a line of hills roughly fifteen miles to the south. Also known in Gaelic as Cailleach na Mointeach (pronounced "callach na montach," with the *ch* as in *loch*)—the Old Woman of the Moors—this range of hills, we were soon to learn, plays an important part in the design of the Callanish complex of megalithic sites, particularly with regard to lunar alignments. Achmore is the farthest site in the Callanish numbering system (Callanish XXII by the Curtises), so Sleeping Beauty has a somewhat different appearance there than from the sites closer to ground zero at Callanish I, seven miles away. From

Achmore, Sleeping Beauty appears to be pregnant, a genuine Earth Mother.

After the short but important visit at Achmore, we continued on to Olcote, the Curtises' cottage on the edge of Callanish village, and we met Margaret again after twelve years. A little grayer and perhaps not quite as sprightly, Margaret had lost none of her wit and enthusiasm, as the next two days proved time and again.

CALLANISH: NAMES AND NUMERALS

To most people, Callanish means Callanish I, the largest and most enigmatic of the many megalithic sites in the area. Arguably, it is also the most beautiful stone circle in the British Isles, and is certainly the best known and most photographed in Scot-

Ron Curtis shows us an ancient standing stone, now recumbent, with its packing stones.

land. Before Margaret takes us on a tour of the site, I must mention a couple of points of nomenclature. There is a movement to rename the place *calanais,* the Gaelic form. The new Ordnance Survey Maps of the Western Islands are changing over to Gaelic, in honor of Celtic roots. Patrick Ashmore, the archaeologist in charge of the 1980 and 1981 digs, wrote a fine illustrated guide to the site called *Calanais: The Standing Stones* (Urras nan Tursachan, Ltd., 1995). However, in virtually every reference work, the monument will be found under *Callanish,* though in very old works, *Callernish, Classerniss,* and even *Tursachan* may be encountered. While I respect any and all efforts to preserve cultures and languages, I will stay with Callanish in the present work to avoid confusion. Visitors to Lewis will not go astray. Callanish and Calanais have the same pronunciation, and almost everyone on the island speaks English (although quite a few do not speak Gaelic).

The other issue of labeling is the numbering system for the various sites, using Roman numerals. Alexander Thom started this system, and was responsible for the first VII in the series. A team of researchers from Glasgow University took the sites up to XII, and Gerald Ponting and Margaret and Ron have taken it up to the XXII, Achmore being the last to be listed. Of the numbered sites, six are stone circles, while some sites consist of just a single standing stone. There are other stone circles on Lewis, such as Druim Dubh and one at Shawbost (discovered in 1983), but they are too far away to be considered part of the Callanish complex. It amazes me that stone circles are discovered every few years in the Outer Hebrides, in much the same way that "new" ancient Mayan cities keep popping up in the Yucatan. In fact, in 1994 an acquaintance of the Curtises discovered yet another circle about seventy-five miles south of Callanish. The best stone circles to visit, are those numbered I, II, and III, all within about a mile of each other, but we'll also look in at site X and, in particular, the wonderful and curious site VIII.

CALLANISH I

Margaret has taken thousands of tourists on tours of Callanish I, and yet her enthusiasm for the subject has not waned. I think she welcomed the opportunity to show the stones to someone who knew a little more about them than the typical tourist, although she was to find out soon enough that I had a lot to learn as well. *Sure, that's why I'm here.*

Knowing what is needed to make certain points later on, she weaves the visitor through the village and the stones in just the right order to build the story of Callanish. She showed us a place in the village where the main A858 road was widened, revealing an ancient cairn, that precipitated an emergency excavation that still occupies much of their time. Then we moved to site XVI, a single three-foot-high standing stone perched next to an old tractor chassis not far from Olcote, their home. She didn't tell us why we stopped to look at this stone—we didn't even get out of the car—she just wanted to make sure that we were aware of it. Then we continued on another half mile and parked the car near the great circle itself. Even Darin, not quite twelve years old and jaded by a dozen recent visits to stone circles, was taken aback by the tall, thin, tightly packed, wonderfully shaped menhirs. Margaret led us into the site as if it were her own home, to introduce us to her family. Indeed, if the stones are able to sense a kindred spirit, it would be Margaret's. I wondered if anyone else had ever spent as much time in this special space, or—in the past three thousand years at least—knew as much about it. There is a wonderful photograph of Callanish I in the color section.

The Stones

The Callanish stones are a hard and beautiful Lewisian gneiss. Their defining characteristic is the curved striations of color evident on almost every stone. Streaks of white and even red run riot on the predominantly gray backgrounds, as if applied by an artist with wide sweeping strokes. The stones are generally narrow of shape, allowing the use of tall stones that are only a fraction of the weight of stones of comparable height at most other circles. At the same time, there's nothing flimsy about the stones, as evidenced by the fact that Callanish I is one of the most complete of the ancient stone circles in the world. Construction dates to about 2500 B.C., Margaret told me, with another flurry of activity around 1800 B.C.

Another reason for the monument's wonderful state of preservation, in addition to the natural toughness of the stone (and its obviously remote location), is the fact that the circle was partially buried in a thick bed of peat for a large part of its existence. By 1854, villagers from Callanish (settled about 1790) had cleared the roughly five feet of peat from all but the circle itself, and in 1857, Sir James Matheson, who made a fortune in the opium trade and bought the island of Lewis with the proceeds, ordered the rest of the peat removed from the site. This very unscientific "excavation" revealed the ancient burial cairn within the stone circle, forgotten for two thousand years. It also left a curious

two-tone color scheme on the stones that persisted for several decades: the portion of each stone above the peat line was gray and weathered, whereas the five feet or so that had been immersed in the acidic peat were bleached a bright white.

Margaret gave us a few minutes to take in the stones, before sharing some of her insights.

"Geologically, the stones could have been brought from virtually anywhere on Lewis," she told us. "I think a lot of them were selected for their shape or their color, and some of them for these things." She indicated a black intrusion on one of the stones, about three inches in diameter. "You see these crystals of green and black? That's called hornblende. It's perfectly natural to the rock, and a sufficient number of the stones have it to suggest that they were selected for it. If you picked up fifty stones at random, you wouldn't get as many with hornblende."

The North Alignment

The northern avenue, a double row of stones, does not align with true north. In fact, its center line runs about eleven degrees east of north. Other alignments at Callanish are so precise—the south row is

Engraving of Callanish published in 1867 by Sir Henry James, Director of the Ordnance Survey. The marks left on the stones when the peat was cleared ten years earlier were still very clear. (Caption by Margaret Curtis, used with permission.)

virtually perfect—that this is undoubtedly not an error. Enigmatically, the newly discovered cairn alongside the highway is aligned with the avenue. Many writers suggest that the so-called north avenue aligns with certain major star risings in ancient times, although theories vary as to which star and when. But it turns out that there *is* a true north alignment at the monument. Margaret took us to the end of the south row and up onto a natural rock outcrop called Cnoc an Tursa. From this vantage point, we could see the similarity of the overall shape of Callanish to a Celtic cross, a feature that several writers have pointed out, although the Celts could not have been here within a thousand years of the time of construction. We could sight along the south row of stones through the circle's center and . . .

"Can you pick out the stone away off at the other end by the tractor chassis? It's just left of the trees and just right of the tall stone in the circle."

I spotted a small stone, the very one we'd stopped at on the way to the circle, site XVI.

"If laying out accurately is of interest to you, stand here. That stone is within a tenth of a degree of true north. And there was no accurate north star to go by when the circle was built, which impressed Professor Thom."

Lunar Alignments

Having had a good overall view from Cnoc an Tursa, we returned to stone 9, just outside of the central circle to the southwest. This stone, along with its companion, stone 34 on the opposite side of the center point, are the only two true outliers in the immediate monument, the only known authentic stones that do not belong to the circle or to any of the stone rows.

"These two stones align with the northernmost moon rising. Rear Admiral [Boyle T.] Somerville spotted this about 1909. He published a paper in 1913, and Professor Thom read it as a young man. In 1933 Thom sailed up here and anchored where you see that float in the loch. His son Archie told us how they anchored there and had their evening meal. Then they had to wash up before they were allowed out, and they clambered up here in the evening. Thom recalled what Somerville had suggested and said something along the lines of, 'Good grief, of course it is a lunar site.' So it was actually right here that Professor Thom first got interested in standing stones. And he spent the rest of his life at it, surveyed all the sites in Britain and at Carnac in France, and proposed the megalithic yard. This is a historic spot, Rob, a spot to inspire."

I resisted the brief urge to kiss the ground, and concentrated on the alignment. The stones were only about seventy feet apart, fairly close together to get an accurate alignment with something a quarter million miles away. Nevertheless, surveys taken through the centers of stones 9 and 34 have indicated an accuracy of one-half of a degree, which is pretty good. Personally, I find that round-topped stones within a hundred feet of each other are hard to use as precision instruments; however, Margaret was about to show me lunar events with foresights several miles distant.

But something more basic was bothering me than short alignments, and here I had to blow my cover and reveal to Margaret that I had a less than clear understanding of the moon's risings and settings. I knew that the moon went through an 18.61 year cycle, and I'd heard of terms such as *major standstill*, but I had no good idea of what happened during these cycles, although I thought I did. St. Oran's mantra came back to me: *The way you think it is may not be the way it is at all.* It took Margaret a couple of days to straighten me out, so stuck was my thinking. The Scots could have had me in mind when they coined the metaphor "thick as two short planks." The reader, unencumbered by prior notions (I hope) and aided by the sidebar on page 170, will twink a lot faster than I. Read the sidebar now.

How the Sun and Moon Rise and Set

The sun always rises on the eastern horizon and sets on the western horizon. The rising and setting positions mirror each other about a north–south axis.

On the long summer days, the sun rises in the northeast and sets in the northwest. On the short winter days, the sun rises in the southeast and sets in the southwest. This pattern repeats itself year in and year out.

The moon also rises on the eastern horizon and sets on the western horizon. The rising and setting positions mirror each other about a north–south axis.

But, whereas the sun takes a year to move from the midsummer NE/NW path to the midwinter SE/SW path, and back; the moon takes only a month to move from its long NE/NW path to its short SE/SW path, and back.

Furthermore, the size of the moon's monthly swing across the horizon gradually varies from a *small swing* (smaller than those of the sun) nearly SSE to SSW, to a *big swing* nearly NNE to NNW.

The minor northerly and southerly *standstills* occur during the small lunar swing. The north and south extremes—the *major standstills*—occur at the the time of the large lunar swing. The word *standstill* alludes to the fact that there is very little change in the moon's extreme rising and setting positions once a month for a period of four or five months. Similarly, the sun's rising and setting positions are virtually the same for several consecutive days either side of the summer and winter solstices.

The complete cycle of maximum lunar swing to minimum and back to maximum takes 18.61 years. Halfway between the maximum extremes—or 9.3 years into the cycle—the moon undergoes less radical changes in its risings and settings over the course of a month. This is the phase of the cycle sometimes called the minor northerly and minor southerly standstills. These minor standstill risings and settings are indicated on the diagram as broken lines.

Finally, the true extreme southerly risings and settings of the moon—the events that seemed to be the most important to the designers of the various monuments in the Callanish complex—can occur at any time of year. In summer, the event would be marked by a full moon; in the spring and fall, a half moon would do the job; in winter, the moon would be invisible or a mere sliver of light.

(This sidebar is a slightly edited and expanded version of page 3 of *Callanish: Stones, Moon & Sacred Landscape,* a booklet by the Curtises [self-published, copyrighted 1990 and 1994] and used with their permission. See sources.)

(This information is based on observations in the Northern Hemisphere. Readers south of the equator will know to change summer for winter, winter for summer. The diagram shows an approximation of observations at Callanish, at latitude 58° North. — R. R.)

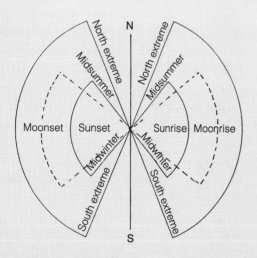

Solar and lunar cycles of rise and set. Solid lines show extremes. The dotted line indicates the moon's minor standstill rising and setting.

I didn't have the advantage of reading the sidebar. (*Stop.* If you haven't read it, do so now. If you have, it won't hurt to read it again.) But Margaret showed extraordinary patience, although at times I feared a swift belt with one of the wooden-backed illustrated placards she was carrying.

"So you don't like short alignments," said Margaret. "Fair enough. Let's look at one a bit longer."

We processed along the avenue to the last stone on the eastern side, stone 8, which seemed to be foreign to the predominantly flat style of most of the Callanish stones. It seemed more blocky, somehow, with a squarish cross-section.

"What about this one?" asked Margaret. She's a great proponent of the Socratic teaching method. "Have a look at it. Is it set crosswise to the row, or what do you reckon?"

"It's set askew," I ventured. "It's set at an angle to the row."

"Uh huh. Well, come around to this side and sight along this flat face. What do you see in the distance?"

"Tea Room—Crafts," I read.

Margaret laughed. "No, I mean away off in the distance."

"I see a cliff. The edge of the stone lines up with an escarpment."

"So we had two stones—9 and 34—that mark the maximum moonrise. The moon is up for twenty hours, drops into the hill there, just above the yellow chimneys, and then it would have reappeared to the right of the cliff, regleamed we say, just before it disappeared below the horizon. And a fortnight later, it plays similar tricks on the southern horizon. So I'm suggesting that's why stone 8 is set askew, and it's up to you to decide for yourself."

We started back toward the circle.

"The problem is that the moon doesn't make the same swing each month," Margaret explained. "At present, it's taking quite small swings across the horizon each month, but these will get greater and greater until, in 2006, the moon is going to rise and set just either side of north and, a fortnight later, it will rise and set just either side of south. This is the major standstill, or whatever words you want to call it. So the sun plods on, doing the same thing each year, while the moon builds up these huge swings, every 18.61 years. The minor standstill is when it's doing the short swings over the course of a month."

I nodded that I understood.

Margaret looked at me suspiciously, but carried on.

"Now, this is where Sleeping Beauty comes into it. Ron showed her to you at Achmore, didn't he? Well, she is not pregnant from here."

Margaret pointed out the range of hills that makes up Sleeping Beauty, but Jaki and I had to allow considerable artistic license to make sense of the shins, knees, torso, breasts, head, and pillow, as they were pointed out to us. I have trouble with constellations like Ursa Major, the Great Bear, too, although I can easily see it as the Big Dipper. Anyway, we could allow that a religious person could easily see a woman—barely—lying on the horizon.

"Now, the point about this range of hills is that when the moon has got to its maximum southern standstill or declination, it will rise from some part of her body. Usually, it appears first among her shins, if you like, and rises up a very shallow angle along her body, silhouetting it, and this is quite dramatic. And this is where, if you like, you can put all the little Earth Mother links between the moon, the menstrual cycle, et cetera. You've got the symbolism built in."

"Imagine how long you'd have to survey the area to find this spot," I said.

"Yeah. And then the moon rises at a very shallow angle until it's due south, and, even then it's only four diameters up, not very high. Then it slides across to that tall mountain over there, Clisham, and disappears. But, at the last minute, it reappears—in the valley. It's dying in Clisham and being reborn or resurrected again in the valley, and this would happen from pretty much all the sites around here.

"Sleeping Beauty," a range of hills seen from all of the Callanish group of stone circles. (Ron and Margaret Curtis, photo and labeling.)

As the basis for an early religion, I must admit, it certainly has something going for it. Do you go to church at all?"

I admitted that I was a born-again pagan and went to stone circles.

Margaret tried to deadpan this comment, but allowed a small grin to gleam under her professorial cap—a purple home-knitted hat to match her anorak.

"Ah," she said, and then continued her description of a 4,500-year-old event. "Now that deep valley in the Clisham hills there, west of true south, no matter what Hawkins and other people have written, they're never going to see that deep valley from here because of the rock outcrop, unless they've got twenty-foot legs or X-ray vision. But what you *would* see, over that car there, would be the moon rolling up over Sleeping Beauty and dropping into

the hilltop. But that wasn't the end of it because a few minutes later you would see . . . this!"

Jaki gasped audibly; I've actually got it on tape. Margaret had whipped out a placard showing the moon's reappearance after clearing Clisham, and framed by the standing stones of Callanish. Very impressive indeed.

"This is the drama," said Margaret, like a director setting the scene. "You look up the avenue, because it draws the eye, and there's the sacred circle, if you like. And you get the moon appearing there, just above the burial cairn. Now, if you've got a priestess or a queen or some person in charge, and if they were inside the circle, they would look quite big. But if you saw a person where we saw the sheep nibbling beyond the circle, they would look like so."

And she whipped out another card, showing the outline of a person standing on the hillside.

"Now, if you combine this picture with that picture, what do you get?"

"A silhouette," said Jaki.

"The man in the moon," said Darin.

Margaret beamed and put her cards away, letting the point sink in.

"So it *is* a lunar temple," I said, my words like a long-lost echo of Thom's sixty years earlier.

"Well, maybe, but look at this," said our guide, enjoying, I think, manipulating our puttylike minds.

A Solar Alignment?

Margaret pointed out that the stone circle was not truly a circle. It is, in fact, what Thom identifies as a "flattened circle." She also asked me if the stones were evenly spaced and I said that they were not.

"Uh huh. So there must be other criteria. Do you see this natural knob on the side of that stone?"

I could see a projection from the stone indicated, at about eye level.

"And this adjacent one has an overhang. Do you see it? Good. Now move your eyes up and down, right and left, until the projection and the overhang form a kind of a window."

"Oh, yeah, you can make a small horizontal window. I can get it really good back here."

"Have you got ground level in it yet?"

"Yup, yup, I've got it!"

Margaret dropped the bombshell.

"A week before midsummer, the sun will rise at that particular spot."

Incredible.

Maybe *too* incredible. I shook my head and smiled. Jaki laughed.

"You know, I have to say, I don't know if these things really . . . I mean, I have to view it with some skepticism." Jaki expressed what I was thinking. Had we been married too long?

"Sure," said Margaret. "You want to know if its real or if . . ."

" . . . we're just looking for things," finished Jaki.

"I don't know!" said Margaret. "Either you like it or you don't."

"I *love* it!" said Jaki, laughing.

"Look, this is all very simple," I said, and proceeded to dig myself a socket hole about six feet deep. "We all agree that there is a small window on the horizon there. That's clear enough. But the sun has got to rise there sometime, it could be a week before midsummer, two weeks, a month. It would be more convincing if it rose on the actual day."

Margaret listened patiently. "But it is so close at midsummer, the sun's standstill, just a tiny change from the week before. If we're right on this, they've selected those stones and put them there extremely carefully, which is related to what you're writing. I don't mind if you reject it."

"I must say that the stones don't look like they've tilted," I admitted. "They're standing straight, so the window *could* have been there four thousand years ago."

"It could act as a warning for someone," suggested Margaret, "a warning that midsummer is coming."

That made sense. After the initial shock of this almost magical alignment had passed, it all made perfect sense.

"Yeah," I said. "I suppose you can't announce a ceremony on the very day. You have to tell them about it a week ahead of time."

Stone 35

The broken stone 35 is the smallest standing stone (squatting, really) on the site. Like the outliers 9 and 34 that form a lunar alignment, it is neither part of the circle nor any of the other rows. But it is very suspicious. Margaret explains:

"This stone never appeared until the path appeared in the past century. The path was built so that Mrs. Matheson could take her friends to see the stones without getting their long dresses soiled. And a story has grown up about this stone. Two local girls both liked the same boy, so one of the girls went to the witch and said, 'I want to get rid of my rival, so I can have the boyfriend.' So the witch gave the girl a belt and said, 'Put it around the other girl and you'll get the boyfriend.' She went home with the belt, but had second thoughts, and, instead, she came up here and put the belt around this stone. Thunder and lightning cracked it in two."

Margaret clapped her hands loudly to simulate a thunder clap.

We laughed at the tale and I asked Margaret why she regarded the stone with skepticism.

"It didn't appear on anyone's plan until the path was made. It looks very suspicious that it's a Matheson placement. It was broken by drunks, not magic. And it's set in concrete! Have you heard of Gerald Hawkins?"

I knew that he had written *Stonehenge Decoded* (Dell, 1965), but Margaret told me that he'd also done extensive analysis of the Callanish alignments in the late 1960s, and that his famous computer analysis had chosen stone 35 as one of the most important, with lots of lines radiating from it.

"And yet it might not even be in the right place?" I asked.

"It *certainly* isn't in the right place! There is no prehistoric hole here. We don't know where it came from. It was re-erected in 1981 simply because it had been here a hundred years, so you can't use it to get meaningful alignments."

Margaret showed us more alignments, solar and lunar, but I cannot take the space here to describe them all. The point is that Jaki and I were absolutely blown away by the astronomical intricacies of this site, stone 35 notwithstanding. Ron and Margaret express it well in *Callanish: Stones, Moon and Sacred Landscape:*

At nearly the center of Callanish I are the remains of a burial cairn, probably erected many generations after the circle was built, and now in ruin.

Dedication and intelligence; planning and organisation; engineering and the economy; the social and political structure—all aspects involved in the Callanish megalithic project—a project that has been likened to the effort by the USA in landing the first men on the moon. Possibly we may never fully understand the nature of the driving force that compelled prehistoric people to put so much effort into the wide range of activities culminating in the Standing Stones of Callanish.

Margaret's Stone

I knew that Margaret had discovered the missing end stone of the east row in 1977, now designated stone 33A. It was a story I'd long been interested in, and to hear it from her was one of my main reasons for returning to Callanish.

This discovery was preceded by careful research. Margaret had seen an indication of the stone on a plan drawn by John Palmer in 1857, the year that Sir James Matheson completed the peat removal at the site.

"I'd laid out a string grid in this grass and used a graduated rod to start prodding through the turf," Margaret told us. Pointing a few feet away, she said, "Out there, the probe just went through the peat with a gritty sound as it hit the till below, but over here it just went *clunk,* and you could draw the outline of the stone. When the official digging was on in 1980, they took the grass off and this—" she tapped stone 33A "—was visible underneath. It looked a bit like a crocodile. We went down lower, through the peat, and found a horseshoe-shaped ring of stones, the packing stones. Softer soil was easily removed and you got a firm hollow that was the shape of the foot of the stone, like a footprint in the snow. So it was quite obvious how it stood."

"I always wondered how archaeologists are so sure about socket holes," I said.

"By the firmness and quality of the soil. It may be a different color. There's no doubt when you get it."

Stone 33A is re-erected with pulleys and a tripod, July 19, 1982. (Photo provided by Ron and Margaret Curtis.)

Margaret explained that the stone was found on croft land, just over the fence from Historic Scotland property.

"There was a mountain of red tape," she explained. "They weren't going to re-erect it on crofting land, so they had to exchange an equal area of Historic Scotland for the space where the stone belongs, which is why you see the bend in the fence. And then Historic Scotland insisted upon setting it in concrete!"

"They did?"

"Yeah!" She smiled and shook her head. "Notice that it's whiter, because it was under the peat. And it's the square 'full stop' stone, like those at the ends of the other stone rows."

Margaret and Stone 33A. They go way back.

"How come it doesn't line up with the others?" I asked. "Which one is out?"

Margaret beamed. "I haven't got *all* the answers!"

Maybe not. But I hadn't realized that Margaret had made another discovery, in 1978. She took us to stone 19, the first one of the avenue's west side row.

"In the old records, individual stones were drawn. At one stage, this one was drawn with a pointed top, and a few years later, it was drawn with a flat top by the Inspector of Ancient Monuments who came up to look at the peat marks on the stones. We wondered why the top was broken, so we got a big stepladder from the local school, and I sort of looked around a nearby stone wall wondering what the . . . *Hey, I've got the missing top!* I passed it up and

it fitted exactly, and there it is, officially glued back on, and it had been off over a hundred years. So I've found a whole standing stone, a tip of a stone, and a circle . . . I guess that's not bad going."

"I guess not!" I agreed.

CALLANISH III

We drove about a mile east on the main road, to Callanish III, with the almost unpronounceable Gaelic name of Cnoc Fillibhir Bheag. (Yanks and Sassenachs who want to have a go can try this approximation: "kroc fill-i-ver veg.") This delightful circle, fairly close to the main road, is the most visited of the minor sites, and has a special place in my heart, because it is the ancient stone circle most like our own at Earthwood. The stones are of similar scale and shape, and it even has a passable Bunny Stone. Actually a flattened circle, like its better-known neighbor, its mean diameter is about fifty feet, a little larger than the Stone Mountain Circle, which is also in the same general scale and appearance as Callanish III. I felt very comfortable here.

Margaret said nothing to change my good feeling, but, as we'd now come to expect, she offered plenty of food for thought. She started with Sleeping Beauty again.

"We wanted to know if there was any piece of land you could stand on to see the circle with the hills beyond. So Ron started up close to the circle here, looking through the aperture of his camera. The stones looked massive, of course, and the far horizon looked small, so he started walking backward until he fell down the slope there, still looking through the eyepiece. He stood up and shouted, 'Look, Margaret, what I'm standing in. A wee ring of stones.' This marked the perfect position to frame Sleeping Beauty by the stones and to observe the rising of the extreme southern moon."

We were getting used to this sort of magic by now, but the next revelation was extraordinary, even by the standards we'd come to expect from Margaret. Inside site III's stone ring are four curious stand-ing stones without any obvious configuration or relationship to the ring. Two of these inner stones coincide with the long axis of the flattened circle, indicating an azimuth of 334°, a symbolic align-ment to the moonset at the northern extreme of the major standstill. But the extraordinary thing is the symbolism of the stones.

"If you think about that hill range there, you've got it," Margaret began, indicating our old friend, S. B. "There are still people who think of the Earth Goddess; it's a very ancient religion." (I thought of the Venus of Willendorf offering that I'd seen at Columcille.) "And the Goddess is not just a single deity, but she is represented in three forms: the maiden, the mother, and the wise old woman, rather like the trinity in Christianity. Does this ring bells with you at all?"

We admitted our ignorance.

"Well, it was new to me, too. And the three forms—maiden, mother and old woman—have got corresponding colors: white, red, and gray. And they've got shapes. The young Goddess is a white triangle, and I'm pointing to one of the four cen-tral stones, which is white and triangular, although it's a bit grayed up with the rain at the moment. We also have the red stone, there, for the mother, either for birth or fertility, and she's actually got breasts. And over there, the truly gray stone, is the third manifestation of the Goddess, which is said to be the wise old crone. People with a more vivid imagi-nation than mine say she's got a cape wrapped around her. And, this 'triple goddess,' if you like, has a consort." Margaret looked around to see if eleven-year-old Darin was out of earshot. "So here, within the circle, is the male. I think he is obvious. And the

The author tries to work out the symbolism of the "consort" at Callanish III.

color of the stone with white quartz streaming is appropriate."

Inwardly I wondered if any other individual in the world was as intimately familiar with a megalithic site as Margaret is with Callanish. Outwardly, I asked if someone had suggested these meanings to her.

"Yes, people come from all over the world to worship here," she said. "And they share their fields of knowledge with me."

Callanish II

This possibly elliptical ring, ten minutes by foot beyond site III, has only five standing stones remaining out of at least ten originals, and it seems likely that other stones or at least sockets remain beneath the turf. It is worth a visit, as long as one is so close, if only to view a wonderful tall triangular stone, the first to greet you as you approach from site III. Darin declared it his favorite standing stone and hugged it, and I don't fault his selection in any way. Margaret suggested that we might be attracted by another beautiful young maiden stone.

We'd gotten pretty wet at sites III and II. We took Margaret back to Olcote and dried ourselves off in her sitting room while she gave me a much-needed crash course on the moon's strange cycle: Lunation 101. Later, after an excellent supper at the nearby Doune Braes Hotel, we returned to Olcote for further discussion about the stones, and to look at a video that Margaret had made about Callanish. We arrived late back at the youth hostel and had to sneak in by flashlight.

CALLANISH X

October 3rd was drier. After a breakfast of fresh rolls and Orkney cheese, we took Darin to see the great broch of Dun Carloway, and had the place to ourselves, nary a tourist in sight. Then we met Ron, who took us on a half-mile hike across peaty moor-

Darin's favorite stone at Callanish II.

land to site X, a ruined circle. It seemed like two miles. Callanish X, or Na Dromannan (na drym-a-nan), is a large ring with roughly twenty fallen megaliths lying around. Ten form a ring, possibly a flattened circle, seven are scattered inside the ring, and three stones may have been outliers. Ron has done extensive work at this site, probing and surveying, and has found alignments to other sites, but he admits that the place is still a great mystery. The main reason that he took us to the site was that he knew my deep interest in the construction of stone circles, and this better than any other site has several excellent examples of fallen stones surrounded by their packers. It would be a lot of fun, I thought, to re-erect some of the better stones where their positions and orientations are obvious—under the guidance of experts like Ron and Margaret, of course.

Travel Tips

I hope that the excitement of the Callanish complex will encourage a few readers, at least, to make the journey to Lewis. If you do, I have four travel tips for you. One, hire Margaret for your own personal tour of the site. Two, visit Dun Carloway broch just seven miles farther on from Callanish village. This ancient broch or dun, round and of exquisite drystone construction, is thought to be a hill fort, one of five hundred or so built around the coast of Scotland about two thousand years ago, although experts disagree about its purpose. Dun Carloway is exceeded in its state of preservation only by the Broch of Mousa, on a tiny island in the Shetland group. Kids love Dun Carloway, because they can explore its circular stairway and chamber, which spirals up between the inner and outer stone walls. Third tip: Stay at the wonderful Garenin Youth Hostel, just two miles beyond Carlo-way. This hostel is a restored "black house," with walls of stone four to five feet thick, and a traditional Highlands and Islands thatch roof, tied down and held in place by boulders hanging off the roof. And the warden is friendly; at least, he was in 1997. When you visit Callanish, Dun Carloway, and Garenin, you'll get glimpses of island life forty-five hundred years ago,

two thousand years ago, and one hundred years ago (although the traditional black houses have been around for several hundred years). My fourth travel tip is to take your supper and have a pint of beer or a malt whisky at the Doune Braes Hotel, midway between Callanish and Carloway. We took our supper there with Ron and Margaret on each of the two nights we spent on Lewis. After a wet slog on the moors in search of stones, it's a great oasis of warmth and cheer. By the second night, we'd all converted to the excellent salmon.

Dun Carloway, near Callanish, Isle of Lewis.

Garenin Youth Hostel and its friendly warden.

The other point of interest at site X is the nearby cliff face that is traditionally thought of as the quarry that supplied the great stones at Callanish I. We all took turns speculating on the possible routes that the ancients might have taken.

Margaret stayed clear of this expedition in order to nurse a sore leg. Good decision. After the rain of the previous day, hillwalking conditions were poor. I slipped on a rocky slope coming back down, bruising the thin skin of my shin, and joined Margaret in the gimpy gambs department.

CALLANISH VIII

Margaret joined us for lunch at the new Callanish Visitor's Centre—worth a visit— and then we drove the ten miles around Loch Roag to site VIII, also known as Bernera Bridge, one of the most fascinating megalithic sites in the British Isles. The Gaelic name for the site, Cleitir, is a little easier to pronounce than some of the others—"cletcher" will do—and it means, simply, cliffs. The site is situated on the south end of the five-mile-long island of Great Bernera, very near to the north end of a modern bridge that joins the island to "mainland" Lewis.

It was a windy, cool, gray, but dry afternoon when we arrived at the site, and we were suitably bundled up for Margaret's tour. She and Ron had already told us that they had re-erected a major stone at this site, without motorized machinery, and they knew that the stone and their methods would both be of the highest interest to me.

A visit to the area by Prince Charles inadvertently precipitated the Curtises' deep involvement in the site:

"The owner of the island decided that it would be nice to stand that stone back up for the royal visit," Margaret told us, indicating the beautiful stone, "which had been lying on the ground for about one hundred years. So the owner sent a local crofter up with a mechanical digger, and this crofter dug a hole for the stone, and pushed it in. Then he decided that as long as he was here, he'd see if there was anything else around. He dug lots of holes and made a horrible mess. A few weeks later, Ron and I happened to be canoeing up from home, which is a lot shorter than coming by car, and we said, 'Hey, that stone's up again, but it looks wrong.' We informed Historic Scotland and in due course, we were given the job of excavating here, to see what information we could gain. It was really quite exciting. And it was quite a challenge as well, given the 11-kv power lines, modern and old field fences, a water supply running through the site, dynamiting that had blown up a good piece of the bedrock, and this crofter running around with a JCB digging holes."

The site is a ledge of a hillside, just above a steep slope down to the sea loch. Four standing stones suggest a half circle, but Margaret pointed out that we were standing at the top of a natural rock cliff that had been smoothed by the glaciers.

"So we don't have half of a full circle up here. Whatever we've got, this is all that there was."

I asked Margaret what she thought the site represented, and her answer would require a chapter of its own, beyond the scope of this book. A few highlights would include a 3200 B.C. carbon date found at the site: her analysis of a stone alignment suggesting that this might be a May Day (Beltane) ceremonial site; and the discovery by dowsing of a patterned pot, an event that Margaret described as "uncanny to me, in the middle of all the destruction."

"Before we dug our trench, the dowser, Jack Fidler, called the pot an anomaly on the site," added Ron.

Margaret, ever observant, interrupted the conversation with, "Look, a gannet!" (A gannet is the solan goose, *Sula bassana,* of the pelican family.) "I've never seen one fly over before. We've had mink walk across the site. We've had otters. We've had seals watch us. It's been super from that point of view."

"So this stone you re-erected, it was in the wrong place?" I asked when the gannet had passed.

"Yes, and luckily the crofter had put it six feet from its correct place, which enabled us to find the socket hold and all the original packers in the *right* place. We were able to empty out the socket hole and find the shape that fitted the stone, so the orientation is right now, as well as the position."

"So you had to lower the stone, drag it six feet, and stand it back up again. It appears to be about four tons, and nearly ten feet high. How many people did it take, and what kind of equipment did you use?"

"Two men helped us, that's all," said Ron. "We used a wooden tripod, a Tirfor winch, and a block and tackle with two or three pulley wheels at each end. No JCB or anything like that." A "Tirfor winch," I learned, is what Americans know as a come-along,

a ratcheted pulling device with powerful mechanical advantage.

"And it's within a few centimeters of its prehistoric position," added Margaret.

The accuracy of the positioning of this stone is critical to the spectacular lunar alignment of the site, as Ron explained.

"You see there, off on the horizon: The moon at the major southern standstill is hidden behind some of those hills, but it appears in a gap. In that short bit of green horizon, there are two small stones, due south. The left-hand one is the bigger of the two and is a cubic block, about a half meter on a side. The one to the right is pointed and fallen over. The distance between those two stones is the moon's diameter, and the height of the horizon gives the moon's position at the major southern standstill. When we came to test this in 1987, we found that we had to go part way down the cliff to get it to work. That may be due to refraction in the atmosphere. Now if you start coming back up the hill—" we followed Ron up the hill "— to here, you now see two more stones on the more distant horizon as they come into view. Again, the left-hand one is cubic, about a meter cube or more, and the right-hand one is pointed and it's fallen over, same configuration as the nearer pair, and same distance apart, about a half degree of arc, equal to the moon's diameter. In some terminology, the cubic stone is the female and the pointed stone is the male, so there seems to be something very definite in the alignments here connected with the moon. We don't claim to have it all worked out yet, but, here, just above a modern wooden peg, the two horizons and the two pairs of stones are coincident."

We looked at the places on the horizon that Ron had indicated, horizons on mainland Lewis across the bridge, roughly a quarter-and-a-half mile away, respectively. How long had it taken the prehistoric people to work out this alignment with such precision, given that 18.61 years would have to go by before the pattern would repeat itself? No wonder

Margaret points out the Beltane alignment.

Two men helped Ron and Margaret re-erect the Site VIII stone. (Photo supplied by the Curtises.)

Ron Curtis and an old friend gaze out across Loch Roag.

that Aubrey Burl, when he visited, told Ron and Margaret that he'd never seen anything else like it in the British Isles.

Margaret stared across the loch to Lewis.

"It's a lovely spot," she mused. "I find it a very elemental spot. Looking south here at noonday, you get the light, the fire of the sun reflecting off the water, so you've got all the ancient elements, really: earth, air, fire, and water. It's a beautiful spot."

Awen, I thought.

WE SAY FAREWELL

Our evening was divided between looking at artifacts and exhibits at the Curtises' little museum, and another fine meal at Doune Braes Hotel. The next morning, we tidied up all unfinished business and questions, and I had a look at Margaret's homebuilt stone circle in her back garden. Margaret presented Darin with a Callanish T-shirt. Our 48-hour visit at Callanish had come to an end, and the time had gone by all too fast. With sadness, we said our farewells to the Curtises and drove to Stornoway. After an hour's wander around the shops, we put our hired car in line for the ferry and waited. Who should appear but Ron, with a sheaf of photocopied papers for us, which he'd promised to send to the States later. He'd met us at the boat when we arrived and now he waved us on our way.

As Lewis faded into the western horizon, I recalled some cryptic passages from the 1st-century (B.C.) Greek historian Diodorus Siculus. In his *Histories,* Book V, Diodorus refers to an account of the 4th century B.C. by another historian, Hecataeus.

Diodorus tells us that in regions beyond the lands of the Celts, there is an island "not smaller than Sicily"—Lewis is disqualified, but not Great Britain—"inhabited by the Hyperboreans, so called because they live beyond the point from which the north wind blows." A magnificent round temple is ascribed to this island, dedicated to Apollo (the sun god). The god is said to visit the island every nineteen years. The moon is described as dancing through the night "from the vernal equinox until the rising of the Pleiades." Some writers interpret one passage as describing the moon clinging close to the horizon. Now, the west line of stones at Callanish marks the equinox, the eastern row, according to some, including Burl, aligns with the rising of the Pleiades star group around 1550 B.C., and the lunar alignments every 18.61 years are, by now, well known to the reader. While many writers, including Atkinson and Chippendale, suggest that Diodorus might be referring to Stonehenge, I tend to side with Margaret and Ron (and Aubrey Burl) who suggest that the curious references to the moon, the equinox, the nineteen-year cycle, and the rising of the Pleiades, all point more likely toward Callanish as the cryptic temple.

As we disembarked in Ullapool, I wondered if we had just returned from Hyperborea.

I like to think so.

— CHAPTER 9 —

Ed Heath,
Megalithic Sculptor

*If a stone doesn't fit, it just screams that it doesn't fit. But if it fits,
and it's proper, and there's harmony, people enjoy it.
They come by and say, "Ah, look at this. There's something
about this that I like." They don't know what, but there's
something there that they like and that they enjoy.*

—ED HEATH

Derek Muirden is director and presenter of the popular PBS-TV series *People Near Here*. My work in cordwood masonry had been the subject of one of his programs and when I mentioned to Derek that I was doing a book on stone circles, he told me that he had just finished work on a *People Near Here* segment about sculptor Edward Heath, builder of dolmens, stone circles, and other megalithic arrangements. I phoned Ed, enjoyed a good long chat about large stones, and arranged to visit him and his wife, Barbara Morris, in November of 1997.

Jaki and I had been home from our megalithic journey for about a month when we set out from Earthwood on the two-hour drive to Ed's home in Sutton, Quebec. It was a gray, bitterly cold day, but we were warmed three ways upon our arrival at Ed and Barbara's 130-year-old brick farmhouse: our hearts by the friendly welcome; our souls by Ed's passionate interest in sculpture and megalithic sculpture in particular; and—not the least important— our bodies by Barbara's excellent cumed squash soup and homemade breads.

Through conversation, and by leafing through a photographic portfolio of his work, we began to learn a little about the sculptor.

Born in Boston's inner city in 1951, Ed has benefited from a cross-cultural and multi-racial background that has made him feel comfortable with all kinds of people and traditions; his outlook is both global and eclectic. Ed's formal education took him to American International College in Springfield, Massachusetts, and Loyola College in Montreal, where he played football. His work as an artist went on hold for a few years while he pursued a career in professional baseball. This didn't surprise Jaki and me. At forty-six years of age when we met him, Ed still has the size, look, and movements of an athlete. And years of wrestling with large lumps of stone have not diminished his strength.

Ed's career in sculpture really began in North America, but blossomed during a three-and-a-half-year work/study program in Europe. There, he added classical techniques to his repertoire, and he was the first sculptor to win the prestigious Grolla d'Oro award in 1983, an honor repeated the very next year. His sculptures found their way into museums in Italy and Malta, and he enjoyed more personal success by winning the First Acquisition prize at the Santa Croce International Competition in Florence. Ed returned from Europe in 1987 and, two years later, settled in the Sutton hills in Quebec's Eastern Townships, where he has set up an indoor studio. In the 1990s, however, his studio seems to have become the planet Earth itself, particularly the 178-acre chunk of it right outside his door. While Ed's more portable work has been bought up for corporate and private collections all over Europe and North America, his most awe-inspiring work was privately commissioned on a 100-acre estate just a few miles from Ed's home, a collection of megalithic monuments on a similar scale to Bill Cohea's megalithic park at Columcille.

Stone sculptor Ed Heath does not play the clown. He has just removed the mask and goggles he uses for protection from stone dust. (Sabine Maher, photo.)

THE FARM

After a long sharing of photographs and stories, and with full bellies, we set out into the sub-zero afternoon to look over the pieces on the farm where Ed has lived since the late 1980s. A large rolling field borders the house and barnyards to the northeast, and several large sculptural groups are scattered carefully over the landscape. (See color section.)

Samurai

Derek Muirden says that it's hard to drive by Ed's place and stay on the road. The piece that probably causes the most near-misses is the first one closest to the house and the road, a sculpture both graceful and massive, which Ed calls *Samurai*. Thirteen feet high, the ten-piece configuration has elements of megalithic construction, oriental grace, and North American Inuit stone building, yet the three influences work harmoniously with each other and with the surrounding environment. Not one of the ten pieces has been shaped. Ed's sculptural expertise manifests itself in the way the natural stones—and a single whalebone—are placed together. The sculpture as a whole weighs some twenty-four tons, but it's perfect balance, both structural and visual, prevents the piece from imposing on the landscape. I was to learn over the next year how important it is to Ed to work with the landscape, to harmonize with it, and never to impose. Further, one of Ed's primary concerns is to improve the environment, and to avoid its destruction. If a tree is damaged or has to be removed for some reason, Ed is sure to more than compensate the land with tree planting or other improvements to the forest, such as providing better drainage where trees are suffering from root rot. Working with nature and not against it is a theme that recurred in our conversations again and again over the next few months.

We had passed near another long whalebone on the way to *Samurai*, and could see yet another one far off in the distance, in the gray November skyline. Jaki's question—"Where do you get these bones?"—was obviously one that Ed had heard before, and his smile was in scale with his football-player size.

"They're government-sanctioned, don't worry. They're mandibles from a blue whale. I got them about sixteen years ago, and it took me fifteen years to find the right use, one fitting the last resting place of a blue whale. A friend knew the minister of fisheries in Newfoundland and I was soon in touch with the person in charge of the last whaling station in Canada. They sent me these bones, by train, by boat, and by truck. This one is seventeen feet long. You couldn't get these today, but they're legal."

I wondered where he found his stones, and Ed's answer brought us back to whales again.

"They're all from the property, from this field, and from the woods."

"Just laying around?" I asked enviously.

"Buried usually, sometimes completely, sometimes with a little bit showing. With the big shovel,

Samurai. (More of Ed Heath's sculpture appears in the color section.)

we pull them out of the ground. Some of them are magnificent. Look at these two. When we pull stones out of the ground, I liken it to the first time I ever saw a whale. When the whale came out of the water—it was a blue whale—it was enormous! And he came up and took a look at us—he didn't spout or anything like that—he just looked at us and his eye was the size of a pie plate. He took his time and then went down and left. It was an exhilarating experience, like taking that stone out."

A few months later, when spring returned to the North Country, I was able to share this exhilarating experience with Ed, at the birth of a new stone circle.

The Farmer's Tomb

We headed farther out into the vast rolling field, a giant sixty-acre bench on the east side of a high hill. Across the valley, toward the equinoctial sunset, a range of hills created a lovely middle-distant view. The ancient mountain, Pinnacle, held sacred by Native Americans, stood high to the southwest. Ed's four husky dogs were enjoying the freedom from their enclosure. One can't be too careful during deer-hunting season, and Ed kept a close eye on all four, as concerned for the deer as for his dogs.

The Farmer's Tomb is a group of seven large unshaped stones, plucked from the ground and used just the way they were found. Ed has arranged them in a diamond-shaped symmetrical group consisting of three standing stones and a dolmen at the diamond's blunt end, itself composed of three support stones capped by a four-ton slab. The pointed end of the diamond is marked by a nine-foot-high menhir, a stone that would be quite at home at Avebury. The 69-foot-long axis of the diamond has a northerly alignment.

Jaki and I particularly enjoyed the dolmen, because it was almost exactly the same size as the one that we had reconfigured with Cliff Osenton a few weeks earlier. But this one had been built with a large "shovel," as Ed calls the huge excavator that he hires for his heavy stone work. The John Deere

The Farmer's Tomb.

892D-LC is capable of digging a 23-foot-deep hole from flat ground with its huge bucket and can excavate and stand up a 25-ton stone.

But, aside from a few scrapes on the stones, which will disappear with a few years' weathering, these ancient stones, rounded by eons of tossing and turning in the earth, look like they might date back to neolithic times. And that's just the effect that Ed wants to create. Quarried stones with drill marks on their edges can be quite beautiful and make very effective stone circles, but they will never look ancient, not in several lifetimes anyway. I asked Ed about the kinds of stones he has on the farm.

"There's schist and a variety of boulderlike stones, possibly glacial. The bedrock is schist. You find a lot of stones in the low-lying depressions of the bedrock. The stones vary in hardness; some of the schist just crumbles apart. There's green serpentine, and pockets of soapstone here as well, which is fascinating, but that's deeper, under the schist. We don't have granite on this side of the mountain, like the ones over at Iron Hill, where I've been working lately. The stones on this farm have been weathered, most of them. So one could imagine this as an exposed mountaintop. In some places, just six inches of earth covers the stones."

"I'm partial to weathered stones myself," I said. "They're more gentle, somehow, than quarried stones. They seem to have the wisdom of age."

"Well, quarried stones, oftentimes, have been violently shaken apart by explosion. But you're right; weathered stones are softer in appearance. They *look* older, whether they are or they aren't."

The Farmer's Tomb looked ancient.

Rejoice

We walked farther along the ridge that runs down through the middle of the great meadow, drawn to a happy, swaying, dancing stone, surrounded by four others, like a jolly mother dancing with her kids and dogs. As with *Samurai,* the defining feature is a balanced giant whale bone, but the net effect is very different, thanks to the character of the stones and Ed's placement of them. Instead of an oriental flavor, *Rejoice* suggests a Native American influence, perhaps the Inuit of the far north, whale-hunters since ancient times.

Ed spent an influential period of time among the Inuit peoples, and became interested in their art and lore, combined in a unique stone sculptural form called the *inuksuk,* stacks of balancing stones that tell a story, a little like totem poles—but different. Ed told us that he has been building *inuksuit*—the plural form—for about seventeen years, long before his work took a decidedly megalithic turn.

"You can speak with the inuksuit in the arctic, or even small stone groupings on the ground. One might tell you, for example, 'Three people have been here. We walk in that direction. A good hunting area for caribou lies two days north.' That sort of thing. You have to be very careful doing this sort of art, because it's sacred. A native woman stopped one day. She'd seen my inuksuit sculptures from the road. She asked if I was the one who did the sculptures, and I told her I was. She then proceeded to ask me all sorts of questions about my work, listening carefully to my answers. It was like taking a test. Finally, she nodded and said, 'Okay, you're doing it right,'

Rejoice.

and departed. If you just stand stones up without studying and without caring, you might do something offensive to someone's culture."

Ed is careful, as the word should be used: *care full*. He respects the sacred beliefs of all peoples, and considers nature to be sacred as well: the birds, the deer, the trees, even the stones. Some spaces are more sacred than others. Ed can *feel* it, literally, in the way that Ivan McBeth feels the energy of stones. And, in his sculpture, Ed can augment a sacred spot, as he was to show us at the very next "piece," just a hundred yards or so from *Rejoice*.

The Sacred Enclosure

Ed wanted to create a place where one could really get in touch with oneself, as well as with the earth spirits. To accomplish this, he knitted together spiritual elements of many different cultures: Aboriginal, Inuit, Japanese, and African. The space is defined by a rectangular mound of small stones, with a small square enclosure at the center. I was reminded of the Clava Cairns near Inverness in Scotland, except that the geometry here was square rather than circular. Four brightly colored wooden posts with snake designs marked the four primary directions, an Australian influence.

"I have some African posts I want to put up as well," Ed told us.

"What's your primary intent with this grouping?" Jaki asked.

"I'm . . . " Ed began, then paused to consider the question more fully. "It's just developing. When I come out here, I get an idea, and, yeah, okay, fine, I'll do it. This is the only piece that I let happen through guidance. I don't know what the guidance is. I just let it happen, and it comes. For example, now I realize that I need to put branches around it so that people can't access it from the field. The entrance has to be from the back, from the woods, from the sunrise—not from the sunset."

I asked about the four decorated posts, each of a different primary color.

"It's connected with rituals from the Native plains people, such as the Hopi in the Southwest. You enter the square box space, there, representing the six directions . . ."

"Six directions?" I asked. Trapped in a two-dimensional mindset, I could only think of north, east, south, and west.

"Down, the earth," explained Ed, "and up, the sky. Now, you might have black and white feathers tied on a string. And whichever way the wind blows, it indicates a color, and each color represents a certain direction, and also life, death, introspection, and illumination. The whole idea is that if the feathers are blowing in that direction, you think in that direction, and it helps give you guidance. And, in the Southwest, the posts are called *pahoes*."

"It obviously works for you," I said.

"It works for me, but the place affects other people, too, in different ways. Some are afraid to walk in here, can't walk in. Some people feel bad bad bad, some start crying, some people take off their clothes."

I do not include a picture of Ed's Sacred Space. It has little to do with stone circles, and much to do with Ed's approach to the sacred realm. Bill Cohea would appreciate the place as a gateway to the Spirit. Ivan McBeth might call it the realm of the transrational. Ed Prynn would say it's mystical stuff. For me, there's a lot of respect at play here, and a photograph cannot add to it, and might detract.

IRON HILL

Daylight was disappearing rapidly, as was any remaining heat in the air, and we had not yet seen Ed's major megalithic creations, over the mountain and a few miles away. Ed put his dogs in their pen, and we piled into our little car and scooted to an area known as Iron Hill. We passed a number of large parcels of land, beautiful even at this bleak season of the year, the properties providing backdrop for some expensive houses. Stones? Everywhere, there were stones. If there was any place in the world where a stone circle contractor might make a living, this was it. Stones and Money, a powerful combination. Ed provided the needed flux: Talent.

For reasons of privacy, the man of means who commissioned Ed to create a forty-acre megalithic wonder will be identified, simply, as Joe. On the way to Joe's place, Ed told us how he landed the commission.

"Joe was down at Art Sutton, in the village, asking if there were any sculptors around. Barbara happened to be just outside the door. The proprietor knew Barbara and told Joe to ask her about a stone sculptor."

"And she just happened to know one."

"She just happened to know one, and pretty soon Joe stops by and sees my work for the first time. He says to me, 'Listen, I got a whole bunch of stones.

What do you think you can do?' So I came over here and he wasn't kidding. He's got stones! Weathered granite, mostly—instant ancient. The stones were mostly in the woods, some in the fields. Most of them had to be dug out and moved to where we wanted to set them. And it was a wet muddy season, that first one. We used two shovels, a bulldozer, a front-end loader, and three trucks in order to move those stones, a major project."

We drove by Joe's property along the road, to get an overview of the site. Most of the settings can be seen from the car. Jaki and I couldn't believe our eyes. With the exception of the Carnac Alignments, we'd never seen so many megaliths in one place! Yet there was plenty of open space between groupings, so that each mini-site had its own character. From the car, Jaki and I spotted a Bunny Stone, almost identical in shape to the ones at Earthwood and Callanish III.

The Saracen Set *at Joe's place, private collection. (Don Davidson, photo.)*

Our initial impression was of a project on a scale similar to Bill Cohea's Megalithic Park at Columcille, the obvious difference being that Bill's stones are more integrated with the forest, so you can't see so many megaliths at one time. We turned around, parked the car, and bundled up as best we could; the air temperature, I'm sure, was rapidly approaching absolute zero. We walked up the snow-covered driveway to the first grouping, called *The Saracen Set,* a circle of Avebury-scale stones. Three pairs of standing stones featured giant lintels. Several other stones stood individually on the circle's circumference. Of all the modern stone circles I have become aware of in the past several years, this one takes the prize for authentic neolithic atmosphere, despite the Arctic conditions under which we viewed it for the first time. The only missing detail that might have improved it as an ancient circle counterfeit would have been a few recumbent stones laying around, as if they'd fallen over. As it is, I'm sure that a lot of tourists have driven by Joe's property, convinced that an ancient civilization once populated the Eastern Townships of Quebec.

The rapidly failing light quickened our steps, as we endeavored to see as much as we could before the starless night sky left us in pitch darkness. Rapid movement, at least, prevented us from freezing solid. We looked at a very large three-legged dolmen, and another group of standing stones. We knew from pictures we'd seen back at the farmhouse that there was a large stone circle on site that we had not yet seen.

Ed led us up a hill at an athlete's pace, and we intersected an avenue of stones near its top end. We turned right, and proceeded up the hill a little farther. A large presence seemed to take form on the edge of the forest. We moved closer and the presence loomed larger and larger.

"I don't believe it!" exclaimed Jaki. Neither did I.

In the diminished light at the edge of the forest, the stone before us seemed as massive as the Devil's Chair at Avebury and of similar shape. At forty tons, it was only slightly smaller. I told Ed that this was certainly the most massive modern megalith that I'd encountered in my journeys—although Manannan at Columcille is taller—and Ed seemed pleased to hear it. After we'd touched the stone and walked around it once or twice in amazement, I asked him how he'd managed to set this giant.

"Big shovel," replied Ed. "I did this one alone. I didn't want anyone around, in case they'd get hurt." The big shovel was the John Deere excavator already mentioned.

We continued down the avenue, aware of ghostly monoliths on either side. At the bottom, we entered the large stone ring called *The Lunar Circle,* thirty-two stones in a true circle almost 300 feet in diameter. The circle's size reminded me of Brodgar in Orkney, although the stones were weathered granite boulders, not sandstone slabs. And unique to any circle, ancient or modern, Ed had placed a large megalith at the center, with a large granite cap on top. From some angles, the centerpiece suggests a giant head with a very Gaulish nose and (to complete the illusion) a French beret on top. The circle is on the slope of a hillside meadow, with woods providing a background for more than 180 degrees of its circumference. The circle encloses a half dozen or so mature trees, another rare feature. The placement on a hillside meadow means that some viewpoints clearly show the circular nature of the group, something that is not always obvious in stone circles built on flat land.

Six inches of snow on the ground reflected enough light to let us find our way back to the relative warmth of the car. Seeing just a part of the work at Joe's place was extremely invigorating and we looked forward to a summertime visit, when we could explore the site in detail and in comfort. We delivered Ed back to his farm, thanked him and Barbara for the great day, and returned to New York and the boys.

A WARM SPRING DAY

In late April of 1998 I learned that Ed was doing some major work at his farm, which would include some stone raising, and asked if he'd give me a shout when he knew he'd be doing megalithic work. On May 1st, a warm and lovely day, Barbara called at lunchtime with a message from Ed, who was out in the field: *Come on up, we're raising stones!* I grabbed a hurried lunch and arrived at Ed's farm just before 4 o'clock.

Ed's main thrust over a planned two-week period was building ponds, part of a water garden creation, again with elements from different cultures. The oriental influence was obvious, tempered harmoniously with the use of indigenous materials: plenty of megaliths. Heavy equipment was everywhere. Three or four large men with large machines were digging, grading, hauling, ditching. Ed had been working with this family for years and had learned how to put his creative ideas into the men's minds, just as they knew how to transfer the ideas into the giant arms and buckets of the heavy equipment. Ed told me that these guys could cut with a four-foot bucket as accurately as most men with a spade. I thought of medieval cathedral construction and how a scupltor didn't necessarily have to take hammer to chisel to create a masterpiece. He simply had to impart his ideas to his skilled stonecutters. But Ed's approach was hands-on. He operated the equipment as necessary, jumped off to manually position a large boulder, and orchestrated two or three other pieces of equipment simultaneously. He sought the advice of the operators because he had a deep respect for their skill and experience, and the feeling was mutual.

The Lunar Circle *at Joe's place, private collection. (Don Davidson, photo.)*

Respect for the Earth Spirits

Ed gave me a tour of work that had been accomplished during the past week. I learned firsthand about the dual purpose of his earth sculptures and megalithic work: a desire to improve the land. I saw that the creation of a large trout pond also helped dry a poorly drained section of the forest, which Ed felt would improve conditions for the trees. And pulling megaliths out of the field made it easier to mow the hay without damaging the blades of the mowers.

Ed took me to a newly erected stone in the woods, a large beautiful slab, nine feet high and nine feet wide, probably weighing fifteen to twenty tons.

"We couldn't pull this one out of the woods. It just seemed to belong here. So we just stood it up, and it's right."

I agreed that it was.

"See that hawk?" Ed asked. "He's been watching us very closely. This is his world and we're imposing on it. But I think he accepts us. Even the deer have been staying near, and watching. Damn, look at that big hemlock, leaning there! That's where the stone came out. It was anchoring the tree. I'll need to bring a machine over here and straighten the tree and put a stone back over the roots."

That's the way Ed works. He knows that he walks a fine line when messing with big stones in the forest. In many ways, the space was already sacred. Would he make it more sacred . . . or less?

"You have to ask permission, you have to honor the earth, the animals, the trees, the stones. If it isn't right, don't do it. When you take a stone out, you lose some trees, so you either put them back up or replant more trees. We give reverence, and, hopefully, the stone stays up. If not, somebody's upset."

In this unstaged snap, Ed points out a hawk, flying nearby. Is the hawk's spirit somehow caught up in the stone? The shadow knows.

I got the sense that Ed wasn't talking about people.

"You never cut down a hawthorne tree," Ed told me. "Over at Joe's there was a hawthorne tree in the way of removing a stone, so I made sure that we took it up and replaced it with great care, took about a half an hour. And everybody's a critic. 'Oh, it's going to die anyway, just cut it down. It's a sticky spikey tree.' No, you put it back. We did, and it's fine. Everybody wanted to know why we took care of that one tree. Well, in ancient times, if you had a hawthorne hedge, it meant that you were welcoming to travelers. If there was no water, you could eat the hawthorne leaves to quench your thirst. They're sacred; you don't cut them down."

It is no coincidence that the stone builders all seem to have a deep respect for the Earth, for ancestors, for tradition and lore, for sacred spaces, for nature. It all ties together. Ivan McBeth has learned time and again that if the stones don't want to stand, they will not stand. Everything has to be right. Ed Heath has had similar experiences. Once or twice,

he has experienced what some might describe as bad energy or bad vibes when things aren't just right. He describes it as solemn: "not a happy feeling." So he tries to keep alert to the impact of everything that he does on the land, and he listens. It's a subject that we revisited later over pizza and beer.

Costs and Commissions

Ed had to get back to the men before they left for the day, to plan for the following morning's activities. Quiet descended over the farm as the machines were shut down and the men departed. I tried to imagine, unsuccessfully, what the daily charges for all this equipment must be. How could Ed generate any income from this? Did it matter?

"Well, it leads to commissions. Eventually, people come and say, 'Can you do this sort of thing for me?' Of course, I don't re-create what I've already done. I create from what they have, and usually they have enough to do something with. If I really need a stone, I can take it from here, but I like to use everything that's at the site itself. If it's small stones, okay, we'll build something with small stones."

"I like small circles, even tiny ones," I said. "In my book I want to tell people that they don't need humungous stones to build a stone circle."

"Absolutely," Ed replied. "A friend of mine on the other side of the mountain, she built a medicine circle of small stones because she didn't have the strength to lift up large ones. And you go inside that circle and—holy smokes! It is *powerful*, more powerful than any stone circle I've been in. It just depends on what you put into it. She put a *lot* of energy into it."

Pizza, Beer, and Stone Energies

We went out for pizza and beer, and brought them back to Ed's large farm kitchen. Barbara was putting their year-old son, Caius, to bed.

I popped "the question," the one Bill Cohea is so curious about. *Why?*

"What draws you to megaliths, Ed?"

The sculptor took a sip of beer and considered the question.

"It happened when I took my first or second large stone out of the ground, when it became more of a ritual than a sculptural thing. It was almost like an alternative to sculpture. The stone became very very sacred as soon as it was removed from its resting place. Now I feel strongly that when a stone is set up in a public space, it has to be in the right harmonic ambience. The stone has to fit in that particular place. If it doesn't—then it doesn't belong there. My early experience was that if a stone fell down, leave it. Don't stand it there again. I experienced that on a few occasions until I learned more about earth currents, water, energies, and things of that nature. I learned that if you work with reverence with those, you *can* put a stone up in an area where it fell down. Most of the time, though, you don't. You get that healthy respect when you take the stone out of the ground."

"I've never actually taken a stone out of the ground. Mine were always laying about on top of the ground."

"There's probably a lot more around you that you don't even know about. Tomorrow, we'll go fishing for stones."

Fishing for stones. Sounded interesting.

"How do you know where to dig?"

"There are telltale signs You might see the irregularity of the landscape, for example, or a different kind of moss growing in the grass. When you put your hand near the ground, there's like a tingly feeling."

"I wish I could feel that. Others have told me about it, but . . ."

"Well, a good way to do it is, first of all, to walk around in circles and loosen up your head space. Then, think of stone and wet your lips, and if your lips start tingling, you may be around a stone."

"Tingling?"

"Yeah. I don't know what happens, there's a *tingle*. It's like when you put alcohol on the skin or something like that—it *tingles*. And if you get that tingling, you've got it! It's right there. You bring that machine over and—whaaaaa!—you bring out a twenty-ton stone!"

I couldn't wait.

"Any plans to stand some up tomorow?" I asked hopefully.

"Oh, yeah. I've had a stone circle in mind to go with my equinox alignment. We'll start it in the morning."

Sunset at the Equinox Group

Through the window, I could see that the sun would soon be setting, so I excused myself to snap a few pictures of Ed's creations. I was curious about one piece in particular that we'd walked within a hundred yards of earlier in the day, one that Ed had only recently set and which he said had an equinox alignment. Equinox sunset alignments are prevalent at Ed's because the slope and orientation of the land is to the west. They're good ones because you can enjoy them twice a year, unlike a solstice alignment.

The grouping consisted of four large stones, but the unique thing about the arrangement was that one of the stones, large and boulder-shaped, was split neatly in two, the pieces set about eighteen inches apart. The equinox sunset splits the stone again, and the vertical band of rays hits a huge flat monolith ten or twelve feet east of the split rock. The fourth stone is recumbent.

Cameras in hand, I arrived at the group just in time to snap a shot. It was May Day, or Beltane in the old terminology, six weeks past the equinox and halfway to the solstice, so the alignment wasn't even close, but I got the idea—and a pretty picture.

A STONE CIRCLE IS BORN

Ed is an early riser. We went for a walk in the woods prior to the arrival of the family of boulder-shaped equipment operators, and looked at dozens of weathered menhirs, just ripe to be plucked from

The Equinox Group.

the forest floor. *Here,* I thought, *is the fabled mega-lithic graveyard.* With great pride and reverence, Ed led me up a wooded hill to what first looked like the edge of a rock escarpment but was, in fact, a hundred-foot-long recumbent stone, broken into three or four pieces. Ed likened the stone to one of the reclining Buddhas in Thailand. I thought of the Grand Menhir Brisé, and figured this one was at least twice as heavy, or seven hundred tons. But who could tell? You couldn't see it all. Sculpture? Tidying up the forest a little was all that was needed here. To just lean flat against the stone giant and ponder the last rays of the equinoctial sun was reverence enough. The place was naturally sacred, and may have been thought of as such for thousands of years.

The sound of machinery starting up took us out of our reverie and we made our way back to the field. Ed explained to Buck, the man on the big John Deere digger, where a water ditch needed to be, and to Buck's father on the bulldozer where other landscaping needed to be done. Ed himself commandeered a small backhoe and prepared for the installation of a large plastic culvert in another drainage ditch, so that the digger could cross to the stone circle site. By 9 A.M. the way across the ditch had been made and Buck, with a large menhir in the maw of the big bucket, joined us at the stone circle site, a cleared square field at the bottom edge of the great meadow.

Setting the Alignment

Ed wasn't building this stone circle for my benefit—it was part of the grand plan for this field—but its birth may have been hastened a bit by my desire to photograph his work.

"I want the center of the circle on the same equinox alignment as that group you photographed last night," Ed told me. "The problem is that from here we can't see the other group or the stick I drove into the ground to the west of it, marking the sunset." We jumped in the small backhoe and zipped up the hill to the wooden stake. Ed positioned me west of the stake, so that I could align it with the split rock group. Then he showed me the spot on the opposite hillside, clear across the valley, that marked the sunset, a spot he'd checked six weeks earlier. To find it, he extended his right arm straight out in front of him and held his thumb horizontally.

"There, you see the yellow schoolbus up on the hillside? Sunset is one thumb's length to the right of that." We compared thumbs. Mine may have been slightly shorter, but so was my arm, so the arc described should be about the same. "Now, I'll go back down. When we're right under that spot, let us know."

Ed returned to the circle site and walked laterally until I signaled him to stop. I ran down the hill to catch the action, and a good thing I did. The five-ton stone, the centerpiece of the circle, was standing in less than five seconds. I barely got there in time to snap pictures. I was expecting that Ed would dig a socket with the small backhoe, but no. Buck simply stood the stone up on its pointy end and pushed it into the ground, like you push a candle into a birthday cake. Five seconds, no more. Ed sized up the location, the orientation, the shape, and pronounced them good. Buck pounded the stone in with two or three whacks of the heavy bucket, and that was it. The stone was probably only a foot into the ground, but I could not move it or perceive the slightest hint of instability. Was that it? Was it done? If so, it had certainly been the neatest job I'd ever seen: No pile of earth from the socket, no damage to the grass.

"Well, it's safe enough, until we get a severe spring frost," said Ed. "Then it could heave. Can't leave it like that. We'll put a few smaller stones around it, covered with earth, to shed the water away. In effect, that deepens the stone without altering its apparent height, and gives it good frost protection."

Fishing for Stones

We only had the one stone with us, and Ed would have to mark out the circle anyway, based upon this center stone, so we turned our attention to stone fishing. This square addition to the field was a virtual Grand Banks of stones, something that Ed suspected from the surface evidence.

"We're at the bottom of the field here, and you can see that the ground is much rougher than up higher. Must be a lot of surface stones. You can see outcroppings, bracken, moss, rough ground. This field has probably never been plowed, too bony. I can just feel the stones."

"Your lips tingling?"

Ed smiled and directed Buck to a bumpy area of the field. With his fingers, he signaled to the operator where he wanted the bucket's teeth to dig in. In no time at all, the giant machine had plucked a clump of schist up to the surface. It looked like a massive stone, and Ed was excited—until the clump broke

The big digger pushes the stone into the ground, like a candle into a birthday cake.

up into a number of smaller lumps of soft schist.

"Too bad," I said. "The stone was just too soft."

"Yeah," said Ed, disappointed. Buck kept one half-way decent stone from the clump, buried the rest back in the hole, and tidied up the spot with his bucket. In three months, vegetation would cover the site again, except that it would be flatter than it had been.

Ed moved a few yards toward the woods, to another spot that he had previously identified with a stick. He was getting animated, you might say excited. "Oh boy, oh boy!" he was saying. He must have been tingling something awful!

In no time at all Buck reeled in a ten-ton piece of beautiful green serpentine, diamond in shape, a trophy megalith! I could see how Ed could get excited about fishing for stones. It was like a whale coming out of the ground.

We fished for another half-hour or so, but didn't catch anything to match the big diamond. We threw most of the little guys back, but kept a few useful boulders. Ed finds great uses for all sorts of stones, and it gave me a new appreciation for my reject pile at Earthwood. Although I'd used up all the good standing stones, I learned from Ed that some wonderful sculptural groups can be created from lesser rocks—if you've got the eye for it.

A Telling Exchange

Ed and the lads had plenty of important work to do with the ponds and landscaping, so he left the stone circle for another day, or until he'd gathered up sufficient standing stones. We returned to the pond area near the house, and, just before I left the work area to return home, I witnessed a small but telling exchange. Buck was down off the machine, and Ed was giving him an overview of what he had in mind for the space between the new trout pond and the road, where a number of smaller stones stood, some capped with balancing stones.

"We'll have to remove those stones, Buck."

"Don't you like those stones, Eddie?" asked the digger driver.

"I like them, but I want a major group over there."

Ed Heath looks at the property as one large canvas, and each part of it has to be in the right relationship to the whole. His plan is grand, in the general

Ed was tingling something awful . . .

. . . and Buck pulled out a ten-ton serpentine!

In 1998, Ed completed his stone circle at the farm.

Ed added a trilithon to Rejoice *during 1998, and the group is now called* The House of Rejoicing.

sense, but he is free to create individual pieces as he goes along. Spontaneity and a unified plan are not mutually exclusive in Ed's world; they are in balance.

In October I returned to Ed's farm. His Japanese garden and pond were finished. He'd added a large trilithon to the *Rejoice* group, and another large monolith for a true north alignment. I was particularly pleased to see the completed stone circle, having attended at its birth. The giant canvas was complete, or very nearly so. In a year's time, all the water pools would be filled, the scars on the earth would be healed with fresh green, and the park would begin to weave its perfect magic.

— CHAPTER 10 —

The Dragon Circle

It was like going from one world to another.

—ED PRYNN

I spent the winter of 1998 in my office, researching stone circles ancient and modern, and drafting the first nine chapters of this book. Frequent communication with most of the people you've met in this part made the work a labor of love.

Each of the "Stone People" had not only invigorated my megalithic spirit, but had extended warmth, love, and friendship. I knew that this was a special group, and, during a conversation with Bill Cohea, the idea of forming Club Meg was born; "meg" for megalithic, of course. We felt that these extraordinary people should be aware of each other's work, and Bill figured that I was the guy to bring the group together. I sent out a first modest newsletter, and Bill sent all of the Club Meggers a letter and a copy of his beautiful Columcille calendar.

I spoke with Ivan McBeth and Cliff Osenton several times over the winter and was delighted to learn that I'd helped to bring the two together. With Cliff, Ivan visited the same quarry where we had helped to build an Irish dolmen a few months earlier. Working without extra help, the pair managed to stand a four-tonne stone up on end in about

three hours. To prevent the stone from toppling over when it made the last swing to the vertical position, small planks were positioned on the quarry floor where the stone's base would land, to absorb the weight of the stone and soften its lateral thrust. This was Cliff's first experiment with actually standing a large stone and Ivan was very impressed with what they'd accomplished.

A NEW CIRCLE IS BORN

Very soon after this two-man stone raising, Ivan told me that the time was ripe for him to realize his dream of building a full-size stone circle without using modern mechanical equipment. Beginning May 22, he would be leading a stone circle building camp on a friend's estate in Surrey, at a place called Dragon Hill. Announcements were being sent to people who had attended druidic camps that Ivan and Julie had conducted in the past, and a notice was placed on the Internet.

One day, I surprised Ivan on his cellular phone. When the call came through, he was at the very center of the future "Dragon Circle" in Surrey, doing layout and alignment. He was tickled that he was standing in the middle of a stone ring to be, speaking to me at a distance of three thousand miles. I told him that I was standing outside of my office, overlooking the Earthwood Circle just sixty feet away.

"I'm afraid my phone cord is too short. Otherwise we could talk from circle center to circle center," I said.

Cliff Osenton and Ivan McBeth spent three hours erecting a four-ton stone at the quarry. Ivan is the bloke on the right. (Cliff Osenton, photo.)

"We shouldn't need phones at all to do that!" Ivan replied.

Ivan was fully charged up for the project. He'd selected stones at a quarry, and they would be delivered soon. He told me that Cliff Osenton and Uri Geller, the famous psychic spoon bender, would be the two featured experts. Cliff would move the stones with his proven megalithic engineering techniques, while Geller would show the group how to apply psychic power to the problem of moving heavy stones.

The stone circle camp—and possibly the first-ever meeting of Club Meg—sounded like something that I could not miss, but another trip to England was not in my budget, and I had already scheduled a building workshop at Earthwood for May 30. I phoned Ron and Margaret Curtis at Callanish and Ed Prynn in Cornwall to let them know about the event and all seemed interested in attending. I *had* to go, if only for a week. Ivan was willing to waive the registration fee and provide a place to stay on site, and the Curtises, Ed Prynn, and I agreed to share a car rental, so my costs would be little more than air fare.

At 7 A.M., Thursday, May 21, 1998, I landed at Heathrow airport. Four hours later, I arrived at the camp on a beautiful estate in the Surrey countryside, just south of London. Out of respect for the owner's privacy, the exact location will not be given. Likewise, some of the names of the participants have been changed to protect their privacy. The events described, however, are entirely true.

THE DRUID CAMP

I found the camp by asking at a bed-and-breakfast that Ed and Glynnis had learned about. Oddly, the lady proprietor had a friend who'd built a stone circle nearby, but she did not know for certain where Ivan's camp was located. She gave me her best guess, however, based on my sketchy clues, and her guess was spot on. I was soon driving up a private estate road, wondering how many millions this place must be worth, and following signs that said, simply, "Camp."

After a quarter mile, I came to a large barn and office, where I met the tall, bearded, and gentle farm dweller known as Balin. Balin had been an active environmentalist, on the front lines of the battle to stop particularly nasty countryside development projects. He had dug himself into tunnels, tied himself to trees, and been arrested for trespass several times; so many times, in fact, that his last judge said that if he was arrested again, he would be locked up for a long time. So Balin was lying low at the farm. He proved to be a solid and helpful fellow over the next week, although he was not strictly a part of the druid group that was to make up the majority population at the camp.

Balin directed me to leave my rental car at the barn and indicated the path to Dragon Hill, where the stone circle was to be built. As I got closer to the site, I began to take in its special qualities. The first outstanding feature can be seen from far afield: a grove of giant sequoia and chestnut trees surrounding the cleared hilltop. When I got to within fifty feet of the nearest trees, I realized that they were far out of the ordinary. With plenty of space between them, the trees grew to maximum girth and developed full crowns. One of the chestnuts had a circumference of thirty-three feet, translating to a ten-foot girth. The sequoias tapered much faster, but were still seven feet in diameter a few feet off the ground.

With my head in the sky for the trees, I was almost upon the stones before I noticed them: huge white blocks of limestone, brilliant in the strong sun and azure blue sky of this perfect English spring day. Then, as soon as I crossed the outer perimeter of giant trees, I saw that I was in a camp, a camp whose appearance conformed with typical perceptions of village life five thousand years ago. Outside of the circle of recumbent stones sat five or six prehistoric-looking hut dwellings called *benders,* described below. I saw cooking and washing facilities, and a firepit. I began to realize that more than stone building was at play here. This was a camp indeed, albeit an apparently empty one at the moment, where people would live, eat, sleep, work, play, make ceremony, love, laugh, and cry. I didn't know that, as the days unfolded, I was to gain an insight into the kind of community that would have been necessary to support a project of this scale in ancient times. But I could not dwell on these things at the moment; I had first to visit the stones.

The Stones

My left-brained analytical side immediately went to work, and I soon counted nineteen of the giant blocks scattered around the site in a ring. In places, two or three blocks were stacked or leaning against each other, and I judged that they'd fallen off a truck and landed in these awkward positions; awkward, I say, because I knew that one of our jobs would be to untangle the mess. Each block had a number on it in red paint, generally a single digit, followed by two decimal places, such as "5.72." I guessed rightly that these numbers indicated the weight of each stone in British tonnes. Most of the stones also sported a large red spiral design, which I learned later was Ivan's mark, by which he'd identified the stones he wanted at the quarry.

The stones varied from 1.5 to about 10 tonnes, with all but two in the 3.2- to 8.3-tonne range. Their shapes varied tremendously: some were pointed, others were long rectangles or trapezoids, one looked like a grand piano. I paced several at twelve feet long or better. My first impression of the stones was that they were . . . big! Much bigger than I was expecting. I wondered if we would be able to move and stand such megaliths without heavy equipment. My next impression was that the stones were extraordinarily beautiful. They were a brilliant white, as if they belonged to a Greek island, not a hilltop in Surrey. They were all limestone, but had different hardnesses and surface qualities that became apparent upon closer inspection. Incredible undersea fossils decorated some of the stones, including one nautilus over a foot in diameter. Scallop and cockle shell fossils abounded.

After the visual impact of the stones subsided, I tried to make sense of the site. It was empty of human habitation, populated only by great trees, great stones, and the odd pheasant running through the circle. Four-foot-long twigs were sticking out of the ground near the stones, with colored flags tied to their tops. I wondered if each flag indicated a stone position, but this theory was not proven by a numerical comparison; there were fewer flags than stones. I was on the right track, though. I learned later that the flags marked the circumference of the circle.

The Benders

Ivan had told me about this unique type of dwelling the previous year, but you really have to visit a bender to understand the construction. There didn't appear to be a soul on site, so I wandered into one of the larger huts and was impressed by its obvious structural strength and quality. Dozens of long hardwood saplings had one of their ends stuck in the ground around the curved perimeter of the building, and their other ends bent in toward the center. Laterally around the building, other saplings were positioned over the radial members. All pieces were roughly an inch in diameter. At many of the intersections, the lateral pieces were tied to the bent saplings with twine. Think of the latitude and longitude lines on

The camp appeared empty.

the Northern Hemisphere of the globe to get a pretty good idea of the geometry of the hut's framework. The largest bender also had four or five larger "posts" made of small trees of perhaps three inches in diameter. These posts branched out naturally toward the top and were tied to the hemispherical framework, lending extra strength to the structure. Canvas tarps were carefully placed over the framework, shingle-fashion, to shed water. Some tarps were used as doorflaps. Door openings were usually about three feet high, and often had recycled windows propped up near the opening for light. Some of the benders even had woodstoves and chimneys.

The buildings were quite dark, even in the day-time, with most of the light coming from around the chimneys and through the doorways. More tarps covered the floor. Replace canvas tarps with ani-mal skins, and the benders could have graced the architecture of a neolithic village.

One of the benders was set up as a kitchen. This was actually the only hut you could walk straight into without stooping, and the only one with plenty of light, thanks to a large south-facing opening. Behind the kitchen bender, I discovered that the little village was served by a very clever hot and cold water system. A long hose approached the site from the direction of a farmhouse a quarter-mile distant. The hose teed into a cold water line and, by way of a shutoff valve, into the hot water system.

"Allo, mate," said a cockney-accented voice from behind me.

Startled, I turned to the voice, which belonged to Des, a lightly but brightly dressed man in his 30s, heavily tattooed and with long ginger hair and beard.

We introduced ourselves. Des was friendly, and accepted me immediately, even volunteering to explain the water system.

"Yeah, that's the cold water line. You can get a nice drink of spring water from this tap any time you like. This branch goes through the water heater there."

"Looks like a propane tank," I observed. "With a door on it."

"Yeah. Calor gas. Two tanks, really, a small one inside a big one. It's Ivan's design. You put a wood fire in the small tank and the water is heated in the manifold between the two tanks."

I noticed that a water line entered the outside tank quite low and exited near the top.

"The hot water goes into this large insulated holding tank," continued Des, "where it can be di-rected either into the shower bender over there, or up into the hot tub, there." He pointed to a large galvanized cylindrical tank on top of a wooden plat-form six feet in the air. A ladder provided access.

"Hot tub?" I asked incredulously. "I gotta see this." From the top of the ladder, I saw that the surface of the four-foot-diameter tub was covered with a sheet of bubble wrap as insulation. The wa-ter was warm.

"Amazing," I said. "You seem to have all the mod cons."

"Too right, mate."

Des, I learned, was one of several people who kept the camp running smoothly, keeping the fire-wood coming, repairing structures or systems, or-ganizing work parties; all the things, in short, that make it possible for a camp like this to function. We talked about stones, the village, my work. I showed him some pictures of other modern stone circles, as well as a couple of my books on alternative build-ing. The books and photos, I found, were of great interest to the people at the camp, and gave me a modicum of quick credibility, which was impor-tant, as I would only be able to stay a week.

Des didn't know where Ivan was, but as the farm van was gone, he assumed that Ivan and John, an-other assistant, had gone out to get wood. My host then excused himself, as he had a bender to put up over the "shit pit," a hundred feet or so downslope to the north. My offer of help was gratefully ac-cepted and Des grabbed his knife, a ball of twine, a small sledge hammer, and a 24-inch pointed iron

The hot tub, water heater, and dishwashing facilities. The kitchen crew takes a break. (Margaret Curtis, photo.)

bar. We loaded up with 8-foot-long saplings from a pile.

The building technique was simple, and easy to do in the firmly packed sand subsoil of this site. Des drove his iron bar into the ground about a foot, then extracted the bar by loosening it and lifting upward, leaving a neat hole in the ground, an inch or two across. Next he jammed a sapling down into the ground as far as he could. We set a dozen of these saplings along the south side of a long pit that had already been dug in the ground, but, thankfully, not yet used. The "shit pit" was six or eight feet long, and about thirty inches deep and eighteen inches wide. We installed saplings along the east and west sides of the pit as well, but left the north side exposed. A heavier stick spanned the long north opening, supported by more substantial cor-

ner posts. None of these components greatly exceeded two inches in diameter. We bent the anchored saplings in toward the larger "beam" and tied them down. We also tied on a number of lateral saplings to the bent framework. Finally, we covered the frame with recycled canvas tarps. The whole job took an hour and resulted in the most pleasant venue for morning contemplation that you could imagine: a long view to the north, overlooking the beautiful Surrey countryside. For people to see you, they would need to be on the next hill, a mile away, and armed with good binoculars.

Soon after we finished, Ivan and John arrived in an old van filled with a variety of short thick wooden planks, some large first-cut slabs from a sawmill, and other assorted chunks of wood. I greeted the gentle giant. "Hullo, Ivan, is this your megalithic kit?"

"Yeah, maybe not quite up to Cliff's standard, but pretty damned good. Oak, mostly. Hey, what do think of the stones?"

"They're big," I said.

Ivan smiled.

"Yeah."

The Camp Comes Alive

Officially, the camp was not to open until the following evening, which is why there weren't many people around Thursday morning. But during the afternoon, lots of regular followers of Ivan and Julie's druid camps began to arrive, a friendly and brightly colored lot on the whole, men, women and children.

Ivan asked if I would help him catalog and name the stones, which I felt honored to do. I acted as his scribe and carried around a large piece of note paper from stone to stone. We measured each stone for length and made a column for estimated finished height. Ivan was inclined to socket the stones more deeply than I would have done, sometimes as much as a third of the stone in the ground. We argued about this good-naturedly, but I always gave in. Ivan was the chief. He looked like the chief, he acted like the chief, he *was* the chief; of this project and of this camp of people. He did not micromanage, though. There was a good chain of command in the camp, good delegation of tasks and responsibilties. But when it came down to it, Ivan was chief.

Necessary aside: I must not give the false impression that decision making was autocratic at the camp. I was soon to learn that decision making was done by consensus, which necessitated many, many meetings. Each day started with a staff meeting. Following a tea break, there was a general meeting of all people present in the camp, including all those who had just attended the staff meeting. Sometimes, this seemed quite redundant. One day, I figured that there were about fifteen people at the staff meeting, and twenty-one at the general meeting. This arrangement was fun, cumbersome, time-consuming, interesting, and very often frustrating, particu-larly to the non-druids at the camp, who were present right from the first day.

Ivan probably didn't need a scribe, or a paper and pencil. He had an uncanny ability to look at a stone, decide its name, and remember its size, shape, and location. Here are the names that Ivan came up with on the first tour of the circle: North, Whale, Titch (the small stone), Blob, Stocky, Stalagmite, Spear, Wedge, Dog, Rusty, Banana, Piano, Peardrop, Amonite, Schist, Boat, Center, Rocket, and Bed. Banana, I think, was my only contribution. These were working names only, and many changed in the next few days. The largest officially weighed stone was Whale, at 8.31 tonnes (9.3 tons). In all, the stones weighed 100 tonnes, all that the budget allowed. Their cost was £40 per tonne, delivered, and they were donated to the project by the mother of Ella, Ivan's friend, for whom the land was held in trust.

By nightfall, about twenty people had arrived in camp, all druids except myself. I sat around the fire for a while but, tired from a poor night's rest on the plane, I was in my sleeping bag in the guest bender by eleven.

THE MEGALITHIC VILLAGE

My second day in camp, a relatively quiet Friday, was really a day of preparation for building the stone circle. These preparations involved some concrete planning about the stones themselves, but just as important, the camp began to tune in to the spiritual aspects of the process in a focused way. It became a megalithic village, united around the collective enterprise of erecting a sacred stone circle.

In the morning, I drove out to Heathrow to pick up Ron and Margaret. After quite a search, I found them sitting patiently in the domestic arrivals lounge, which was not marked in any way. The couple was tired, having flown down from Storno-way with a connection in Glasgow, and accompanied by a crate containing a mother cat and kittens. "I've got to feed the kittens every few hours with a dropper," explained Margaret.

We caught up on things megalithic during the ride to Surrey. Ron and Margaret checked into the B & B and we continued on to the site, where I introduced the Curtises to Ivan, Julie, and several of the other druids that I'd come to know. They were delighted with the village, and with the details of the living systems such as water, sanitation, and food preparation. As archaeologists and students of the neolithic lifestyle, they know more about prehistoric living conditions than I did, and they seemed to be favorably disposed toward the *very* old world atmosphere of the camp.

My view is that the task—in this case, building a stone circle by hand—dictates the process. You need lots of people, so you need shelter for the people. You make the shelters from indigenous materials. You need a place to store and prepare food. You need a special hut for the chief. You need water, fuel, and tools. You need a shit pit.

You also need some rules or guidelines for appropriate behavior. A Gate House bender had been set up on the way up the hill, close to the circle, to serve as an official port of entry into the village. Because the Curtises and I were special guests, and early arrivals, it never occurred to us to stop by the gate, that turned out to be something of a mistake. I do remember passing by the gate and commenting on the sign outside of the Gate House that said, "No alcohol. No drugs. No electronic music." It wasn't until the next day, when Ed Prynn arrived, that I learned a bit more about the gate and its purpose. Like every other visitor, Ed was taken through the gate by one of the druids, a process that took about half an hour. The Gate House was much more than a registration booth. It was a place of cleansing and acclimatization, a halfway house between the outer world and the inner circle. No additional rules were discussed, just the offer of a cup of tea and a smoothening out. "There was no

ritual work," Ed told me later, although passing through the gate does seem to have served a ritual purpose. I probably would have benefited from it myself; the experience might have helped me to understand some of the social patterns that were deeply intertwined with the stone moving over the next few days.

Opening Ceremony

At dusk on Friday, I brought Ron and Margaret to their digs, and found a nice country pub for a pint of bitter. I returned to camp about ten, just in time to participate in the official opening of the camp, led by Ivan and Julie. A couple of dozen people gathered around a large ceremonial fire near the center of the circle. The druids invoked the spirits of the Four Directions and asked their blessings. Julie, the perfect priestess, conducted the ceremony with her usual verve and skill. She told us about the use of the druidic word *awen,* which would be an important part of the next few days' activities.

"We've got a tool for alignment that we're going to be using to align ourselves during the day, and to help move the stones around. It's just a chant we do together. Some of us are used to doing it, and we use the word *awen,* but you could substitute something else, like *amen* or *om.* Awen is a druidic tone that draws on inspiration from the Great Spirit."

"An *om* is basically a Hindu or Sanskrit word," added Ivan, "which is considered to be the basic sound of the universe. All of these chants draw on the Creator or Creatrix. They are all chants that make things happen. When we align and hold hands, we are one. When we chant, we invite the Great Spirit or Goddess to join us. So, it's really good to try to chant something that makes you feel good, and in accordance with your religion, beliefs, culture, whatever."

Julie performed a benediction, and then she led the druid members of the gathering in their awen chant, three long resonant aaaawwwweeeennnns. A pleasing and inspirational energy filled the circle.

Next, Ivan asked us all to give a wish for the camp, anything we thought important. Various wishes were offered up, and people's inflections mirrored their sentiments. "Peace." "Strength!" "Freedom!" "Understanding . . . Courage . . . Laughter . . . Surprise!"

"Teamwork," said a dronelike voice, bringing laughter.

"Unity," said Ivan.

After the final wish had been expressed, Ivan and Julie declared, "This camp is open!"

Julie led the group in singing a joyous song, and then Ivan spoke in a subdued and reverent voice."

"I'd like to have a little word about this fire. This is our fire, *our fire,* and its flames dance at the center of *our circle,* a circle created by our coming together. It reflects our spirit, and our spirit will make us strong. So it's not just the duty of the site crew to keep this fire going. It's the duty of all of us. So if you get up in the night for a pee or something, keep the fire going. Try to keep aware of what's going on here at the center of our camp. Nurture our fire, nurture our spirits. Treat it with a lot of love, because we are one, and this is part of us, too. Now, the hot tub is piping hot! The last one out, please put the cover on."

Ivan's gentle speech was my first indication of the man's skills in bringing together the physical and the spiritual realities, the rational and the transrational. He would be ably assisted by Julie, Des, and the others, but, ultimately, the responsibility to keep both sides going smoothly rested with the chief. Over the next few days, as various situations arose that affected camp life and stone moving, I was fascinated by how Ivan managed to hold everything together. I became convinced that Ivan was giving me yet another glimpse into the prehistoric; there had to have been a strong leader to orchestrate the myriad disciplines required to build even a modest-sized stone circle.

Drumming and fire dancing carried on for an hour or two, and I crashed well after midnight.

STONEBUILDING: DAY ONE

Each day began with a ritual, which consisted of Ivan playing a gentle, clear, and hauntingly beautiful flute fugue at 7 A.M. Ten minutes later, a few campers joined Ivan in the Dance of Life, a Cherokee dance that Ivan had picked up somewhere along the way. The dance took place in a circle and invoked the Four Directions with simple movements and Native American syllables and cadences. I joined in right at the end and vowed that I'd be on time the next morning. Why? *Why not?* It's all part of the experience.

The normal pattern of events following the flute alarm and the Dance of Life continued with breakfast, staff meeting, tea, general meeting, and then getting to work. Work never started before 11 A.M. during my week at camp.

First Meeting

I collected Ron and Margaret from the B & B, and we arrived back at camp just in time for our first communal meeting at the megabender, the largest hut.

The group of thirty or so people was predominantly of druidic persuasion, three-quarters of them, I'd guess. It seemed natural and to some extent functional to adopt the same social and political arrangements as had been used at previous druid camps that Julie and Ivan had organized. The meetings always began with holding hands in a circle and three good awens. They were conducted by talking stick, much as I'd witnessed at Julie's house at the autumnal equinox. Ivan would usually begin by outlining the day's activities. On this first day, there was little to discuss. We were all along for the ride, and individual speeches consisted mostly of introducing ourselves and explaining why we were there.

The most interesting introduction was probably Ivan's.

"I build stone circles because that's what I do, and that's what I believe the land has entrusted me to do," began Ivan. "And so I wait, and from time

Ivan leads an early morning Dance of Life.

to time, the land gets in touch, usually by way of its guardians, the people who actually live there. They come to me and they say, 'Look, I have this dream to build this stone circle. What do you reckon?' And we spend time meditating on the subject, getting in touch with the land, and actually living on the land. This is most important. And over time, you just tune into the place, you're open to the omens, to the guidance. And then, at a certain time, it's like the balance changes, like you've caught a wave and you're surfing. And from that moment onward, things start to manifest, faster and faster. And eventually the stone circle is born. And it's something we take seriously and yet with as much humor and lightness as possible. So I guess that's me in context. I'm just really looking forward to working with everyone here in order to realize that dream. And . . . well, let's go for it!"

We were welcomed to the land by Ella, the pretty young lady for whom the land was held in trust.

She thanked us all for coming. The spirit of the group was very high indeed.

After the meeting, I witnessed a curious but age-old ceremony. Joan and a few of the other druids had collected drums, pans, sticks, and anything they could find to make noise, and were marching around the circle's perimeter—clockwise, of course—making the most raucous disharmonious loud din that they possibly could.

"What's going on?" I asked Des.

"They're chasing off any evil spirits that might be lurkin' about," said Des.

Wooding

Morning meeting was also the place where volunteer groups were organized: wooding, cooking, washing up, the food-run to town, and the like. Ivan's idea this first day was to begin with a task that would involve everyone in camp, so that we would develop a sense of community and teamwork. We needed three large pine poles for stone moving, and firewood. Five hundred feet from the site was a woodlot, part of Ella's farm, and we all set out to do some wooding.

One problematic aspect of the work experience at camp revealed itself very early on. The men, one or two in particular, jumped at what they thought were manly jobs: selecting and chopping down the pine trees, for example, and the women were largely left to watch or drag out smaller pieces of dead-wood for the fires. No one complained, but a little bit of male chauvinism was beginning to make it-self known. Otter, one of Ivan's lieutenants and an experienced woodsman, took over the tree-felling operation. Unfortunately, Otter lacked the ability, the sensitivity, or the inclination to bring everyone into the process. Margaret observed this keenly and mentioned it to me later.

The Stones Prepare to Move

Unlike his Swan Circle, where Ivan set the center stone last to "bring the circle on line," he decided to take the opposite approach here. The Center Stone would be set first, to launch the project and give it a focal point. I advised against it, making the case that a great big stone in the center would be in the way of getting accurate alignments on the other stones. Ivan acknowledged that this was true, but he felt he knew the alignments well enough around the perimeter to position the other stones correctly. And, he intended to shape the top of Center into a point, to improve its usefulness as a backsight.

Ivan and Julie had organized any willing and available people into an awening crew, whose job was to impart psychic energy into the project, in an effort to lighten the load. Typically, a group of people, predominantly but not exclusively women, gath-ered around the stone-moving action in a horse-shoe shape, holding hands and directing their en-ergy to the stone. At the critical moment, Ivan would orchestrate the aweners with a command, such as, "Okay, aweners, gently now, here we go!"

Did the aweners in fact lighten the load? Well, many people on site thought they did, including some who were actually levering the stones. They certainly contributed to good feeling on the site, with the exception that one or two women would have really rather been levering the stones.

Friction

As were we preparing to disentangle the Center Stone from its place in a pile, a new face appeared on site: Angus from Scotland, a contractor experi-enced in moving stones. Angus had the knowledge to direct the site crew, probably more than Ivan, Otter, or myself. Of all the people present, Ron Curtis was probably the only one with a better sense of heavy-lift engineering than Angus, but Ron was quite content to stand back and observe, only oc-casionally taking Ivan aside for a quiet suggestion. Unfortunately, Angus brought a very disagreeable manner with him, which introduced serious nega-tive vibes to the site. He began to shout orders, demand changes in technique, even swear angrily—something that was not done in the camp. I ex-pected Ivan to tell Angus to shut up, but that wasn't Ivan's style, and the situation deteriorated. "Who is this guy?" I wondered.

My vantage point was as an observer and pho-tographer, and it allowed me to spot several prob-lems, the most serious of which I would relate to Ivan. I saw lots of what I considered to be unsafe practices: barefoot workers, too many people near the stone being moved, and levers being used in-correctly. Otter was setting up with too long of a lever arm between the fulcrum and the load, which was causing the levers to creak, a sign of impending failure. Nonetheless, despite the chaotic scene, Cen-ter was levered up onto blocks that afternoon, and would be ready to transport to its honored position the next day.

During the height of this hectic afternoon, Ed Prynn arrived. It was great to see my old friend again, but we barely had time for a hug, as the ac-

As Center is raised onto blocks, prior to transport, the aweners endeavor to smooth the way by imparting high energy into the scene.

tion was so intense. Archdruid Ed, in his traveling clothes, jumped right in and helped, hauling on ropes or awening with the various teams, as required. When work stopped, I introduced Ed to Ron and Margaret, Ivan and Julie. Wherever he goes, Ed is an instant celebrity. With his long white hair, pink face, and thick glasses, he looks the part of a Cornish eccentric, and within a few minutes of hearing him speak, his listeners know that Ed's the genuine article.

At the evening campfire, I enjoyed Balin's stories about his encounters with the authorities while trying to save the planet. Angus arrived and joined the group. After a while, he called me aside and we had an odd conversation about the afternoon's events. Angus was explaining to me all the things he saw that were wrong: safety issues, organizational diffi-culties, improper use of equipment. Despite my initial negative reaction to the man, I found myself in agreement with him on almost every point. He seemed to respect my views, so I decided to take a chance.

"Angus," I said, "I know that you know your way around stones . . ."

"But?"

"May I be perfectly candid with you?"

"I wish you would."

"The way you came on today, arriving on site and just sort of shouting orders. It . . . well, it made for some very bad vibrations. Did you not feel it yourself?"

"Oh, aye, Rob, I did. I did. I felt it, and I was the cause of it. I felt bad after, but I just couldn't believe what I was seeing. I couldnae help myself!"

"These people are new at this. They could benefit from your experience. But they're not like your contracting crew. And there can be only one boss, or it's chaos. Talk to Ivan. If you spot something, go through Ivan. He's smart, he listens, and he makes good decisions. If you have specific concerns, speak up at meeting tomorrow. Give your truth in a quiet, gentle way. That's what will work with these people."

"Aye, and will you speak as well, Rob?" asked Angus.

"I will."

STONEBUILDING: DAY TWO

The next morning I joined in again at the Dance of Life, but I wasn't getting any better at the movements. In fact, I was getting worse. On the way to the village to pick up Ron and Margaret, I ran into Ed and his friend, Alan Connett, in Alan's car. Ed had decided to move out of the B & B in favor of staying in camp.

"I thought you'd paid for two nights, Ed," I said.

"I have, Rob, but that's not the place for me. This is where I want to be. The camp is magic, Rob. I've never seen anything like it. I have to stay at the camp. Look, you take my room tonight, get a bath and such like. It's all paid for."

That sounded good.

Sunday's Meeting

The camp's population had risen considerably. I counted fifty-five heads at meeting. It was Sunday, a beautiful day, and Uri Geller was supposed to be the big drawing card. After the communal awening, Ella's uncle, an Anglican minister and a chartered engineer, delivered an invocation, a blessing on the site and its people. It was a well-received prayer, non-denominational, and the reverend concluded with advice that came from his experience wearing two different hats.

"Act as if it all depends on you," he said, "wait as if it all depends on the other." This wisdom was greeted by many a solemn "Oh!" from the druids, their accepted way of expressing agreement with the person speaking.

Ivan started the talking stick.

"Thank you for that lovely benediction," said Ivan. "My friends, I am sorry to say that it does not appear that Uri Geller will be coming to help us move the stones today. I know that this is a disappointment, and I am sorry that it has not worked out. But we have a beautiful day, lots of beautiful people, and we can move the stones without Uri Geller!"

Again, several in the gathering offered "Oh's." I was disappointed that Geller would not appear, but not surprised. I didn't really expect him to levitate any large stones, but I was curious to witness his modus operandi.

I deliberately positioned myself across the circle from Angus, clockwise with respect to the passing of the talking stick. I wanted to hear his comments before I spoke.

Angus was calm and collected, almost contrite. He spoke of the need for a positive chain of command and suggested that those with a little more knowledge about moving the stones should meet together to plan the next move before the fact. He pointed out some of the specific safety issues that we had discussed, particularly that too many people were too close to the stones. I think everyone was pleased to see and hear a different side of Angus.

The talking stick moved to Ella's uncle, the minister/engineer, who also spoke of safety, as did Ron Curtis and Ed Prynn. The younger people's comments had more to do with the joy of the camp and the stone moving, except for a few women who expressed their desire to be a part of the stone moving, and not aweners exclusively.

The talking stick came to me. Much of what I'd wanted to say had already been said, so I decided on a parable.

"In about 1320, an itinerant barber or surgeon happened to be traveling through Avebury, and came across a curious scene. We know the date by the coins in his leather pouch. The local people were busy tearing down the great stones of the outer circle. We know he was a barber or surgeon by the scissors and lancet found in his pocket. The surgeon must have decided to join in the festivities on a lark, perhaps on a fine day like today. We know all these things about him because his skeleton and possessions were recovered in 1938 during excavation. His pelvis had been crushed and his neck broken when the thirteen-tonne block collapsed upon him.

"Yesterday, I had very little to do with the actual moving of the stones, but, through my camera lens, I was able to focus on things that others might have missed. Fulcrums must be placed much closer to the load, for safety and to prevent broken levers. We don't have any surplus levers on site and we've broken one on the small first stone.

"In work like this, following the chain of command is of the utmost importance, as Angus has said. Anarchy in these circumstances can result in death or serious injury. If someone spots something that doesn't look right, tell Ivan immediately."

The stick came around full circle to the chief, who thanked everyone for their comments.

"Communication is of the greatest importance," he said. "Try, stop, and re-evaluate. We need an organic chain of command. We need to move at the speed of the stones."

I thought it had been a good meeting. I asked Ron Curtis what he had thought of it.

"It was long," said the laconic Scotsman.

Down in the Valley

By now, I was aware of a lot of politics and power games going on at camp. I don't think Ivan got too caught up in this. I sensed that he sympathized with this druid order, as opposed to being religiously intertwined with it. Ivan found spiritual value in a variety of teachings and creeds, and he really would have been quite happy without all the innuendo and hierarchical distractions. It seemed as if some of the druids—not many—were disturbed about

there being nondruids at camp. Perhaps the presence of outsiders made it difficult to practice the camp way of life that they'd become accustomed to, although some people were not at all modest about their nudity.

Of greater import than internal politics, however, was a threat from outside. From my first day at camp, I'd heard rumors that the people in the village and on nearby estates were a bit concerned about the nature of the activities up on the hill. This was not entirely unexpected. After all, this was ultraconservative English countryside, stockbroker country, old money, new money, lots of money. By Sunday, the rumors had become stronger. Ella and her parents had strong local ties, and they were a respected family. They got on well with neighbors and wanted to keep it that way. The word was that (in some local eyes, at least) this business of putting up a "pagan circle" on a hilltop that could be seen for miles, of having who knows how many unwashed, long-haired, naked dropouts running around up there for who knows how long, of imagined and unimagined health and safety problems . . . well! I had a vision of a wizened old landowner in his manor house two miles away, his telescope trained on the hilltop. "I say, Lovie, we must get this lens cleaned. I can't see clearly the things that I object to!"

A rumor circulated that the mayor had been on site, on behalf of the local council. This turned out to be true, and the mayor turned out to be a great guy, very supportive of the camp, and a valuable ally in keeping the camp from being closed down by the ill-informed locals. But he was feeling pressure from his fellows on the council, and he let Ella know that, while he, personally, thought the activity on the hill was upbeat and positive, the camp was, nevertheless, on very precarious ground.

Center

Communication on site was infinitely better on Sunday. Ivan took command, but called together Ron, Otter, Angus, and myself before deciding on the next move. Then he supervised the work, keeping a careful eye on the load points, keeping extra people away from the critical areas. When lifting or hauling was ready to begin, he orchestrated the aweners. "It's all magic," said Ed Prynn. And the speed of the stones actually moved up a gear.

The six-tonne Center was soon levered onto twelve-inch-diameter wooden rollers made from the pine tree trunks cut the previous day on the farm's woodlot. As the stone had a good flat bottom side, it was not necessary to make a sledge to transport it. Also, there were over a hundred people on site, and I once counted thirty-three men and women pulling the stone by rope. Even a right angle turn presented no difficulty; angling the rollers enabled us to steer Center to its destination.

A Festive Lunch

We broke for lunch at 3 P.M. John Michell, the great megalithic writer and expert in sacred spaces (*Megalithomania, The New View Over Atlantis,* etc.) had arrived on site, and joined Ed, Ron and Margaret, and me for lunch. He was fascinated by what was going on at the camp, the whole scene, not just the stone moving.

Our attention was drawn toward the center of the circle, where a horseshoe of aweners had gathered around Ivan, who was seated on a wooden chopping block. Julie and another druidess were joined by two druid men of average size, and the four formed a square around the chief, alternating man, woman, man, woman. We were wondering

Large rollers and plenty of pullers made it easy to roll Center around the site.

The druids apply their lightening technique to Ivan . . .

. . . and the chief is elevated like a sack of feathers.

what was going on, and someone who'd seen this before said, "They're going to lift him right up in the air." *Not likely,* I thought. *They'd need to average better than eighty pounds of load apiece.* Julie couldn't have weighed a lot more than that herself, and the other three were not particularly musclebound either. The aweners began their chant, and the four levitators joined their hands on top of Ivan's head. They, too, were chanting. After a while, when someone decided that Ivan was charged up and ready to launch, the four moved their hands to his posterior, and, slowly, lifted the chief straight up off his stool, about two feet. I was amazed, and the event brought incredulous smiles to all who witnessed it. Ivan was giggling wildly and his four bearers were not struggling in the slightest. Uri Geller could not have put on a more impressive show.

I asked Julie and Ivan what they'd felt during the experience. Ivan said it was like he was floating in air, an incredible feeling of joy and lightness. Julie said something like that he seemed as light as a feather; it took no effort whatever to lift him.

"And you weigh . . . ?"

"Twenty-three stone," Ivan reminded me. Three hundred twenty-two pounds!

Was something more than muscular strength involved in raising Ivan into the sky? I don't know. I wish I'd been one of the four bearers. They could have lifted him, I suppose, with great muscular effort, but there was no evidence of this. I was very impressed, along with most everyone else.

Inspired, perhaps, by his levitation, Ivan disappeared into his bender and emerged in a new costume, consisting of nothing more than a large multi-colored pair of shorts. He *looked* like a chief now, a Polynesian chief.

Standing the First Stone

After an hour's break, we resumed work. Des and a few others had already excavated a large socket hole to receive the stone, an almost perfect four-foot cube of air in the ground, with a 60° sloped ramp carved into one side for Center to slide into. The hard-packed sand soil was a joy to work in. Normally, sand collapses as you dig, but this sand was halfway to sandstone in density; it carved easily with

Didgeridooers bless Center's socket hole.

a spade, but could support vertical sidewalls. The Curtises and I were a little concerned about the lack of packing stones on site, but Ed Prynn was used to this type of material. "You'll pack 'em with sand, Rob. Couldn't be better."

Anyone who wanted to, even children, took turns removing a ceremonial spadeful of earth. Next, Ivan called for ceremonial gifts to be placed in the hole. A partial list of offerings includes: lots of crystals (some from other sacred sites), a squirrel's tail, a rabbit skin (prey from Wizard, the hunting dog who kept the camp supplied with meat), a poem, an egg, bark from the pine rollers, a 1998 penny, and oranges carved into the megalithic ball shapes found in Scotland. It looked like a fresh midden heap. I kicked myself for not having brought a stone fragment from the Earthwood circle, but I threw in a picture of the circle instead, and Ivan assured me that the energy would still be there.

With more than thirty people pulling the stone, a few pushing, and a couple of dozen aweners lightening the load, the stone moved very quickly across the flat ground to the entrance ramp into the socket. Based on archaeological evidence at Stonehenge, we decided to place a couple of planks upright on the far side of the socket to lessen the impact of the stone on the socket walls. The six-tonne stone was rolled to the edge of the socket hole, then Ivan orchestrated the final pulls.

"Okay, we're going for it now. To establish a rhythm, I'll shout, 'One, two, three, *four!* One, two, three, *four!*' And on the third time, 'One, two, three, *four!* I want you to pull for all you're worth!'"

Ron and Margaret were perched on a nearby stone for the best view. John Michell was standing close to the action, absorbed. Ed Prynn was in among the young people on the rope. I was hoping my video camera battery would last another couple of minutes.

Center gathers speed, and topples into its socket hole. (Margaret Curtis, photo.)

At 6 p.m. GMT, Sunday, May 24, 1998, the Dragon Circle was grounded with a ceremony around Center.

"Ready now," said Ivan. "Aweners, begin … good. Okay, people on the rope. One, two, three, *four*." The stone moved slightly, but the rhythm was the important thing. "One, two, three, *four*. One, two, three, *four!*"

The stone took off like a Titan rocket: slow, steady, then faster. Too fast, really. Its forward momentum kept it going after it had reached the critical balance point over the socket. It fell into the hole, all right, but did not take the desired 60° angle as the base of the stone got caught up on the far side of the hole. The whole event, which took about four seconds, was at the same time thrilling and a bit anticlimactic. Nonetheless, all present let out a great cheer. The stone rested at a 30° angle.

The next hour was spent levering and pulling the stone into its final position, and the stone was fully vertical by 5:40 P.M. Ivan gathered the hundred-odd people around the stone in a great circle for a group awen. The circle was founded now with its Center Stone, and the spirit of the crowd was festive. After the group broke up, I congratulated Ivan on a job done very well indeed. It had been an amazing day. Allowing for the late start and the long and entertaining lunch break, I reckon that Center had been rolled to its place, stood up, plumbed, and steadied in not more than three hours of actual work.

I took Ed up on his offer and spent the night down at the B & B.

The Chieftain

The owner of the B & B had arranged for several of us to visit her friends who built a stone circle just a few miles away. Thus, on Monday morning, Ron and Margaret, Ed and Alan Connett, and I traveled to Cranleigh to visit the jolly little circle of Dominic and Abigail Ropner, which the reader can visit with us in chapter 15.

We got back to the camp just after noon. The crew had already moved one stone out of the way, and were just breaking for lunch. The afternoon's project was to move a large pointy stone into its equinox sunrise position, due east because of the distant horizon. Ivan told me that a chieftain was always in charge of building the stone circles, and that the Chieftain always stood in the east.

Our new Chieftain stone, at 6.50 tonnes, was 10 percent heavier than Center. Worse, at 12'10" in total length, it was more than three feet longer. And, it was far more awkwardly shaped; it didn't have a flat side. In short, I thought it was the most difficult stone left on the list, although three were heavier. Ivan chose it deliberately.

"If we can do this one, we can do anything," he said.

I told him the story of the fellow who began each day by eating a live frog. After that, the rest of the day was easy.

Ivan laughed. "Yeah, that's it," he said.

Chieftain moved very slowly indeed. His awkward shape (definitely a male stone) made it impossible to use rollers, not without a sledge or stone boat, at any rate. As we did not possess sledge-making materials, and Chieftain was only thirty feet or so from its socket, we decided on another method of transportation: levering on tracks.

We laid out a heavy wooden track made up of two long pine tree trunks laid parallel to each other and about three feet apart. This forty-foot track would provide a means for transporting the stone from where it fell off the truck right to its socket hole. The idea is that you maneuver the stone onto the track, its length perpendicular to the track, and centered so that half of the weight is distributed on either side of the imaginary center line running between the wooden rails. Then, in theory, you simply lift and lever first one end of the stone forward, then the other. This is very similar to the method that Cliff Osenton uses to raise stones, as described in chapter 7. The stone's own considerable weight offsets the weight of the end you're lifting, and you make a few inches progress on each lift and heave.

It's also a bit like rowing the stones along the quarry floor, and may have been the method used to deliver the huge Easter Island statues from the quarries to their platforms. Island legend says that the stones "walked" to their platforms.

Chieftain did not cooperate, not one little bit. His awkward shape made it hard to judge his center of gravity and hard to move along the tracks because of protrusions and valleys on his underside. A flat regular stone would have been a piece of cake with this method. But we struggled and persevered, and by 7 P.M., we had him within six feet of his keyhole-shaped socket. At 7:20 P.M., it began to rain and Ivan called it a day.

After a vegetable curry supper we all sat around to hear Margaret's excellent illustrated lecture about Callanish in the megabender. She kept the group enthralled for two hours.

RAINY DAY

My diary sets the tone:

Tuesday was the pits. The morning was lovely, but frustrating. Ivan asked me to work out a plan of action to prepare work for Cliff, due to arrive tomorrow morning. One problem of logistics follows another, which thwarts my efforts. Ivan, who has the final word on all positioning of the stones, is not available for questions; he is locked up in a staff meeting that seems to go on forever. I set up a lever with a fulcrum close to Bed, the 3.2-tonne stone. I invite the beautiful Gretchen from Germany to lift the stone off the ground, and she laughs with the ease of her effort. "I am lifting 3000 kilograms!" she squeals.

I pick up R & M at 10. The regular meeting doesn't start until 11:30 (supposed to be at nine). The non-druids are very frustrated, probably most of the druids are, too. About 12:30 we start work. Goes well until

"Walking" the 6.5-tonne Chieftain along a wooden trackway, first one side, then the other. Balin provides a little counterweight. (Ron Curtis, photo.)

Chieftain's center of gravity wanders too far east and we get jammed. Rain stops work about 1. I am writing this in my bender at 1:30. Should have gone to Birmingham to visit Aubrey Burl, but weather looked okay this morning. Now don't know what to do.

From 3 to 5 P.M., we move Chieftain three feet on tracks. Very tough, wrong equipment. At 5:15, we stop again for rain. Discouraging. I worry about tomorrow, because rain is pervasive. I hope Cliff will be able to work.

Conditions did improve. My bedtime entry:

At 8 P.M., after supper and Ivan's hot tub, we have a very productive session, bringing Chieftain easily right over the socket. Use slippery bark on pine log to decrease sliding friction. John, the local mayor, trips and gets slightly hurt. I tell Ivan, "It's getting dark." He calls it a day. We'll raise it tomorrow with Cliff. Ed, Jeremy and I go to The Crown, Chiddingford (J's village) for a couple of pints. Great old 14th-century pub. Great conversation about moving stones. I get back to camp about 11:30.

ENGINEERING AND THE SPIRIT

Cliff arrived at 6:30 A.M. and the engineer was ready to work. He only had three days to spare, and wanted to waste no time. This site gave him an opportunity to test some of his ideas about standing stones. Over porridge, I explained that things didn't get going very fast around here in the morning.

"Well, we could start, couldn't we?" asked Cliff.

"Well, no, it doesn't quite work like that. It takes a village to raise a stone, you see, and there are meetings . . ."

He just looked at me.

Cliff unloaded his megalithic kit on site while I motored down to the village for the Curtises; we arrived back at 10:15, just in time for meeting. Cliff introduced himself and told a little of his background. Ed regretted that this would be his last day in camp, saying that being here "was one of the greatest experiences of my life."

The Departure of Ed Prynn

At 11:30, with Ivan and Cliff working side by side, Chieftain was moved to the very edge of its socket and everything seemed to be going well. Ed Prynn had stayed as long as he could, wanting to see Cliff in action, but now it was time for him to catch his train back to Cornwall, and he would have to miss Chieftain's plunge. The Club Meggers joined together for a group photo and then wished Ed safe journey. Ed made his way out of the circle, but could not rush his departure, as he had lots of new friends to hug and to share farewells.

On the opposite side of the circle, the crew was ready to resume work on Chieftain, almost chomping at the bit to drop him into his socket. Ivan stood near the center of the circle like one of the giant sequoias, looking pensively around the site. Nothing was happening, as far as I could see, and people were waiting for work to resume. Cliff was standing by not knowing quite what to do, wondering what was going to happen next. It was a very strange situation, I thought. *Why is Ivan just standing there when we're so close with this stone?* A few more minutes went by, and the big guy was still looking around the site like a hawk. Impatiently, I approached him and said the crew was wondering when we'd be getting back to work.

"Cool it, Rob!" said Ivan. It was as close to angry as I ever saw him get. Then he composed himself and said, with extreme patience, "Look, Ed's leaving, but he hasn't left the camp quite yet. This is a very sacred time, when someone is leaving the camp. We have to respect his leaving. We'll return to the work after Ed has left."

I felt about half a megalithic yard tall.

A few minutes later, we levered Chieftain into his socket.

Ivan, who had been in charge of everything to do with stones for several days, asked Cliff to supervise of the rest of the day's work. "I'm tired,"

The Club Meggers gather at Chieftain: Margaret and Ron Curtis, Ivan McBeth, Edward Prynn, Cliff Osenton, Rob Roy. (Kieron Sibley, photo.)

Ivan and Cliff direct the positioning of Chieftain over its socket hole.

said the chief. "I need to rest." Cliff was glad to oblige and Ivan disappeared into his bender to catch up on some much-needed rest.

Titch and the Megalithic Railroad

The lunch break was rather long, as was the custom, and Cliff got up to work, or rather, in his view, to *prepare* for work.

Chieftain had fallen into its socket at about a 40° angle, not much better than Center. And, because of the stone's extra height and awkward shape, Cliff knew it would be difficult to make it vertical. Plus, our rope-pulling crew was down to a fraction of Sunday's number. Cliff needed a "thrust block" or backstop, something substantial to lever Chieftain against, and the only candidate on site was little Titch, a 1.50-tonne triangular lump, perched eighty

feet across the circle. Cliff knew that he could move Titch next to Chieftain with minimal help, but several guys were keen to get involved and before long six or eight people had helped Cliff set up what I call a megalithic railroad: a track made up of 3 x 9 blocks. For rolling stock, Cliff used five of his levers, each ten to twelve feet long and about four inches in diameter. For a sledge, he used a couple of long 3 x 10s that were already on site. The weight and friction of the stone was quite enough to keep the timbers together. Titch rolled along this track with the greatest of ease. I thought back to a time over ten years earlier, when eight of us had struggled with the South Stone at Earthwood, only slightly heavier than Titch. The major difference: ground preparation (the wooden trackway) and the quality of the rollers.

Chieftain is levered into its socket. He landed at a 40° angle.

Cliff Osenton, with hard hat, and his megalithic railway.

Titch fairly flew across the circle. Everyone involved seemed to be carried away with the sheer joy of the project. Six or eight men were moving track forward as fast as they could. In fact, as the ground began to slope downward toward Chieftain, our biggest problem was slowing the stone, so that it wouldn't run right off the end of the track.

In a matter of minutes, Titch was positioned behind the leaning Chieftain, but it needed to be flipped off of its makeshift sledge and onto firm ground, in order to be used as a backstop for levering the big stone.

Meanwhile none of us working on this project had the slightest idea that a serious socio/political problem was brewing in the background.

Dissension in the Ranks

I noticed that Julie was trying to organize some aweners to come in on the project. Joan, the druidess introduced in chapter 6, was already positioned well behind Chieftain, and was trying to impart some spirituality into a scene that, I learned later, she thought was decidedly lacking in that quality. Cliff was unaware of all this. For his part, all he was doing was setting up a large fulcrum, which happened to be a 3,300-pound stone. Titch was flipped off of its stone boat with levers, and it fell into perfect position for its intended purpose. Unfortunately, it bumped Chieftain as it fell, not real hard, not enough to do any damage to either stone, but Titch bumped Chieftain.

For some reason that I never clearly understood, this bumping was the last straw for two or three of the druids, and they began to complain that the project had lost all of its cohesiveness, that the camp was no longer a team, that the aweners weren't called in to help early enough and had been relegated to second-class status, that this bigshot engineer had come into the camp and changed the whole character of the proceedings. Julie intervened. Cliff was, after all, an invited guest, and was simply doing what he'd been asked to do.

Cliff smiled and kind of threw his hands in the air and said, "I'm just moving a fulcrum into place. Thought I'd do it during lunch, so we'd be ready to stand this stone when everyone was ready."

A small minority didn't see it this way. They saw a stone moving, without sanction, without blessing, and moving awfully damned fast, too. Someone saw its bumping into Chieftain as a kind of sacrilege. Julie tried to calm down the minority, and tried to organize some awening.

Work resumed. The drummer set the beat, and those inclined to work with Cliff did so, and those who preferred to work in the transrational realm were encouraged to do that. Cliff was so caught up in the engineering task at hand that I don't think he fully appreciated the importance of everyone's needs being met. Ivan could, but Ivan was sleeping the sleep of the weary.

Work progressed for a while. The stone would be levered upward off of Titch and a few blocks of wood would be placed between the stones to con-solidate the gain. The levers would be reset, another gain would be made, and more blocks of wood installed. Chieftain moved slowly upward, but finally seemed to say, "No more. I will rest here now." The method had reached its maximum gain. The angle of the levers was becoming too high for the lever people to reach, even with ropes. Cliff tried levering from the side, even from both sides at once, but Titch could not foot this type of thrust, and simply moved backward, instead of Chieftain moving forward. Everyone was getting frustrated now, and the rift between the rational and the transrational opened even wider. We had reached a crisis point and Julie knew it. She called a meeting to be held around the Center Stone in twenty minutes.

Cliff looked at me as if to say, "What's happening? It's just a minor setback. We'll work it out." I was better prepared for this sort of thing than he, having been in camp five days longer, but I, too, was somewhat taken aback by the current state of affairs. I can't be sure of this, but I suspect that the

Even with Titch as a backstop, levering Chieftain was very awkward.

problem was partly a function of the strong importance placed on doing things by consensus in this particular druid order. The group bends over backward to make sure that not a single individual feels railroaded, left out, or even overruled. This approach has some fine qualities, but is not conducive to things getting done quickly or smoothly. Thus, the interminable meetings.

A Pivotal Meeting

Ivan, blissfully oblivious to all that was going on just 150 feet away from his bender, was wakened, briefed by Julie as to the nature of the crisis, and given a few minutes to ready himself for chairing what could be a pivotal meeting.

It must have been about 5 o'clock when we gathered together. The day had been cold, damp, and gray, but skies brightened somewhat just before the meeting was called to order.

Ivan asked that everyone express what they were feeling, and, if they had any kind of suggestion for a solution, to please share it. With that, he started the talking stick on its way. It was a powerful meeting. The two or three druids who had been most offended by the seemingly new work process since Cliff had arrived expressed themselves calmly and from the heart. They took care to impress upon Cliff that they had nothing against him, that they were sure that he was good at moving stones, and that he was probably an okay guy, but that this method of working without due consideration for the spiritual side was very disturbing, so disturbing that something must be done about it. Others, druid and non-druid alike, tried equally to avoid stepping on the beliefs and values of the disgruntled campers, but pointed out that Cliff was doing a great job and, essentially, that we should not get caught up in squabbling, which took energy away from the project. But the person who brought the issue to clarity was Ella, by whose grace we were all here this week.

"I can't believe this bickering that's going on!" Ella began, teary-eyed. "We spend all this time at meetings arguing about silly things, and the stones don't move. Here we've got a guy who can move the stones. Let him get on with it! You can't believe the problems I'm having trying to hold this camp together. We all have to be off of this hill in a week. I've been given notice. And here we are bickering over nothing. I just can't take it anymore."

Ella, normally strong and calm, was crying now, and I believe that every single person in the circle empathized with her.

The stick passed on, and the general sense of the group was to move stones by whatever means suited you, physically or metaphysically. The stick came to me.

"Before he left, Ed Prynn described Ivan as both a stone doctor and a mystic," I began. "For Ivan, the two disciplines are not mutually exclusive. He understands both sides. Because of that, he is the perfect leader for this kind of project, and to harmonize the different viewpoints of a group as diverse as this. You may not realize it, but you've opened a Pandora's Box with this camp. For the first time, you've opened your camp to the whole world, and they are here now; some are helping to support this camp as paying customers, and their needs, too, have to be considered. There are people here who have come to learn to move the stones, others to teach, still others to share ancient lore. Cliff Osenton is an extraordinary man, but he could no more coordinate the aweners than he could dance the Swan Lake Ballet. Ivan has asked for suggestions. Mine is this: Let Cliff work in his field, and assign another person to coordinate the other side of the endeavor. And respect each other."

Sometimes I tend to run off at the mouth.

Cliff was one of the last to speak. He took the stick and held it for a few thoughtful seconds, looking at it, as if for guidance.

"I am not insensitive to others, as has been suggested. I am a medical engineer. I have worked in hospitals in Nicaragua, where I saw little children like these here, literally torn to pieces by the ravages of war. I am not insensitive. I have deep feeling, particularly for our ancestors who built these circles. I was asked to come here to do a job. I have to stay focused on that job. And there are lots of people here who are desperate to get on with it. I don't come here with all the answers. I come here to try and to learn, like the rest of you. Anyone who wants to join in with ceremony, they are welcome. I do not mean to show disrespect to anyone, to any beliefs. Please . . . carry on . . . and let me do the same."

Several "oh's" could be heard from around the circle. Truth be told, I might have sneaked one in myself.

A few others spoke and then the stick returned to Ivan.

"I hope that this meeting has made us all aware of some things that we perhaps were not aware of before. Thank you for your comments. I hope that work will resume with a new understanding. Cliff is the engineer for the next session; he is at the top of the chain of command. Listen to him and learn from him. Julie will coordinate with Cliff and with those not involved in the heavy lifting to bring back the sense of unity to the camp that was lost for a while."

The sky brightened, as did the mood of the camp. Only Chieftain remained awkward and stubborn, and, after a very slow-moving work session, we secured him for the night at about a 65° angle.

Joy and Melancholy

Wizard the wolfhound provided a rabbit supper yet again. About 8:45, Des marched around the camp blowing a conch, and shouting, "Rob Roy will speak about modern stone circles in the megabender at 9 o'clock. Rob Roy! Modern stone circles! Nine o'clock." I have never had such an introduction.

I read excerpts from an early manuscript version of this book, and told anecdotes about people well known to the camp, such as Ivan, Ron and Margaret, and Ed Prynn, and some who were not known, such as Bill Cohea and Ed Heath. I tried to describe the motivations behind the modern stone

builders, how most of them had a strong spiritual reason for building the circles.

"I, myself, am perhaps the most analytical of the group, but I'm trying to find my way in the spiritual realm, and I'm sure that you people and this week at your camp—our camp—will help me along this path." This was greeted with hurrahs.

Many people, druid and nondruid alike, stayed on to chat after my hour's reading was finished, and we shared hugs and farewells. I left the camp with Ron and Margaret, feeling both melancholy and joy. I had decided to spend my last night at the B & B, so that I could pack early and get off to Heathrow on time. I was sad that Ivan and Julie had not attended my reading, but I knew that they had a lot on their minds. Cliff, of course, had been up at 3 A.M. to travel to the camp, and took an early night in his van. Ron and Margaret told me that my readings and comments had been a hit with the twenty-odd people who'd attended.

Farewells

I was up at 6:30, bathed, and packed up. Ron and Margaret would be staying on for another day or two, so we said our sad farewells: tears, hugs, the lot. We'd become good friends during this incredible week together.

"How are you going to write it?" asked Margaret.

"I don't know. I have to digest it all. Right now, it's overwhelming," I said.

I returned to the site at 7:45 to see Ivan and Cliff, but only two of the campers were up at this hour. Normally, the Dance of Life would be in full swing. Ivan had not yet appeared. Not wanting to disturb the chief, I walked back down the hill to Cliff's van and joined him for coffee.

"Sorry I didn't stay on for your talk, Rob. I had to get away and be alone for a while. You know, yesterday, I came *this close* to packing up and going back home."

"I wouldn't have blamed you. What changed your mind?"

"I realized that there's a lot of good people here, and a lot who are really enthusiastic about raising the stones. This is an incredible opportunity to experiment, so I guess I'll stay through the weekend."

We spoke for a while, vowed to stay in close touch, and said our farewells. As I started toward my car, I looked up to the camp on Dragon Hill, and there I saw Ivan's unmistakable silhouette against the clear morning sky. He made a single wide sweeping wave of his right hand. I bowed to the chief and he bowed back. Not a word was spoken, but it was the most touching farewell of all.

EPILOGUE

Thursday's journey home was uneventful, but seeing Jaki and Darin at the airport was very emotional for me. I felt like I'd been away a year—five thousand years—not just a week.

A few days later, I called Cliff, Ron and Margaret, and Alan Connett, to find out how things had gone after I left.

On the day I left, the crew returned to the stubborn Chieftain stone. Its base was awkwardly jammed in its socket, and the regular kind of levering could not handle the situation. Two new methods were successfully employed to stand the stone up straight, and they are described and illustrated in chapter 14.

The next day, Friday, was slowed down by a long meeting about rabbits. One of the druids, new to the camp apparently, said that rabbits have a spirit and a consciousness and shouldn't be hunted. So work didn't start until after two o'clock.

"I really thought about packing up again, but things got better," Cliff told me. "Using rollers, we moved two stones of about 5½ tonnes each. On Saturday, we got two stones up in a day. On Sunday, we did an 8-tonne stone, making five up altogether, the center and four corner stones. Efficiency picked up a lot. On my last day there, there were no meetings and a lot of work got done. I guess it dawned on

them that nothing was happening during meetings."

Once, a stone was reluctant to move and everyone came up to touch the stones with their hands to make it easier to move. Meanwhile, Cliff rearranged the fulcrum and the blocks, and, lo, the stone moved easily.

"The aweners reckoned they'd succeeded!" Cliff laughed. "You believe what you want to believe."

The Curtises (who had return tickets to Lewis) and Cliff Osenton left camp before it was closed down. The arrangement with the local authorities was that stone movers could meet one weekend a month until the circle was completed. Alan, who lived fairly close by, continued to attend work sessions over the summer and autumn. He and Ivan—and others—kept me posted.

By August, efficiency had improved even further. One Saturday, a crew of Ivan, Alan, and five others managed to move a five-tonne stone several feet and stand it up in its socket, all in less than four hours! I commented to Alan that fewer people seemed to make it easier, not harder, to move the stones. He laughed and pointed out that experience was the key. The following day, Ivan was unavailable, and the crew was down to five. Nevertheless, another five-tonner was transported thirty yards to its location . . . *uphill!* I asked Alan how they'd managed that.

"First, we elevated the stone quite high up on blocks. Then we built a downhill runway using the large tree trunks and levered it to its location. This took about three hours. It was incredible, but we figured we didn't have quite enough manpower on hand to stand it up."

The story continues. As I write, in April of 1999, Ivan stays full-time at the farm and fourteen stones are in place; the circle should be completed by the time this book is published. As for Cliff, he has conducted some very successful experiments in the Preseli hills of Wales, moving three-tonne stones along the old Stonehenge route. "It's an axe job," Cliff told me. "We're getting closer to the ancient techniques all the time." See chapter 14.

Over the summer of 1998, I came into contact with Jean-Pierre Mohen's excellent book *The World of Megaliths* (Facts on File, 1990), in which he describes the moving of a 32-tonne block at Bougon, France, by two hundred men, in July of 1979. Coincidentally, Cliff has started up a correspondence with a French megalithic engineer, Bertrand Poissonnier, who has been conducting similar experiments in Europe and in Ethiopia. In June of 1997, Poissonnier moved the same 32-tonne block at Bougon, in front of Jean-Pierre Mohen, using only ten men! What progress has been made in eighteen years! "The idea is not to push and not to pull the stone," says Poissonnier, "but only to turn the wood under it."

When I think of the ancients, who raised the 350-ton Grand Menhir Brisé in Brittany five thousand years ago, I shake my head in wonder and despair. Yes, we're making progress, and I expect that a lot more will be made in the next few years as the Cliff Osentons, Ivan McBeths, and Bertrand Poissonniers of the world meet and work together. An interesting story is shaping up for the future. But will the future ever intersect with the past? Will we ever shape, transport, and erect anything like the Grand Menhir? And without modern shortcuts? I hope so, and I want to be there when it happens.

In the meantime, I am pleased with what we've done so far, and feel honored to have been present at Ivan's stone raising at the Dragon Circle, to have witnessed the organization, the society, the bickering, the politics, the logistics, the people who all made it happen. Indeed, for that magical week, the present *did* intersect with the past.

And I learned a thing or two about stone building . . .

Turn to the color section for a photograph of the nearly completed Dragon Circle.

— PART 3 —

HOW TO BUILD A
STONE CIRCLE

— CHAPTER 11 —

Designing a Stone Circle

You know, all these alignments, mixed with the sacred geometry,
mixed with the megalithic yard, mixed with the shapes and sizes
of the stones—it's just an incredible weaving of all sorts of different
things to try to build the perfect circle, which you'll never get.

—IVAN MCBETH

In part 3, we depart from narrative storytelling in favor of a more technical, explanatory style. I'll try to maintain a light tone, but some readers may find some sections a little on the heavy side. Other readers, those just chomping at the bit to build a stone circle, may have jumped straight here, three-quarters of the way into the book, for a crash course on all the technical info, without "bothering" about the human side. If you resemble that remark, stop! I urge you to start this book at the beginning, for two important reasons.

First, stone circles, both ancient and modern, cannot be separated from the people who built them. They are what they are because of the human input: the intent, the aesthetics, the trials. Any circle built hurriedly, and without a care for the kinds of considerations detailed in the first ten chapters, is doomed to failure. Oh, sure, anyone who can operate a backhoe can stand a few stones up in a circle, and probably in such a way that they won't fall down real soon. You don't need a book of this size to tell you how to do that. The failure that I speak of is that the circle will not be grounded in

these human qualities that, frankly, allow the circles to work their magic, whatever you perceive that magic to be.

Second, a lot of lore, anecdotes, and, yes, important building and design tips are salted throughout the first ten chapters. Don't miss them. And sometimes I'll be referring to a site or a situation that you should already know about. So git back there and read!

THE SITE

You can't design a house—or a stone circle—without knowing what sort of a site it will be built upon. You need to know the size of the available space, the qualities of the soil, the availability of materials, whether or not you have good horizons for alignments, so many things. If someone wants to build an underground house, at least in northern climes, I generally advise them to try to find a gentle south-facing slope to take advantage of passive solar heating. The orientation of the home, in this case, is a very practical and pragmatic consideration. Easy. But stone circles have been built successfully on a variety of sites, and they always seem to fit. Why? Because they have been designed with the site in mind, as well as with the greatest concern for the intent and purpose of the circle.

In ancient times, the circles were used for ceremonial purposes, and as gathering places to observe various social customs. In short, they were used the same way as early Christian parish churches or cathedrals. The architecture and scale of the great temple of Stonehenge suits the Salisbury Plain, and the Salisbury Plain suits Stonehenge. Boscawen-Un, in Cornwall, is tucked away in a cozy little depression on the moor, a little parish church by comparison with the great cathedral of Stonehenge. Each is in its correct site. In addition, it must (by now) be blatantly obvious to even the stodgiest archaeologist that the sky and the movement of the heavenly bodies were important factors in the siting and design of most of these ancient open-air gathering places.

The Use of the Circle

Before deciding on a site, one must be clear about the intended use of the circle. Will it be used for public or private gatherings? For large or small groups? Will it be a small personal space in a private garden or wooded glade? Will there be a fire circle at the center? Is the circle strictly for aesthetics, a sculpture on the landscape? Is the intent to tap into the Earth's energy or to create sacred space? Is the circle to be used as an astronomical observatory? The answers to these kinds of questions will impact site selection, so let's examine them in more detail.

Large gatherings require large spaces. The space can be open, as at Stonehenge and Castlerigg, or in a more private wooded situation, such as at Rollright or at the stone circle being built by the Church of the Four Quarters in Pennsylvania (illustrated in chapter 15). If alignments are important to the group, such as for celebrating solstice or equinox observa-

tions, or the Celtic cross-quarter days between these events, then the view of the horizon will also be important, particularly in certain specific directions. Nearby trees and buildings have a tremendous impact on horizon alignments, sometimes in a very frustrating way. Imagine building a circle with a summer sunrise alignment, only to have someone build a house on the next lot to the northeast, spoiling the observation. More on alignments below.

Small gatherings are nice in small, cozy places. Earthwood is like that. The little druid circle in Julie's back garden in Bristol is like that. An added plus is that a smaller and presumably less expensive piece of land is required. Nearby woods or even nearby houses may not be a drawback, particularly if alignments are only meant to be symbolic and not accurate to an astronomer's standard. A small fire circle or firepit can be included at the center, or at one of the two foci of an ellipse, if so desired.

Sculpture or earth art. Stone circles or groupings are often built purely as art. People can wander through, stop and enjoy, contemplate, meditate, whatever they like. Aesthetics should be a consideration with *all* stone circle design, of course, but if a group is purely sculptural, such as *Knowledge, Innocence, and Growth* (chapter 15), the sculptor will be very concerned about how his or her piece integrates with the surroundings. We cannot all be great sculptors like Eddie Heath, but we can all *try* to be, and, thanks to the very nature of stone circles, there's a good chance that our efforts will be rewarded with uncommon success, like the work of Eddie Prynn.

Sacred space. Ivan McBeth and others are very concerned that stone circles be built on sacred spaces. Many believe that the Earth is criss-crossed with energy lines, known as leys, and that it is good to build on these lines, or, better, on the intersection of these lines. Ivan's Swan Circle, you may recall, was built on a ley connecting the ancient sacred sites of Stonehenge and Glastonbury Tor. If you are into dowsing for energy, or know someone who is, you will want to dowse a prospective site carefully before choosing it. To describe dowsing techniques and to advise how to find and tap into the Earth's sacred energy lines are beyond the scope of this book, but I have listed several sources in the annotated bibliography and in the sources, to help you track down this kind of information. Several anecdotes describing what might be called transrational experiences are reported in the first ten chapters, and some readers, I fear, may just dismiss them out of hand, which I think would be a mistake. Personally, I continue to approach these matters with a healthy and open-minded skepticism, believing that this is the way that I will be receptive to learning.

With regard to siting, I tend to side with those who feel that you can bring magic or energy or sacredness or whatever you choose to call it to a site by building a circle and using it.

Geomancy, and the ancient Chinese science of placement called *feng shui,* are each concerned with the integration of sacred spaces and building with regard to various Earth energies. Most of the feng shui principles or rules are soundly based in common sense, for example, that buildings should be constructed on high ground instead of in a valley, or that trees on the northwest side of the house will bring happiness to its occupants. Obviously, we do not want to build in a flood pain. Less obvious is that trees on the northwest side of the house, at least in temperate climates, protect the house from the most severe winter storms.

I do not claim expertise in geomancy, but I find it interesting that very often a stone circle builder will dowse a piece of property and determine that the best site for the ring happens to be the most practical: having good drainage, high ground for

A Question of Balance

In a conversation at Columcille in late 1997, I explained to Bill Cohea that I didn't have a ready answer when people asked me why I build stone circles, a question of great interest to Bill, given the unique path that he has followed. I told him that I thought of the stone circle as a work of art, a sculpture, really. I'm not good at drawing things, but I have a sense of architectural balance and proportion, which I am able to express in buildings I design, and stone circles. Bill nodded, and picked up on the question of balance. We agreed that each stone has to be set just right: spacing, direction, height, everything.

Although I had only a cursory understanding of *feng shui,* the ancient Chinese science of space and placement, I'd discovered its importance at both the Earthwood and the Stone Mountain circles. Without exception, all of the modern stone builders have exhibited a Zen-like feel for proportion in their work. Why? Maybe one has to have this sense of balance to really appreciate stone circles in the first place. If

someone's heart and soul are not in it, the project won't come out right. Suppose a wealthy property owner said to a contractor, "I'd like a stone circle on my front lawn. Go get some stones and build it." Well, unless the contractor himself is into the spirit of the project (for it is obvious in this hypothetical case that the landowner is not), the project will not come out right, will not be in balance and harmony with the site.

Suppose the property owner has the required vision and enthusiasm. Now, the choice of heavy equipment contractor is of the utmost importance. Disaster looms if a contractor is chosen whose heart is not in the project, who thinks that the whole thing is a silly waste of time. On the other hand, stones can do things to people. Bill Cohea and Ivan McBeth have each told me about operators who have never done anything like standing a menhir in place, getting caught up in the excitement and becoming converts. This may be the most fun they've ever had on the job.

views and alignments, good access, whatever. Magic may be nothing more than common sense, and vice versa.

Astronomical alignments require a special site to really work well. I like astronomical alignments, and it is a source of frustration to me that the Earthwood circle is poorly placed to fully take advantage of them. The circle is in a low spot on the property, and our acre of cleared land is surrounded by forest, so I am constantly having to maintain my summer sunrise and winter sunset alignments by keeping paths cleared through the forest. On the positive side, these tunnels in the trees introduce a dramatic element into the observation. The sun suddenly bursts into view as its shallow sunrise or sunset arc moves into the clearway. The sun's orb, then, on the point of the standing stone, is doubly dramatic. And I get a lot of firewood from maintaining the alignment paths. (See photograph in color section.)

Given a choice, though, it is better to select a site with a distant horizon, if accurate astronomical alignments are important. The alternative is to be satisfied with symbolic alignments, discussed when we get to alignment design and layout.

Marginal Land

In my previous book, *Mortgage-Free!*, I advise prospective owner-builders to seek out what I call "marginal land" to build upon, land that has been devastated by forest fire, perhaps, or scraped clean of its topsoil and gravel, as was the case at Earthwood. Every house that we can design and build has a negative impact on nature. You can't improve upon nature by building a house in it. So let's really do something positive for the planet and build on wasteland, land stripped bare by thoughtless people, and bring it back to useful, green, oxygen-producing landscape, instead of lifeless moonscape. Similarly, our stone circle, now in a sea of grass where only hardscrabble subsoil was exposed twenty years ago, has actually improved both the aesthetics and—by the care we have taken—the ecology of the land.

So-called wasteland may be the perfect site for a stone circle. If you're really lucky, it might even provide the stones themselves. In virtually every case, such land will be a whole lot cheaper than any other land you can buy.

"But I've Already Got My Site!"

I know, for many, there will not be a great deal of choice about site. You've got your property already, and you want to build a circle on it, near your house, perhaps, so that you can enjoy it whenever you like. This was our situation at Earthwood, and I touch upon some of our specific siting and design considerations in chapter 2. You still need to design with the site, no matter how large or small it might be. Ed Prynn did not know he was going to build a megalithic complex in his relatively small garden, but he did; and it works, too, on both the physical and metaphysical planes, and has given pleasure to thousands of visitors over the years.

If you've already got your site, I can't give you any site specific recommendations without actually visiting, and I don't do house calls unless they're in places like the South Seas and requests are accompanied by a pair of round-trip air tickets. What I will do, however, is to list a few common-sense feng shui–type considerations to keep in mind.

Good drainage is important! Don't build into the water table. In the north, your stones will heave with the winter frost; in the south they will sink out of sight (or site, same thing). If space is really tight, and you absolutely must build where the ground is sometimes wet, bring in enough clean earth and topsoil to build up the circle site above surrounding grade. Yes, this will cost money, and then the new earth will have to be hand-graded, planted, mulched, et cetera. But the alternative is almost certainly a failed circle that will not grace your life or your equanimity.

Is bedrock a possibility? If so, you'd best dig a few test holes at least as deep as the stone sockets, before finally laying out your circle.

Leave plenty of "air" around the circle. Don't try to crowd too large a circle into a small space. Build to scale. You need to be able to walk around the perimeter of the stones with space to spare.

Access. Is there a natural approach to the site? Accent it with an avenue or a pair of gate stones. Draw the visitor into the circle.

Take a year to plan! Do you *really* know your site? Do you know where the sun rises and sets on March 20th, June 21st, September 22nd, and December 21st, and the cross-quarter days in between? Do you know where true north is? Maybe there's a magical way that the sun penetrates between those two high-rises across the road on your birthday or wedding anniversary. Do you know how the trees on site impact the sun and shade during the various seasons? You should. What's more, you'll need a year (or more) to gather stones and get to know them. By taking your time, you'll create a more carefully thought out space.

Get to know the site before you build a stone circle. "Live on it," says Ivan McBeth.

THE STONES

I worked as a stone mason's laborer in the north of Scotland for several years. The master mason for whom I most often labored would remind me frequently—particularly when he was running out of building material on the scaffold—"You can't build without the stones, Robbie. No, you can't build without the stones!" This piece of standard wisdom may not be as self-evident as it first appears. In Scotland, there seemed no shortage of stone, and this has been my experience in northern New York as well. But I knew a fellow in Georgia who desperately wanted a stone fireplace, although there was no stone in the county, only sand. My friend made his own stones by filling a large form with sand mortar, colored with cement dye. When the slab cured, he'd break it up into building stone. He made several different colors of stone that way, then built his fire-

place. In Florida, another place without stone, I am amazed at how Disney's "imagineers" construct beautiful enduring "stone" of concrete.

We have said that design needs to be site specific. It should also be *stone specific;* that is to say, the size and shape of the ring may depend a great deal upon the size and shape of the stones you are able to procure. You wouldn't lay out a hundred-foot-diameter circle if the only stones you can find are two or three feet in length. In this case, build a small circle. Alternatively, build a Native American medicine wheel or a labyrinth of small stones. These constructions can be very powerful, and don't require huge stones or heavy lifting. "Mystery of the Medicine Wheels," in *National Geographic* (January, 1977), is a good article on the subject. See the sources in the back of the book for information about labyrinths.

People ask me what kind of stones are best. There is a simple answer: *Use what you've got!* Or what you can easily get. Unless you're being commissioned by wealthy patrons who want a certain stone—Vermont white marble, for example—I have to presume that cost is important to you. If the overlying strata is sandstone, then build of sandstone. If there is an abandoned granite quarry down the road, build of granite. If the area is peppered with glacial erratics, build an erratic circle. The only stones that should be avoided are soft stones, stones that flake off easily, and stones with fractures running right through them. I have one red sandstone in my circle with lots of vertical fractures. Eventually, water freezing in the fissures will break up the stone.

It almost goes without saying that long and narrow stones are preferred over short and stubby ones. They have a superior height-to-weight ratio and are generally easier to handle. The central monolith at Callanish extends 15'9" above grade and weighs almost 8 tons. Burl says it "could comfortably have been set upright by a gang of thirty workers" (Burl, *A Guide to the Stone Circles,* p. 149). I say, without bravado, that I could install such a playing card stone vertically in its socket with an enthusiastic team of ten able-bodied helpers, *without machinery,* thanks to the stone's foot-thick playing card shape and skills I've recently learned. However, I am equally certain that I could not engineer the manual raising of the Swindon Stone at Avebury with one hundred strong-backed helpers. Despite being a foot shorter than the central monolith at Callanish, it is roughly eight times heavier and much more awkward of shape. Modern megalithic engineers, even Cliff and Ivan, still lag far behind our ancient forebears, although the gap is being closed quite rapidly, and I predict astounding feats will be reported before the 21st century's first decade is complete. Some of the people mentioned in this book are likely to be involved.

Procurement Suggestions

I was brought up in New England, lived in northwest Scotland for several years, and settled in northern New York, so it's easy for me to forget that not everyone is surrounded by stones. There are over a thousand known ancient stone ring sites in the British Isles, but virtually none east of Rollright near Oxford; none, in other words, in the most populated part of England. (Until Ivan built the Dragon Circle, of course!) The most likely reason? Lack of suitable stones.

Megalithic construction throughout the world is practiced where stones are plentiful. Luckily, in most of North America and the British Isles, stones are commonly available, many of them having been delivered to site by successive ice ages. Still, although you might live in an area of stones, you might not see any laying around, or not enough to build a circle with, anyway. Here are some suggestions.

Stone walls. Farmers have spent centuries on both sides of the Atlantic clearing fields for cultivation. They are among the world's most skilled heavy-lift engineers, without taking college courses, either. Cliff Osenton learned many of his techniques working on the farm. In the past, these farmers transported large stones to the edge of the fields by stoneboats pulled by draft horses or oxen; today, a tractor's front-end loader is generally the preferred method. Take a walk along country roads, or around the perimeter of a cultivated field—with permission, of course. Don't be surprised to find several good candidates for a stone circle. Autumn is the best time to conduct a stone search, after the leaves have fallen and the ground vegetation is gone. Each year I discover several new standing stone possibilities that I'd walked or driven by dozens of times before.

And what constitutes a "good candidate?" For our purposes, an ideal stone is one that is long and not too big around. If it is long and also fat, it will be awkward to transport and to lift, even with heavy equipment. This general rule scales right down to small stone circles. It's easy enough, for example, to build a ring of small round boulders, but, somehow, it just doesn't satisfy our aesthetic sense of a stone circle. Build a little ring with longer and narrower stones, however, even though they may weigh less than the boulders, and the circle feels right.

A stone's weight is probably the most serious obstacle to moving it, although a particularly awkward shape can also present problems. With the stone called Chieftain, in Surrey, shape was even more troublesome than weight. Long narrow stones usually have reasonably low weight and favorable shape. Calculating a stone's weight is explained in the sidebar opposite.

Girth has a great deal to do with volume, and, therefore, weight. The Earthwood stones were relatively slender compared with, for example, the stones we worked with in Surrey. Relatively flat "playing card" stones, like those at Callanish, Brodgar, Stenness, and at the Four Quarters circle in Pennsylvania, are relatively lightweight for their height, whereas stout stones, like many standing at Avebury, Castlerigg, and Columcille, are quite heavy by the foot. A good example of how an inaccurate estimate can really mess up one's perception of a stone is a new monolith we recently erected at Earthwood, to replace the rather feeble summer sunrise outlier.

Calculating the Weight of a Stone

A bathroom scale will not be much help in determining the weight of that six-foot long piece of granite in your backyard. We'll need to work it out by measurement.

A stone's weight is the product of the stone's volume and its density, which, for our purposes, is most conveniently measured in pounds per cubic foot.

Let's use the South Stone at Earthwood for our example.

The base of the stone was quite flat, so the height (length) of the stone was fairly easy to determine. Along one edge of the stone, I measured 77 inches, while the opposite edge came in at 83 inches, giving an average height of 80 inches, which can be expressed as 6'8" or 6.667 feet. Next, we need to know the average cross-sectional area of the stone. Luckily, this one was fairly regular and a good approximation can be made by averaging. I took width and thickness dimensions for all four sides of the stone, measuring about every sixteen inches along the stone's length. This resulted in the following chart, with figures to the nearest half inch:

Height off ground (inches)	Width		Breadth	
	North (inches)	South (inches)	East (inches)	West (inches)
8.0	26.0	24.5	18.0	22.5
24.0	26.0	27.0	17.0	21.5
40.0	23.0	24.5	17.0	21.0
56.0	26.5	24.5	18.0	19.5
72.0	27.0	27.0	19.0	20.0
Average:	25.60" width		19.35" breadth	
	(2.133 feet)		(1.613 feet)	

The average width of 2.133 feet times the average breadth of 1.613 feet gives an average cross-sectional area of 3.441 feet, which is then multiplied by the average height of 6.667 feet, resulting in a volume of 22.941 cubic feet. Using the density of anorthosite as 165 pounds per cubic foot (from chart below), we get a weight in pounds of 22.941 times 165, or 3,785 pounds. If you want the answer in tons, simply divide 3,785 pounds by 2,000 (pounds per ton), which results in about 1.9 tons. If you're British, you'll want to divide by 2,240 (pounds per tonne), resulting in about 1.7 tonnes.

These calculations are easy to do with a pocket calculator. Note: If you are easily fooled by numbers, you might find stones to be slightly easier to move in the British Isles.

Here are typical weights of various stone types measured in pounds per cubic foot. These should be considered as approximate values only. Sandstone, for example, can vary from 127 to 159 pounds per cubic foot, depending on its quality. Lightest stones are listed first. Other stones, such as quartz and sarsen, hardly vary from the figures given.

George Barber surveys the South Stone.

millstone grit	138	granite, anorthosite	165
sandstone	142	shale, slate, diorite	172
sarsen	154	gneiss	176
feldspar, marble,		trap rock, dolomite	180
rhyolite	160	basalt, dolerite	190
limestone, quartz	165		

The great heart-shaped lump was delivered by the town highway department in a large front-end loader. I'd figured a weight of 6 to 6.5 tons, based on a guessed thickness of about 3 feet. In reality, the average thickness of the stone was closer to 4.5 feet, and the stone weighs just a tad over 9 tons! The loader was maxed out, with hardly any weight on the rear tires as it rolled along. But I digress. I want to share more ideas of where to find stones.

Roadsides! It's worked for me, and I haven't had to actually pay for highway stones either, at least not yet. When country roads in hilly or glacial moraine areas are being widened, lots of large stones get dug up, and they have to be got rid of somehow. This can be a real pain for the highway crew, as the stones might not fit easily where the small material is being dumped. Ask and ye shall receive.

House excavations. I built my first stone circle with large boulders that came out of the excavation of our underground house. It was easier and more pleasing to put them in a circle than to roll them into the woods or try to bury them somewhere. Recently, some friends built a new addition to a cliffside house. Excavation of sandstone bedrock was necessary and there are enough wonderful playing card-type standing stones in the huge rubble heap to build a half-scale model of the Standing Stones of Stenness. Some are nine feet long, yet only weigh a ton or so, as they are just four to six inches thick. I've arranged to trade a copy of this book for some of the stones. I brought the first installment to Earthwood on my little pickup and Darin and three of his friends set it up outside of Darin's Littlewood house, on November 29, 1998, his 13th birthday. It is now known as the Birthday Stone.

Stone quarries. Well, duh! But won't these stones be expensive? Yes, if you want to buy stones of high commercial value, stone types used for building, for example, like granite and good dense sandstone. Four Quarters recently spent close to $4,000 for four lovely sandstones slabs, delivered. The stones for the circle described in the previous chapter cost

This nine-ton stone, found during road widening, maxed out the town's large front-end loader.

The nine-ton stone was flipped into its shallow socket with an excavator. It now marks the summer solstice sunrise.

about £4,000 ($6,600) delivered, not a bad price for such large and lovely stones. But you might find a real bargain if you go to the quarry and look over the "scrap heap" for rejects. Many quarries have such a place. Stones there might not be perfectly flat, so a quarry specializing in paving stones might set them aside. I scored three beauties in this way. A friend who used to do blasting for the quarry took me over one day, just to show me around, and I ended up with three long and beautiful "rainbow" striated stones, for free!

The author extracts stenness-like standing stones from a pile of excavation material.

Darin plumbs the Birthday Stone.

Abandoned quarries. There are actually quite a few of these around. You may recall that less than two miles from Earthwood, there is an abandoned anorthosite quarry belonging to my friend George Barber. He is a man "of the stones" and is quite happy to make donations to a good cause, such as a stone circle. You may have to track down the owner of an abandoned quarry by searching through tax maps at the county real property office, but the effort might be worth the trouble. Extracting the stones from an old quarry may be another inconvenience (as well as an expense), involving heavy equipment hire, and signing liability waivers. But if the stones are free, or if you can barter something for them, well, a little inconvenience is not so bad.

Farm auctions. Farm auctions? Yes, although this procurement suggestion may only be useful for my British readers. Sometimes, you can find a consignment of old granite fence posts at farm auctions. These posts can be up to ten feet in length. Roy Dutch of St. Austell, Cornwall, built a lovely stone circle this way, for about five hundred pounds (eight hundred dollars).

Keep your eyes peeled! I am constantly spotting stones I hadn't seen before, just while out walking along the road. Late autumn and early spring are the best times to spot them, when the leaves are off the trees. Maybe they are in the town's right of way. Maybe they are on a neighbor's property. Knock on doors. Nothing ventured, nothing gained. I've never been turned down, and I've never had to pay for a stone.

Relative Sizes: Keeping to Scale

Have a goal as to the size of stones you want in your circle, and express this goal in terms of parameters; for example: *I'd like the stones to stand between four and six feet above ground.* The search for stones may satisfy your parameters, or it may cause you to alter them somewhat. Maybe you just can't find enough of the stones in your range as you thought you could. You could put fewer stones in the circle, of course, or you might need to adjust your parameters to three to five feet high, for example. In general, it's good to keep your standing stones in similar scale. It's okay for some to be taller than others; this is the case at almost every stone circle, ancient and modern. But they should be in the same scale or magnitude. Eight-footers intermingled with two-footers tend to look disproportionate. Perhaps the smaller ones can serve as a separate feature, an inner circle or avenue, for example.

What? You can't find *any* large stones? And there are plenty of nicely shaped small ones available? Don't be afraid of building a small circle with small stones. They can be magical. And they are easier to build. Chapter 12 is for you.

No stones *at all,* large *or* small? Well, one guy built *Carhenge* in Alliance, Nebraska: Cadillacs and Continentals sticking out of the ground. (See chapter 15.) Adam Jonas Horowitz, in Santa Fe, New Mexico, is currently building *Stonefridge: A Fridgehenge,* at the city landfill. Mark Klein, a sculptor in Almond, Wisconsin, wanted a certain shape, so he poured his own six-foot-tall concrete obelisks. How about *Strawhenge*? For a harvest celebration, simply stack straw bales in the field. For something more permanent, fasten the bales together with reinforcing rod, and cover them with chicken wire and two coats of cement stucco. I'm serious. People are building beautiful straw bale houses all over North America. Straw is plentiful, inexpensive, and easy to handle.

From the Stone Age to the Appliance Age: Stonefridge: A Fridgehenge, *art by Adam Jonas Horowitz. (Photo courtesy of Adam Jonas Horowitz.)*

Builder and sculptor Mark Klein cast his own menhirs in concrete.

Stones Before Design

It is usually better and easier—and always cheaper— to design the circle after you have a pretty good idea of the stones available. One of the rules for building a low-cost house is equally valid with regard to stone circles: *Design around the available materials*. Once you've gathered the candidate stones together on site, cataloged them, and become familiar with them on a first-name basis, you can let the stones themselves design the circle. How can this be? Well, let's look at just one hypothetical example. Suppose that despite all your best efforts, you can only come up with four really substantial standing stones. An obvious choice would be to place them at north, east, south, and west, which will give you both a North Star alignment, as well as sunrise and sunset equinox alignments for both spring and fall. Other, lesser stones can be spaced

between these corner stones. Got five stones? How about a center stone, as at the Dragon Circle or Boscawen-Un? Or use the fifth stone as an outlier for midsummer sunrise. Only three are available? Leave out the south stone in favor of a recumbent or sitting stone, to give a place for the North Star observation. Got the idea?

These are hypothetical cases only, just to illustrate the value of flexibility. Stone circle design is a numbers game, and this is why I say that the pool of stones itself will help to design the circle. The alternative is to get locked into a rigid design that you *absolutely must have,* and then to go out and try to procure the stones you need to satisfy this plan. The quarry owner will love you, but you'd better have deep pockets.

UNIT OF MEASURE

Before we can start drawing stone circles or laying them out on the ground, we need to discuss units of measure, a matter that was important to the builders of the ancient circles (sometimes fastidiously so, as will be seen) and should be important to us for practical reasons.

When I build a house, be it round or rectilinear in shape, I always use a tape measure delineated in feet and inches. Lumber and sheet goods are available in whole units of feet in the United States, so we do not use a metric rule. A house built in Europe, on the other hand, should be laid out in meters, for the same reason: Always use the unit in currency. This common-sense approach to measure is self-evident.

On a hilltop near the center of Newport, Rhode Island, stands a circular tower considered by many to be the most enigmatic building in America. One of the strongest pieces of evidence pointing to the Viking origin of the tower is that both the inner and outer diameters of the wall are in exact units of the old Icelandic *fathmon,* a word closely related to our fathom. But, while a fathom is exactly six feet,

The Newport Tower, Newport, Rhode Island, is laid out in Norse measure.

the Icelandic fathmon translates to an English measure of about 6'1¾". It happens that the inner diameter of the tower is almost exactly 3 fathmons (18'5¼"), while the exterior diameter is almost exactly 4 fathmons (24'7"). As a builder of many round buildings, I can assure the reader that had this building been built by the English, the diameters would have been 18 feet and 24 feet, exactly. Other architectural and historical evidence also suggests that the Newport Tower is of Viking origin, but the dimensions alone are compelling.

The Megalithic Yard

Alexander Thom (1894–1985), a native Scotsman, retired as professor of engineering science at Oxford in 1961, at age sixty-seven, and began full-time devotion to his surveying work in the megalithic field, which he had begun quietly before World War Two. His work led to three important theories, largely substantiated by what most leading experts acknowledge were the most meticulous and accurate surveys of stone circles and other megalithic

sites ever undertaken. First, Thom determined that the "circles" were not all circular, nor were they just bad attempts at circles. He discovered complex geometrical shapes: ellipses, eggs, deliberately "flattened" circles, and others, discussed later in this chapter. Second, working with his son and grandson, Thom was able to determine previously unknown astronomical alignments in many of the great sites, including Callanish, already discussed in chapter 8. Third, Thom found compelling evidence for the use of a common unit of measure used in sites from Brittany to Orkney: his famous megalithic yard (MY.).

Chapter 5 of Thom's fascinating *Megalithic Sites in Britain* (see bibliography) goes into great detail to prove the existence of the MY, which he fixes at 2.72 feet. To arrive at the unit, Thom surveyed hundreds of stone circles in Great Britain (later extending his work to megalithic sites outside of Britain) and used basic statistical mathematics to find the common unit of his surveys. Readers interested in Thom's proof should go directly to the source.

One of the most extraordinary pieces of documentation for the megalithic yard has already been mentioned in chapter 1: Four of the largest circles in the British Isles, at three of the most important sites (Brodgar, two at Avebury, and Newgrange), all have the same diameter of 125 MY. The skeptic might say—and remember, skeptics are useful people—that what is significant is that the four circles are the same, and, having established that fact, they can be divided up into any whole number of fictitious units. This would be a great argument if not for the fact that the MY pops up in so many other whole-number measurements, for ring diameters, perimeters, spacing of stones, and more. No other discovered unit seems to serve as a common denominator quite as well as the megalithic yard. Furthermore, Thom's precise surveying led him to the conclusion that the ancients maintained a very high degree of accuracy throughout Britain and Brittany, and that this would have required an administration center somewhere, a repository for the sacred measure. He postulated that actual megalithic yardsticks would have been carried to at least the major important sites over the length and breadth of the land. However, Dr. Aubrey Burl, another leading expert, remains skeptical of the central repository theory, pointing out that "in each region of megalithic rings the standard lengths seem to differ slightly" (Burl, *Stone Circles of the British Isles*, p. 73).

At least two actual "megalithic yardsticks" may exist. One is a hazel rod recovered from an early Bronze Age burial mound at Borum Eshøj in East Jutland, Denmark, measuring 79 centimeters in length and divided by hashmarks into proportions of ⅕, ⅖, ⅕, and ⅕. Thom fixes a MY at 82.9 centimeters, so there appears to be about a 5 percent discrepancy. However, in a published paper, Margaret Curtis says, "Euan Mackie gives reasons for preferring an overall length of 81.3 cm (for the Borum Eshøj rod)." Margaret points out that there exists "the possibility of the rod having altered length

Margaret and Ron Curtis show their models of mega-lithic yardsticks to Cliff Osenton.

during three thousand years in a waterlogged grave" ("Megalithic Callanish," published in *Records in Stone: Papers in Memory of Alexander Thom*, p. 439.)

The second possible megalithic yardstick is an oak rod found at an Iron Age stronghold in Borre Fen, Denmark. It measures 135 cm in length and, after a 3 cm "button" at the top of the stick is subtracted, is divided into eight demarcations of about 16.5 cm each. Five of these units, then, total 82.5 cm, very close to Thom's MY of 82.9 cm. Margaret and Ron brought models of these two yardsticks to share at the Dragon Circle. The originals are in the Danish National Museum.

My own theory on the megalithic yard is that, yes, it does exist, and that it is based upon a human pace. Cliff Osenton has shown me that great accuracy can be incorporated into stone circle layout with no more tools than a pair of human legs. My own pace is close to 1 MY of 2'8⅝". I have to stretch my pace to come up with the English yard of 3 feet. But everyone is different. I am—or used to be—5'11" tall. In Orkney, about the time of megalithic construction, men averaged about 5'7" in height, within a general range of 5'3" to 5'10" (John W. Hedges, *Tomb of the Eagles*, p. 190). One's pace is partly a function of height, but also depends on how the individual walks. A variation of 10 percent or more from "average" can be reasonably expected among, say, the adult males in a tribe. My thought is that one person (perhaps the chief or shaman) lays out the circle, and his unit is considered standard. It would be a simple matter to pattern the average pace of this individual into wooden pace sticks. If Burl's idea of regionalized measure is correct, this standard pace could be used from circle to circle. And, in all regions, the standardized pace measure would not be far off of Thom's megalithic yard.

And what finally happened to the megalithic yard? Thom himself says, "We might speculate that this unit was left in the Iberian Peninsula by Megalithic people to become the vara of recent times and to be taken to America by Spain."Thom quotes several values of the vara in comparatively recent times, from 2.7425 feet in Madrid to 2.75 feet in Peru to 2.778 feet in Texas and California (*Megalithic Sites in Britain,* p. 34). The Madrid vara varies from the MY by less than 1 percent, while the New World vara has become 2 percent larger than the ancient measure.

My natural pace, as I've said, is about 1 MY. Try your own pace, for fun. Measure the length of ten "normal" walking paces. Divide by ten to get your own personal yard.

I'll always use feet and inches for a house, but if I'm going to build a stone circle, I feel compelled to stick with the original unit of the ancients. Use the table in chapter 2 to help convert megalithic yards to feet, or to meters if necessary.

DESIGN

At last, halfway into the chapter, we get to design. But if you've got your site, your stones, and a unit of measure, the most important part of the design work is already done. Now, you can gather your paper, pencil, and compass and sketch some ideas.

Geometry

A stone ring can have numerous shapes. In the Norse world, they even liked to arrange standing stones to resemble the hull of a longboat. Let's go through the more common configurations.

The Circle. The circle is the predominant shape among both ancient and modern stone rings, probably because it is the easiest to lay out on the ground. Drive a stake firmly in the ground where you'd like the center of the circle to be. Clip one end of a tape on a nail at the center of the stake, and, holding the tape at the desired radius (half the desired diameter), stretch the tape tight and walk clockwise in a circle around the center stake, marking the ground as you go. (Counterclockwise or *widder-*

shins is considered to be bad luck!) What could be easier? You can mark the ground with white chalk, small pebbles, sand, stakes, whatever suits you. Your circle is laid out.

If you don't like the position or size of the circle you've created, move the center, change the radius, or do whatever you need to do to make it right. You could work out the size and location on a scaled site plan and transpose your most pleasing effort to the actual ground. We drew such a plan at Earthwood, because we wanted to see how the stone circle would relate to the other sizes and positions of nearby round buildings. See "Designing the Earthwood Circle" in chapter 2 for other specific design details we adopted.

The only slight drawback of the circle shape is that it has only one focal point, the center. What would you like to have occupy this unique point? A standing stone? (Boscawen-Un. Callanish. Ivan's Dragon Circle.) A ceremonial stone? (Stone Mountain Stone Circle. Columcille. Stonehenge, if the Altar Stone theory has any validity.) A firepit or fire circle? (Earthwood.) Empty space? (A majority of stone circles, both ancient and modern.) Only you can choose. My only comment is: *Keep your center!* It's a magic spot, so don't lose it; it's hard to get back. Here are some tips about preserving your center point:

With the open space or firepit concepts, you can adopt my little trick of burying a female threaded pipe fitting just below grade at the true center. This fitting can be driven into the ground or set in concrete. A flat stone cover, level with grade, allows access to the pipe fitting and protects it, even in the case of a firepit. I keep a four-foot-long galvanized pipe with a male end on site, usually leaning against the North Stone, which can be threaded into the buried fitting. At the top end, I have fastened a termination cap and painted a little white dot at the center of the cap. With the pipe installed, the white dot corresponds to the exact center of the circle, where it is useful as an accurate backsight, either for observing celestial events, or for adding new outliers in the future. I use it frequently.

If you put a stone in the center, try to place it so that the point of the stone is precisely at true center. If the stone is flat on top, such as an altar stone, mark the exact center in some permanent fashion, such as setting a crystal or special stone into a chiseled socket. The crystal can be held in place with epoxy resin. If putting steel to your stones is philosophically unacceptable, how about enameling a little cross, star, or rune?

Ron and Margaret like the idea of a goat laying out the circle. Simply tether the goat to a firm stake. The goat will eat a circle of grass. (Could this have been the precurser of the modern crop circles?)

Laying Out a Circle Without Using the Center

Let's suppose for a moment that the ancients didn't have any rope or string, an unlikely scenario, but worth considering. Or that *you* don't have any string. Or that you just want to have a little fun on the beach. Try the Cliff Osenton method of laying out a circle. I've done this on beaches in Scotland and was amazed at the accuracy that is possible. The key is to keep a steady pace, which takes a little practice.

First, walk a straight baseline along the beach (AB on the drawing). Now choose a point (1) as the location of the first stone, and mark it with a little circle or a flat stone. From stone 1, walk three paces along AB and mark the spot with an X. Now, turn your body to a right angle with AB—you can sight along your extended arms to help in this regard—and take one pace forward. This is the location of stone 2. Mark it as before. Now extend a line through stones 1 and 2 so that it becomes the new baseline CD on the drawing.

Return to stone 2 and walk three paces toward D. Stop and mark the spot with an X. As before, set your body at right angles to the new baseline (now CD) and step one pace forward. Mark the spot with a flat stone or circle. This is the location of stone 3. Now, as before, extend a new line through stones 2 and 3 so that it becomes the a new baseline EF. Continue repeating the process in this fashion, and after a while, you will be sneaking up on stone 1 from the other side. If your pace is consistent, you will find that you have traveled a remarkably true circumference. You may or may not come exactly to the point of beginning. But I'll guarantee this: The exercise will put a smile on your face.

On the beach, and on paper at home with a compass, I have found that a circle created as described would have nineteen stones in it, a number common among ancient circles in the Land's End district of Cornwall. If only two paces are moved forward instead of three, the resulting circle is quite a bit smaller and has twelve stones, the most common number found in ancient stone circles. Coincidence? Probably. But, remember, once you've got the circle itself, stones can be placed along the perimeter with respect to sunrises, sunsets, cardinal directions, whatever.

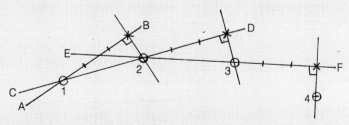

Pacing a circle. Three steps forward and one step to the right produces a circle with nineteen stones. Two steps forward and one to the right results in a twelve-stone circle.

The Ellipse. The ellipse is sometimes confused with the oval or egg shape. Strictly speaking, an ellipse is the path of a point that moves so that that the sum of its distances from two fixed points (called *foci*) is constant. This definition suggests lots of good things, the most obvious of which is that the shape has two focal points instead of one. So, if you're torn between a firepit and a special standing stone, you can have both! The *implication,* if you think about it, is that the ellipse is almost as easy to lay out as the circle. Here's how you do it with a piece of thread, two push pins, a pencil, a piece of scrap wood, and a sheet of paper:

Refer to the diagram below. Using a foot-long piece of thread, tie two small loops into the thread, so that the loops are about six inches apart. Draw a baseline on the paper and pin the paper to the board. Put a push pin through one of the loops and press it through the paper into the wood at a point A on the baseline. Put the push pin into the other loop and stick it into the paper at B on the baseline, four inches from A. Use the pencil point (movable point C) to keep the thread tight as you draw the ellipse. Because AC plus BC always totals six inches, the definition is satisfied. Draw the top half of the ellipse, then the bottom half. Experiment with focal points at different distances from each other: two

inches, three inches, four inches. As the distance between the foci (A and B) increases, the resultant ellipse becomes longer and narrower. When the two foci are very close, the ellipse approximates a circle.

One of the advantages of an elliptical ring over a circle is that it can be made to fit certain sites better. Consider a situation, for example, where there is more building space in one direction than in the other. Gaining familiarity with ellipses is much easier to do with tacks, thread, and a board than it is in the field, where it is much more difficult to mark the perimeter, so start with the model. Similarly, trial and error layout is easier on a scaled plan than at the site. Once the most favorable size and geometry is worked out on paper, the focal points can be scaled off onto the actual ground.

On site, simply drive two stakes into the ground at the desired foci. Tie a loop into each end of a rope of the required length, which can be scaled up from the paper plan. With a loop around each focal stake, scribe your ellipse onto the ground using a sharp stick, like drawing with a giant pencil. The goat method requires a ring that can slide along the rope, and a short tether tied from the goat to the ring. If he doesn't get too tangled (watch closely!), the goat will eat a nearly perfect ellipse out of a grassy field.

We have listed two sensible reasons for choosing an elliptical rather than a circular design: the availability of two focal points and the opportunity to elongate the ring to fit the site. I have encountered no evidence in my research to suggest that either of these considerations was important to the ancients. In *Megalithic Sites in Britain,* Professor Thom makes a strong case that neolithic stone ring designers were very concerned with the numerology of the geometry. In particular, they seemed to be fixated with ring designs where both the diameter and the circumference could be expressed in whole numbers, or integers. The abstract number pi, at 3.1416, was a great inconvenience to them. For example, while any ring meeting the definition

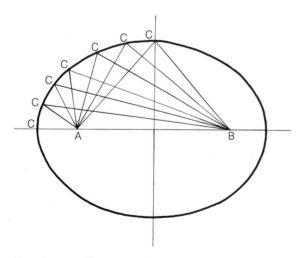

Drawing an ellipse.

The Geometry of Megalithic Ellipses

Consider the ellipse below. The two foci are labeled F and F'. The shorter axis is defined as the line EE', and the longer axis as DD'. Triangle FOE is a right triangle, and if its sides a, b, and c are whole integers, it is also a Pythagorean triangle, such as 3-4-5, illustrated. The line from F to E to F', in this case, is 10. But the definition of an ellipse is that a line from F to any point P and on to F' is always the same, so the broken line F to P to F' is also 10 in this case. D, D', E, and E' are simply special points, in that they fall on the major and minor axes. As FF' is 6, and F to D and then back through F to F' is 10 by the definition, then DF has to be 2. In letters, which hold for any case: (FDF' – FF')/2 = DF. In numbers, for this specific case: (10-6)/2 = 2. All of the major dimensions, except the perimeter, are in whole numbers, just the way the ancients seemed to like it.

But these stone circle designers wanted the perimeter to be integral as well, and with all their designs (circle, ellipse, egg, and flattened circle), they seemed to employ mathematical tricks to try to effect this desired result. One trick that helped was to use another unit called the megalithic rod, equal to 2.5 MY. This unit gave them more opportunities to wind up with whole numbers in the perimeter. Another trick, useful

with circles, was to use diameters such as 7 or 8, which gave close approximations of integers in the circumference 22 and 25, respectively. These numbers produce close but not exact values of pi at $22/7$ (3.143) and $25/8$ (3.125), respectively, not far off the more accurate 3.142, taken to the third decimal.

According to Thom, about twenty existing stone rings in Britain are definite ellipses and "another dozen or so" are less certain, the uncertainty resulting from the ruinous condition of these sites. In chapter 6 of *Megalithic Sites in Britain,* Thom includes a detailed table of all of these ellipses, with the major and minor axes and the perimeter in megalithic yards. Most of the rings of the chart have perimeters that are very close multiples of either the megalithic yard or the megalithic rod. The unique mathematical considerations of several of the sites are examined in detail, including several of his actual survey plans, extremely interesting by themselves.

It is beyond the scope of this book to go into greater detail on the geometry of ancient sites. Nor will it be of compelling interest to any but the most hardcore aficionados of ancient geometrical knowledge, who should try to track down Thom's hard-to-find books, listed in the bibliography.

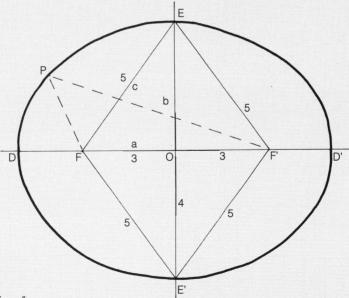

A "Pythagorean Ellipse."

given above (the path of a point so that the combined distances from two foci is constant) qualifies as an ellipse, the ancient stone ring designers wanted something more. Thom's research indicates that they would design ellipses based on Pythagorean triangles, such as the 3-4-5 and 5-12-13 right triangles, which were known to them. By sticking with these Pythagorean-based ellipses, it naturally happened that both the length and the width of the ellipses could be measured in whole units, which seemed to please the designers no end. All this occurred a millennium or two before Pythagoras came up with his famous theorem. "It is remarkable," says Thom in the introduction to *Megalithic Sites in Britain,* "that 1,000 years before the earliest mathematicians of classical Greece, people in these islands not only had a practical knowledge of geometry . . . but could also set out ellipses based on Pythagorean triangles." Later he adds: "We do not know if Megalithic man knew the theorem. Perhaps not, but he was feeling his way towards it" (p. 27).

If you are interested in following the patterns of the ancients, you may wish to employ the 5-12-13, 8-15-17, or the 12-35-37 right triangles, all of which, according to Thom, were known to prehistoric stone ring builders. The sidebar on page 247 will tell you how the triangles are incorporated into the ellipses.

The Egg. A normal egg is *not* elliptical, in case you were wondering. An ellipse is symmetrical about both axes; an egg is a mirror image of itself only along the longitudinal axis. An egg is, well, *egg-shaped,* like the breakfast fruit that comes out of the business part of a hen. One end is blunt, the other end kind of pointy.

Egg shapes are rare among the ancient circles, but are quite real, nonetheless. Thom's surveys and analysis have revealed two distinct types of eggs, which he calls type I (six examples known) and type II (four examples). Thom claims that both types incorporate Pythagorean or near-Pythagorean triangles.

The diagram below, after Thom, shows the geometrical construction of an ancient Type I egg-shaped ring. To draw your own, follow these steps:

Draw a semicircular arc, occupying the entire left side of the diagram, with radius r based upon a center at A. Triangles ABC and ABD are right triangles and the shape of the egg depends on the size and orientation of the triangle. The example is drawn using a 3-4-5 Pythagorean triangle, with AB as the leg of 4 units. Next, draw arc EF of radius r', the radius based upon an assumed center at D. F is also where DB extended intersects with the perimeter. Similarly, draw arc GH with an center at C. G indicates the point where CB extended intersects the perimeter. Finally, draw arc FG, the pointed end of the egg, with a center at B, and a radius (r") equal to BF.

To construct the egg on paper is relatively simple. To create the shape on the land requires the same criteria as for any of the geometrical shapes: a well-prepared site, good measuring sticks (ancient) or tape (modern), and careful workmanship. Ivan

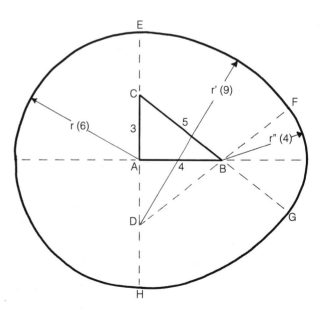

A Type I egg, after Thom. I have used a 3-4-5 triangle to draw this one, but other right triangles will do.

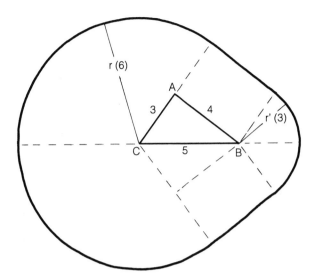

A Type II egg, after Thom. Again, I have used a 3-4-5 triangle to aid construction, but others will serve as well. The hypotenuse lies along the axis of symmetry. This time, there are only two different radii employed, with centers at C and B. The straight parts of the egg's perimeter are tangent to the circle arcs already drawn. This egg is not as elegant as the one in the previous drawing, nor, according to Thom, as common.

McBeth laid out his Swan Circle—actually an egg quite similar to the type I design described above—using a string, a tape measure, wooden stakes, and a pocket compass.

The Type II egg is quite rare—only four were known to Thom—but is included here so that all of the major ancient stone ring shapes are presented. The construction is shown above. Of all of Thom's stone ring geometries, this one is the least pleasing to me aesthetically, and the least convincing mathematically.

A drawback with both egg shapes is that they cannot be easily created by a goat on a tether, particularly a goat lacking in basic high school geometry.

Flattened Circles. Thom identified thirty-three ancient rings having a geometry that he calls a flattened circle. All but two of these fall under the subdivisions of Type A and Type B flattened circles, the odd couple being variations that Thom calls Type

D, and which I will ignore because they are so very close to the Type A variety. And, although there are many more true circles than flattened circles, the flattened group includes some of the most important stone rings: Castlerigg and Callanish among the Type As, and Long Meg and her Daughters among the Type Bs.

A Type A flattened circle is actually a true circle for exactly two-thirds of its perimeter, or 240° of arc, before it "flattens." To construct one, follow the diagram below. Using O as the center, describe a 240° arc CMANG. The 120° angle COA is easily made by joining two equilateral triangles as shown in the diagram. Bisect OC at E and OG at F. Extend AE to D and AF to H. E and F become the centers for arcs CD and HG, respectively. Finally, draw arc DBH using A as the center.

A Type B flattened circle is composed of a true hemisphere below the base line MN and a shape above the base line that looks very much like half of an elongated ellipse, although, strictly speaking, it is not. To construct a Type B, trisect MN at C and E. Extend AC to F and AE to G. C and E are the centers for the small arcs MF and GN. Complete the ring by describing arc FBG using A as a center.

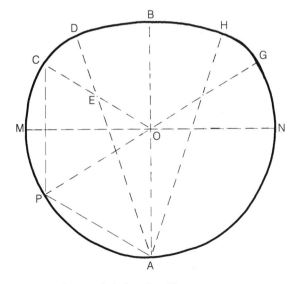

A Type A flattened circle, after Thom.

Flattened circles have a cozy feel to them. The entrance to the circle is nicely located at B on either diagram, and the small diameter AB (which is the only axis of symmetry), can be used for an important alignment, such as midsummer sunrise.

It is interesting to speculate about why the ancients developed the flattened circle, when a regular circle was so much easier to lay out. Mathematical analysis by Thom may give a clue. We know that the number pi expresses the relationship between the circumference of a circle and its diameter, or 3.1416. A Type A flattened circle has the number 3.0591 as the relationship of its "circumference" to its long diameter (MN on the diagrams). A Type B construction has a circumference/diameter ratio of 2.9572. Each of the circumferencial ratios for flattened circles are very much closer to unit 3 than is our old friend, pi:

	Pi	Δ3	% error
True circle	3.1416	.1416	.0472
Type A f.c.	3.0591	.0591	.0197
Type B f.c.	2.9572	.0428	.0143

Where "pi" is perimeter/greatest diameter, "Δ3" is the difference between pi and integer 3, and "% error" is the percentage error or deviation of Δ3 from integer 3 (Δ3/3).

Remember that the designers seemed to favor whole numbers for both diameters and perimeters. They probably wished that pi was three, but as wishing didn't help, they tried to do something practical about it, like inventing more favorable geometric figures.

Before leaving this discussion of stone ring geometries, I must mention that some of the ancient circles—not many—did not follow any of the patterns described above. But even these were not just poorly laid out on the ground, according to Thom. Rather, they incorporated very complex geometries, and more specific purposes. In fact, the greatest of all stones circles, Avebury, falls in this category. How-

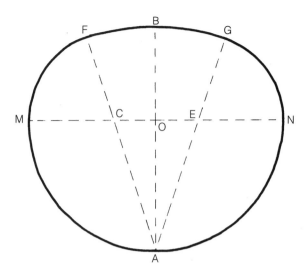

A Type B flattened circle, after Thom

ever, it must be said that Thom's analysis of the Avebury geometry has many detractors among archaeologists, and even among mathematicians.

Stones: Shape, Number, and Symmetry

There is no single "right shape" for a stone circle, although there may be one that is right for you. Choose the shape you like. As for units of measure, if you like the megalithic yard, use it. If you don't like it, use the unit that's on your tape measure. Or pace the circle, like Gaudry Normand did with his wonderful white quartz circle in Bonsecouers, Quebec. (See chapter 15.)

Once you have decided on a shape, you can draw your individual stones onto the diagram. It's fun to try stones in different patterns and places. Shape, color, and height of stones are all important to design. Chapter 2 shows how we made use of these considerations at Earthwood.

Some or all of the stones can be keyed to certain astronomical or geographical alignments, such as at Stone Mountain and the Swan Circle. If alignments are important to you, read the next section, then pencil these alignment stones onto your diagram where they belong.

Symmetry is a matter of personal taste. Some like it, some don't. If you do, decide on one or more axes of symmetry. Our Earthwood Circle is symmetrical, but the outliers follow alignments, which are not necessarily symmetrical, particularly if horizons are irregular.

Alexander Thom thought that the use of whole integers in layout was very important to the ancients, but it probably won't be very important to you, unless you are in to some form of serious sacred geometry. If you are, see the bibliography for sources of additional information. The feet on a tape measure are conveniently subdivided into inches, plenty accurate for stone circle work.

To transpose your design from paper to the land, you will need a wooden stake or two and a good flexible tape measure long enough for the job. If precise orientation and astronomical alignments are not important considerations, you can lay out your circle now. Otherwise, finish this chapter.

ORIENTATION AND ALIGNMENTS

Many stone circles, such as Stonehenge, Callanish, and Castlerigg, have obvious entrances. Many don't. Most of the modern stone circles do not have obvious entrances, including my own, but some do. If your plan calls for an entrance, the circle needs to be aligned accordingly. Perhaps the entrance is from a specific compass direction or along a certain celestial alignment; maybe it's just the shortest distance from the door of your house.

There are many ways to indicate an entrance to the circle. Two stones can be set closer together or farther apart than any of the others. Two stones can be larger, or a different color, or a different shape. You might include an avenue of stones leading to the circle, as at Callanish and Avebury. Perhaps the entrance stones can be capped with a lintel stone, forming a trilithon, and indicating an obvious doorway.

Precise or Symbolic Alignment?

A bunch of lumpy stones arranged in a circle make a pretty blunt instrument for measuring accurate astronomical alignments. Much more sensible designs could be suggested for celestial observation, including stone rows and single stones in conjunction with distant natural features. Nonetheless, many of the designers of ancient circles (if not most) undoubtedly did include alignments that were important to them: solstice and equinox observations, for example, and even the days midway between those celestial events: the cross-quarter days known later to the Celts as Imbolc, Beltane, Lugnasad, and Samhain. But were the alignments very precise, or merely symbolic? In small circles with rude stones, we almost have to say *symbolic;* for better accuracy, one needs a distant outlier stone. For *precision* accuracy, use a distant mountain peak or a cleft in the hills several miles away. Of course this limits the possible places that you can site your circle, but, in ancient times, before individual land ownership ideas, that is exactly what the designers did. In places like Callanish, designers must have taken many years, perhaps several generations, to find the right sites and to design a structure that worked with the landscape. To me, the best compromise is to use outlier stones at least a hundred feet from the center of the circle to get halfway decent alignments.

There are two different ways to get astronomical alignments: extremely complicated mathematics, or visual observation. I strongly recommend the latter. To calculate even a fairly simple alignment, such as the rising sun on the longest day, the following must be considered: the latitude of the site, the location of true north, the azimuth on the horizon where the event takes place (measured in degrees east of true north for sunrise), the altitude of the apparent horizon over the true horizon, refraction of light in the atmosphere close to the horizon, and something called *parallax,* the apparent movement of an object when seen from different positions. (To get a good sense of parallax, albeit exaggerated, hold a pencil two feet in front of your nose and then look at an object across the room first with one eye, then with the other eye.) Until early 1999, I would have said that any reader who wants to learn how to do the calculations had better have a good understanding of mathematics, a scientific calculator, and a copy of Thom's *Megalithic Sites in Britain,* which has the necessary instructions, tables and formulae. But now there is another option: Sig Lonegren's Web site has excellent graphics that explain how to figure sunrise and sunset alignments on various days of the year. See sources.

There is no doubt in my mind that the ancients used observation, not calculation, to get their alignments. For certain lunar alignments, post holes along the avenue at Stonehenge indicate that observations took place over ninety years or more, several generations. Such long-term observations were necessitated by the moon's 18.61-year cycle, already discussed in chapter 8.

We have the advantage of calendars, and we might as well use them.

Solstice Alignments

As we approach either the winter or the summer solstice, the sun's rising and setting pattern changes little from day to day. The sun can be thought of as in a virtual standstill position. This is why we can observe a solstice sunrise or sunset alignment quite accurately for two or three days either side of the true solstice date.

The summer solstice falls on June 21, and the winter solstice on December 21. These dates could vary by a day as a result of the leap year cycle, so always check a current and reliable calendar that includes this type of information. Aren't all calendars reliable? They are not. Based on a faulty calendar, I assumed that the moon would be full during our visit to Callanish in 1997. In fact, the moon was new at the time!

Be fully prepared to record the sunrise or sunset (or both) on the solstice and near-solstice days, so that you can make use of it in your stone circle alignments, or, better, with outlier stones away from the circle. To do this, you must know where the center of the circle will be. Set up a vertical stick at the center, and know how to find the center again in case the stick falls over. The easiest way to record the desired alignment involves two people. One stays at the center and sights over the vertical stick. At the desired distance of an appropriate outlier stone, the other person stations near to the anticipated alignment, carrying another stick in hand. At the moment of sunset (or sunrise) the person siting from the center directs the other person to set his or her stick vertically so that the sun appears to be balanced right on the point of the stick, as seen by the center observer. The assistant can set the stick in the ground with a sledgehammer, or mark the ground with a stone or wooden stake. I made a stack of concrete blocks at my winter sunset alignment, and, a few years later, when the right stone finally came along, I replaced the blocks with a lovely pointed menhir.

Be prepared for the event. If there are trees on the horizon, as at Earthwood, you may have to go out two or three days before the event and trim branches and undergrowth to let the sunlight through. Take an alignment a day or two ahead of the correct calendar date, in case of cloudy skies on the actual day. Your markers may have to stay up for a year or more, so make sure they are secure.

Equinox Alignments

The situation is quite a bit different at the times of the equinox, the dates of which are normally March 20 or 21 and September 22 or 23. Check your calendar. For one thing, at these times of the year the azimuth of the sunrise and sunset changes substantially. The azimuth changes by about .7° of arc—equivalent to the sun's diameter on the horizon—each day. Another major difference from the solstice events is that the sun does not approach a limit, and then start its way back. A few days before the autumnal equinox, for example, the sun rises to the north of its equinoctial azimuth of September 22 or 23. By early October, the sun is rising a few degrees south of its azimuth on the equinox. One advantage of this fact is that observations can be interpolated, useful in case of cloudy skies on the actual day. For example, if you bisect the angle of the sunrise azimuth on September 21 with the azimuth on September 25, you will get an accurate interpolation of the sunrise on the 23rd. If deciduous trees are a factor, consider doing your alignments in March, before the leaves form, instead of September.

Given a true horizon, such as in a Kansas cornfield or on a small island in a large body of water, the equinoctial sun rises at exactly east and sets at exactly west. However, the horizon's actual elevation at your site is likely to throw a spanner in the works, as will trees, so you really need to be out there observing. If a symbolic alignment is okay, the east-west line will serve double duty as a directional finder and an equinox alignment.

Interestingly, the equinox dates do not fall exactly halfway between the solstice dates, because Earth's orbit around the sun is elliptical, not circular. There are about 88 full days between the winter solstice and the equinox dates, whereas the number of days between the summer solstice and the equinox varies between 91 and 93.

Cross-Quarter Days

The traditional dates of the Irish Celtic cross-quarter days (roughly between the solstice and the equinox) are: February 1, Imbolc; May 1, Beltane; August 1, Lugnasad; and November 1, Samhain (pronounced "Sahwane" or "Saw-wen"). Note that these dates do not fall exactly between the solstices and the equinoxes.

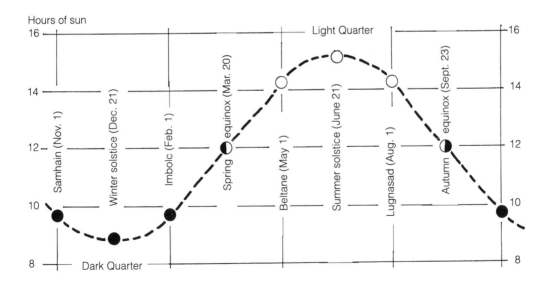

Daylight hours through the Celtic year, at Earthwood, about 45° north latitude. The Celtic year starts at Samhain, which also begins what I call the Dark Quarter, characterized by long nights. Imbolc ends the long darkness and ushers in a three-month period of rapid transition until Beltane, which begins the Light Quarter. Lugnasad begins another period of rapid transition, culminating at Samhain. Note that around the time of the solstices, the change in daylight hours is very slow. This is known as the sun's standstill. At Earthwood, the sun rises in the NE and sets in the NW on June 21; and rises in the SE and sets in the SW on December 21.

How to Find True North

Here are four fairly easy methods for finding true north:

1. The North Star. We are fortunate in the present day that Polaris is a fairly accurate indicator of true north. In fact, the star rotates ever so slightly around true north, being about 1° from the pole at present. (And it's getting better. By the middle of the 21st century, Polaris will only be ½° from the pole. Unfortunately, it will get no closer and will begin to deteriorate after midcentury.) For most purposes, including laying out a stone circle, we can consider it to be "spot on." All we need to do is (a) find it, and (b) record it on the land.

Finding the North Star is easy, at least out in the country where the night sky is not contaminated by a lot of artificial light and pollution. Everyone in the north temperate regions knows the prominent and familiar group of stars called the Big Dipper, in the constellation Ursa Major (the Great Bear). The two stars that make up the side of the dipper away from the handle are known as the *pointer stars,* because they point to the North Star. In fact, if you estimate a distance of about five times the spacing between these pointer stars, in the direction of the top of the dipper itself, you will come to Polaris. Polaris is also the end star of the handle of the star group called the Little Dipper, part of Ursa Minor.

For those of you who want the best possible accuracy from Polaris, use another star in the Big Dipper as an indicator: Alioth (or Epsilon Ursa Major). When Alioth is directly above or below Polaris on a perpendicular axis known as the true meridian, then Polaris itself is at its most accurate position for determining north. A plumb line will cover both stars at this time. If you shoot Polaris at its worst possible time, indicated by turning this diagram 90° right or left, your alignment will be a degree out, and if you put the point of a stone on that alignment, at times the stone will appear to be 2° in error.

The tricky part is to drop a perpendicular line from Polaris to the ground. Again, stand at the proposed center of your circle and have a pair of assistants position themselves under Polaris as far away from you as possible. The assistants should be armed with a long straight white pole, an accurate plumb bubble on a 4-foot level, and a flashlight. Your job is to position the pole so that it is right under the North Star and pointing at it. Your assistants' jobs are to hold the pole exactly vertical, as viewed from your position at the center, and to shine a light on the pole so that you can see it easily. The flashlight will also be useful in checking the plumb bubble. If the pole does not aim at Polaris when the pole is plumb (as seen from the center), direct your assistants to move right or left and try again. When everything is right, mark the position of the bottom of the pole accurately on the ground. A line from this point to the center is a true north line, or meridian. It is difficult for a single

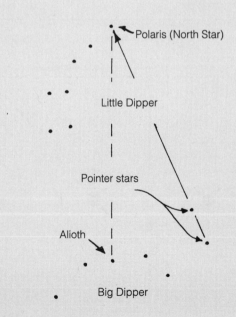

Finding north by the stars.

254

assistant to handle a pole, a level, and a flashlight all at once, although taping the level to a long straight piece of lumber might make it possible.

2. **Magnetic correction.** Borrow or buy a good magnetic compass, not one dug out of a breakfast cereal package. The needle on a magnetic compass points to the north magnetic pole, a point in the Canadian arctic about a thousand miles from the true North Pole (and, curiously, about seventy miles below the surface!). Magnetic north fluctuates over the years. Picture a bar magnet inside Earth with an eccentric wobble. The true North Pole, on the other hand, is the hypothetical extension of Earth's axis of rotation and has been quite firmly fixed for thousands of years. Following the instructions that come with the compass, correct the instrument according to the present-day magnetic deflection. You can get the magnetic deflection for your area off of the key at the bottom of a United States Geophysical Survey (USGS) map, or, in Britain, from the Ordnance Survey map. Other countries have comparable mapping systems. To give an idea, at Earthwood, near Montreal, magnetic north is about 15° west of true north. In Devon, in the south of England, magnetic north was about 5° west of true north in 1999, but the difference is decreasing by about ½ every five years. When using a compass to find north, check your results with a quick look at Polaris, to make sure you've made your correction, well, correctly.

3. **Sunrise and sunset.** Paul Simon sings, "I get all the news I need on the weather report." The evening weather report, or the weather section of the daily paper, might be all you need to find true north. Our TV weatherperson gives the time of sunrise and sunset each day. Halfway between those two times, the sun will be in true south. On a sunny day, plumb a long straight stick (called a gnomon) in the ground. Halfway between the published times of sunrise and sunset, the sun is at true south. Extend the shadow formed by the gnomon and use it as the your north-south

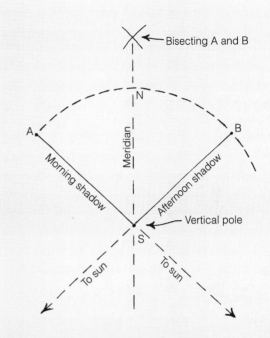

Finding north by the sun.

axis. For accuracy, it is important that you use sunrise and sunset figures *for your area,* not for some city quite a ways east or west of you, or, worse, in a different time zone. This method—and the next—is most accurate at the time of the summer solstice.

4. **Equal Shadows of the Sun.** Toward the south side of any flat level ground area or surface (such as a flat and level sheet of plywood), erect a vertical upright pole at point S. About 10 A.M., mark the extremity, A, of the pole's shadow on the surface. Now describe an arc on the ground or flat surface, using S as the center, and SA as the radius of the arc. In the afternoon, keep a close eye on the pole's shadow. When it again intersects the arc you have drawn, mark the intersection as point B. Simply bisect arc AB at point N. NS is the true north-south meridian. This elegant method of finding north could easily have been used by people four thousand years ago, when there was no useful North Star.

Take your alignments as described above for the other important solar days. As with the equinox dates, you can interpolate the cross-quarter days. By including these alignments, the stone circle can divide the year into eight parts. Beltane, celebrated as May Day in much of the world today, brings in what I call the Light Quarter of the year in northern latitudes. Conversely, Samhain in early November ushers in the Dark Quarter.

These cross-quarter days were originally festival days in the Irish Celtic calendar. There is no reason why you shouldn't incorporate other holidays or even personal days, such as family birthdays or wedding anniversaries into your stone circle. Should you remarry, try to have the second wedding at about the same day of the year, so that you do not have to remove and re-erect a large stone. (Just kidding; better to do a new stone.)

The Four Directions

Druids pay homage to the spirits of the four primary directions: north, east, south, and west. Modern druid circles will have stones positioned at those points, as was done in Surrey, and as we have done at Earthwood and Stone Mountain. The key to laying out the cardinal points is to know true north (or true south). How to determine true north is decribed in the sidebar on pages 254–55.

The east-west line is simply at right angles to the north-south line. To construct this right angle in the field, look at the diagram below, and follow these steps: First, drive a stake into the ground at the center point A of the circle, along the north-south line NS established by one of the methods in the sidebar. Set a finishing nail into the stake, leaving an inch exposed. Next, clip one end of a tape measure (or

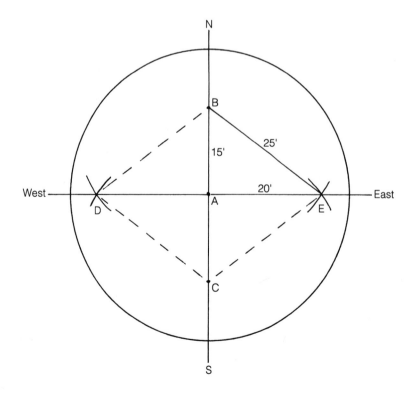

Finding west and east, once north and south are known.

string) on the finish nail. Using a certain known length, 15 feet for example, intersect SN at points B and C, north and south of A. Now, using a greater length, 25 feet for example, and using B and C as centers, draw intersecting arcs at E (east of A) and D (west of A). Line DE is the true east-west line. The length of string or tape you use is not important in constructing this line, as long as BE is greater than AB, but if you choose the numbers suggested (15 and 25 feet), DA and AE will each be exactly 20 feet and you will have created four Pythagorean triangles, ABD, ABE, ACE, and ACD, useful if you want to make a megalithic ellipse, as suggested by Thom.

Our use of the various solar alignments as well as true north at Earthwood is discussed in chapter 2. Other alignments that might interest you are lunar (the moon), stellar (bright stars), and terrestrial (other stone circles, sacred sites, or natural land features). Planetary alignments are not useful, as the planets follow patterns unrelated to our year, and, therefore, do not rise at particular places on the horizon with any kind of regularity.

Lunar Alignments

As I learned from the Curtises, it takes the moon 18.61 years to do what the sun does in one year: establish its extreme northerly and southerly rising and setting. For a variety of very technical reasons way beyond the scope of this book, it would have taken the ancients at least ninety-three years to fully verify the cycle. The interested reader will need to find a copy of Alexander Thom's *Megalithic Lunar Observatories* for a very detailed (and highly technical) discussion. Evidence of these long and continued observations can be found at Stonehenge in the form of post holes, as well as at other ancient sites in Scotland.

Today, the stone circle designer has four choices for determining lunar alignments:

(1) Wait for the next major lunar standstill in June 2006, and record the alignments at that time. Record the extreme northerly rising and setting of the moon. Two weeks later, you'll be able to observe the extreme southerly rising and setting. It would be a good idea to take observations for two or three months before and after that time, as well, in case of bad weather. Your next opportunity will be during the winter of 2025.

(2) Consult *Megalithic Lunar Observatories* (see bibliography) and try to figure the alignments mathematically. Be prepared to make a lot of neuron-crunching calculations, with almost unlimited potential for error.

(3) Check Sig Lonegren's Web site listed in the sources. Sig tells me he is working on the lunar alignment problem and hopes to have clear instructions on his site by the year 2000.

(4) Forget lunar alignments altogether. This is the approach that I've taken at Earthwood. (I will, however, be out there in June of 2006 checking the extreme southern rising and setting, if I'm not at Callanish with Ron and Margaret watching the real thing!) I do not know of any modern stone circle designer who has incorporated lunar alignments in their design. Perhaps the reader will be the first.

Stellar Alignments

If you are fortunate enough to have a good long unobstructed horizon, and don't live near a city, you will find that recording the rising and setting of major bright stars is much easier than taking lunar, or even solar, alignments. You can make the necessary observations (for some stars) any clear night of the year, although midwinter is best, when you've got more hours of darkness to play with, and a clearer sky, as well.

Remember that any particular star always rises and sets at the same place on the horizon, every day of the year. [Note: The previous statement assumes that the particular star rises and sets *at all;* stars that rotate close about Polaris, obviously, *never* rise or set, at least not as observed from most populated areas in North America and Europe.] But most stars do not start to show up clearly until they get by the *extinction angle,* which Thom defines as "the smallest apparent altitude at which, in perfectly clear weather, [a star] can be seen." Below the extinction angle, the light from most stars is absorbed by the atmosphere. Only the very brightest of stars, such as Sirius, can be seen rising close to the horizon.

Besides Sirius, two other good candidates for star rises are Rigel and Aldebaran. Two of the best-known Native American medicine wheels, those at Bighorn, Wyoming, and Moose Mountain, Saskatchewan, incorporate these stellar alignments. Some archaeoastronomers speculate that megalithic sites feature alignments with the Pleiades group of stars, as well as Capella (both near the constellation Taurus the Bull). Other bright winter stars you might want to consider are Betelgeuse and Rigel (in Orion), Pollux (in Gemini), and Procyon (in Canis Minor). A star chart will help you to find them. You will also need to know the hour of the star's rising, according to the date of observation.

Terrestrial Alignments

Ivan's Swan Circle is on an alignment or ley between Stonehenge and Glastonbury Tor, so he has stones to mark both. It happens that this is also the equinoctial or east-west alignment. Another stone aligns with Smalldown Knoll, a local geographical feature.

For local landscape features, usually mountains, hills, or notches, simply eyeball the feature. For sacred sites some distance away, you will have to transpose angles from an appropriate map to your site. This is easy to do if you have a map that includes both the stone circle location and the desired feature ... or a globe.

Alignments at Earthwood

Really impressive alignments require a site with a distant horizon in the required direction. Trees, buildings, and nearby hills can spoil the effect. However, at Earthwood, I maintain my summer sunrise and winter sunset alignments by keeping paths cleared through the woods. If anything, the paths make the alignments even more impressive, as the sun suddenly appears from behind trees, and yet is still on the apparent horizon. Because the stones at Earthwood, even the outliers, are in a bit of a valley compared to the surrounding trees and terrain, the use of a single stone to mark the alignment works great. The sun's orb seems to be sitting right on the point of the stone, or just above it. If you've got a really spectacular site, on a hilltop, or with a long view to the horizon, you may wish to frame your sunrise or sunset with a pair of stones. Consider Stonehenge, a spectacular site for alignments. Most people think it was designed so that the sun rises over the Heel Stone at sunrise on the longest day, but this doesn't make sense. Why block out the very event you want to observe by placing a great hulking stone in the way of it? Originally, the Heel Stone had a mate, now only evidenced by a socket hole, so that the sunrise was perfectly framed by two great stones.

Why bother with alignments? Well, they're fun. And the observation event provides a great excuse for a party, gathering, or sacred celebration on certain pivotal days in the year. Recently, we've incorporated a new alignment at Earthwood that makes a lot of people smile, sort of like a magic trick. It's so outrageous that I just have to tell you about it.

The round Earthwood home is primarily heated by a round 23-ton stone masonry stove located at the center of the house. In 1982, when we built the house, and long before we built the circle, I deliberately included a 24-inch-high "standing stone" on the southwest side of the stove, right at the focal point of the open plan living/kitchen/dining room. You also need to know that the walls of the house are sixteen inches thick of cordwood masonry, individual short logs laid transversely in the wall, much as a rank of firewood is stacked. An insulated mortar matrix fills the spaces between the logs.

At Earthwood, I inserted a plastic tube through just the right log-end . . .

. . . so that the winter sunset lights up a crystal placed at the center of a standing stone in the masonry stove.

Fifteen years went by. For several years, neighbors had been coming to the stone circle on December 21 to watch the beginning of winter over the Sunset Stone, and to have a beer. Each December, I thought of completing my original plan, which was to figure out which log along the southwest arc of the home needed to be removed so that the last horizontal rays of the setting sun on the shortest day would hit the center of the wee "standing stone," positioned so many years earlier. The complexity of the project always caused me to table it for another year. From observation, I could tell the general area of the wall where the log would have to be removed to admit the desired shaft of light. But I couldn't figure out how to visually line up the stone, the setting sun, and the critical point in the wall, to find that *unique* point that had to be exactly on the sunset-to-stone alignment. The right log had to be removed the first time, otherwise the project could deteriorate into mass destruction. Finally, in 1997, I thought of a way to work it out.

Next to the cordwood wall in question, and a couple of feet east of the suspected location of the magic point along this 93,000,000-mile alignment, there is a window; a window that lets in a lot of light, light that hits a cast-iron cookstove next to the masonry stove at sunset on the shortest day. I covered the window with a piece of 2-inch-thick Styrofoam, in which I'd drilled three 2-inch-round holes, one in the center and one each a foot above and a foot below center. On December 19, two days before the solstice, the sky was clear, and the setting sun projected three tidy disks of sunlight on a white drawing board that I'd leaned against the cookstove. One of the bright spots was at the same elevation as the center of the standing stone that had been laid up in the masonry stove long ago. I reasoned that the correct log to remove from the wall would be the same distance west of the hole cut in the Styrofoam as the center of the bright disk of light was from the center of the standing stone.

I was pleased that there was, in fact, a log at just the right location. This was a stroke of luck, as the chances were much greater that the magic spot would end up on the edge of a log or on a mortar joint. My luck got even better as I discovered that the log in question had a slight taper to it, and I was able to drive it out with a sledgehammer, causing almost no collateral damage to the wall; a little mortar broke off the exterior surface. I replaced the log with a 17-inch-long piece of 1½-inch-diameter white plastic pipe, jammed into the much larger log space filled with fiberglass insulation. On December 20, I was able to eyeball from the surface of the standing stone, through the pipe, right to the setting sun on the horizon outside the house. This required adjusting the angle of the pipe somewhat, by rearranging the fiberglass packing. I cut the log into four pieces, and, after transposing measurements, drilled 2-inch holes in the inner and outer pieces, in just the right places. The innermost and outermost of the little log pieces were returned to the wall in their original places, and there was just enough play in the holes to get an exact alignment on December 21, the first day of winter.

Jaki and friends observe the winter solstice sunset from the backsight stone.

Guests were amazed when, at 3:21 P.M., the shaft of light appeared on the left edge of the standing stone as a small hemisphere of light, traversed the stone diagonally up and to the right, and became a full two inches in diameter at the exact center, very bright in contrast to the general background light. After a minute, the bright spot disappeared as the sun set. I'd waited fifteen years for this event, and it was perfect!

Next, we took our guests down to the stone circle to catch the setting sun over the Sunset Stone, which happens each year at 3:33 P.M. So now we have two events to watch each winter solstice, and the one in the house gives a ten- to twelve-minute warning of the event down in the circle.

In 1998, a friend and I chiseled a semicircular 2-inch-diameter hole into the standing stone at just the right place for the winter solstice sunset. We

Winter solstice sunset at Earthwood over the foresight stone.

have two interchangeable crystal spheres, like glass eyes, one of quartz and one of labradorite, which we can set into the hemisphere to receive and refract the last rays of the setting sun.

On the longest day, June 21st, we have a sunrise alignment. The pipe fastened to a receiving socket at the exact center serves as a backsight. At 5:22 A.M., the sun rises over the Bunny Stone and a more accurate foresight stone at the edge of the woods.

DESIGN SUMMATION AND SUGGESTIONS

I hope this chapter has given you sufficient information to design a stone circle that fits your goals and personality. Whole books have been written about some of the specific design considerations, such as sacred geometry and the use of sacred space. If greater depth in these matters is required, the bibliography will be helpful.

I don't like to speak of rules for stone circle design, which tend to constrain the imagination. You can be as rigid or as free as you like with your design. However, I do have three generalized *suggestions,* which you may accept or reject as you like.

(1)   Know your site, your stones, and your capabilities before you design. Don't try to do the undoable.

(2)   Stone circle design includes choices about its use or purpose, shape, size, unit of measure, integration with the site, making use of Earth energy, and astronomical or other alignments. Some of these considerations may not be important to you, and you might have others that I have not thought of. Everybody's different.

(3)   The last suggestion is one I give to my students at Earthwood Building School, with regard to housing. *Build a small structure or model first.* You'll learn a lot and get lots of useful practice. Make little mistakes on the small project instead of great ones on the large project. In fact, small circles are the subject of the very next chapter!

— CHAPTER 12 —

Small Circles

*A friend built a medicine circle of small stones because she didn't
have the strength to lift up large ones. And you go inside that
circle and—holy smokes! It is powerful, more powerful than any
stone circle I've been in. It just depends on what you put into it.
She put a lot of energy into it.*

—ED HEATH

If *megalith* means "large stone," then I submit *minilith* for "small stone," and *microlith* for "very small stone." Let's start with microliths, and have some fun with models, window arrangements, and very small circles out in the countryside.

MICROLITHIC CIRCLES

You might like to try out some design ideas with a little model before you tackle the larger one in the back garden or out on that lovely grassy knoll. Or perhaps you're a city dweller. Well, you can enjoy a stone circle just like someone in the country. Build it in a shallow ceramic vessel or other suitable container. No stones or spare space at all? My neighbor suggests *virtualithic,* making stone circles on your computer. Another option is to use chapter 15 as a guide and go visit other people's stone circles.

Models can be made of extruded polystyrene, clay, cement, or concrete. But little stones work as well as anything and have their own individual characters. Sometimes people come to the office, see the Earthwood circle, and learn of my continued involvement with stone circles. "Well, I couldn't do that," some say. I grab a handful of #2 crushed stone from the pathway outside, scatter them nonchalantly on the tabletop, and arrange them in a ring.

"What have I just done?" I ask.

"You've made a little stone circle," they observe.

"That's right," I reply, tossing the stones back out on the path with practiced drama. "From here on, it's just a matter of degree."

Actually, these #2 crushed stones yield excellent microlith material. These are the stones that get sorted between a 1-inch grid and a 1½-inch grid. While most of the stones are rather nondescript, one in a hundred (or so), make great little standing stones. Our local crushed stone is a uniform blue-gray limestone, but another quarry a few miles away offers crushed stone in a variety of brown tones. In Vermont, crushed white marble chips are available, and a friend in Quebec has access to crushed white quartz. All make wonderful little standing stones. Maybe the people at the quarry or the local building supply yard will allow you to pick through a pile. Other sizes work equally well. Number 1 stone

When Darin wasn't helping Cliff, Jaki, and me to reconfigure a five-tonne dolmen, he built little dolmens and microlithic stone circles by standing stones in soft wet clay. The stones came from the beach near Merlin's Cave at Tintagel Castle, Cornwall.

(left) We built this charming little circle, complete with dolmen and stone avenues, near Wapta Falls in the Yoho National Park, British Columbia. The stones are two to three inches high.

There were no stones near St. Augustine Beach, Florida, so we built this little circle of oyster shells.

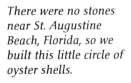

is caught between a ½-inch grid and a 1-inch grid, #3s are up to 2 inches, and #4s are up to 3 inches or more. Longer stones sometimes slip through the gridwork, as they come vibrating down a slope, but they won't be very big around, so they are perfect for a microlithic stone circle.

The other great sources for lovely little standing stones are the seashore and riverbeds. Stones from these sources will be polished smooth, and have a different character from the crushed stone, but they can have equally intriguing shapes.

Let's look at a few microlithic stone circles that the Roy family has "built" over the years. (There are photographs in the color section, too.)

Models

Ivan McBeth has been asked to submit design ideas for several stone circle projects in the British Isles. He once made a scale model of a proposed "Mil-

lennium Circle," a project that never materialized. Ivan shaped the "stones" from extruded polystyrene, and mounted them on a plywood base covered with modeling grass.

You can put a stone circle in a window alcove or outdoor window box. You can rotate it to get different alignments and various shadow effects. You can change its configuration when the fancy suits you. In the pictures on the next page, Darin demonstrates the construction of such a circle. If you can push candles into a birthday cake, you can build a minilithic stone circle.

A number of years ago, a very accurate model of Stonehenge was built by Albert Dowdell, now deceased, just five miles from the real thing. Jaki, Darin, and I managed to track it down from an old book. The new owners, a young couple who had only owned the place for a year or two, were delighted to show us the model, and have us in for a cup of tea.

Ivan's model for a proposed Millennium Circle.

Darin pushes stones into the slightly damp sand like candles in a birthday cake.

Darin adds more stones until the model is finished. Darin's circle features an avenue, a dolmen at the center, and outliers, all in a twelve-inch planter.

Stonehenge model, by the late Albert Dowdell, built just five miles from the original.

We were very surprised to hear that we were the first people to stop and ask about Dowdell's model. The "stones" were cast in concrete in just the right shapes, the tallest not much over six inches in height.

INTUITIVE CONSIDERATIONS

Those modern builders who are particularly concerned with the sacred space aspect of stone circle creation—and they make up a substantial percentage of all circle builders—know that size is not necessarily a critical factor in a circle's power, energy, or usefulness as a place of healing or for meditation. Ed Heath's epigraph at the beginning of this chapter makes the point very well. In *The Swan Circle,* Ivan McBeth shares some of the intuitive considerations that he applies to stone circle design and construction, many of which are recounted in chapter 6.

As the reader knows by now, I tend to be rather left-brained or analytical in my approach to creating stone circles. In an effort to balance the discussion, I sought out help from Selena Fox of Circle Sanctuary in Mt. Horeb, Wisconsin, whose style is well grounded in the right-brained or intuitive mode. A priestess in earth-based religion, Selena has been building and blessing small stone circles for ritual purposes since 1975, in Scotland, the United States, and elsewhere. (In Antigua, she built a modern stone circle and found an ancient one!) Her best-known creation is the Stone Circle at Circle Sanctuary Nature Preserve, described in chapter 15. During her quarter century of work with sacred spaces, she has distilled five important considerations

Note: These five considerations were first published in a copyrighted article by Selena Fox in issue 68 (summer, 1998) of *Circle Network News,* P.O. Box 219, Mt. Horeb, WI 53572, and are reproduced here with permission.

to keep in mind while creating stone circles, large or small. Here they are, in her own words:

1. *Intention & Plan.* Before you begin, carefully consider the reason(s) for wanting to create a stone circle and what function(s) it is to serve. What types of rituals, meditations, and other spiritual activities will take place there? Who and how many will use it? Who will help create and maintain it? What is the anticipated lifetime of the circle? Meditate, consult your dreams, review your journal, and do other inner work to clarify intention and develop a plan for the creation and upkeep process.

2. *Site Selection & Design.* Keep intention and plan in mind as you consider possible sites, and then develop the design for the site selected. Choose the site carefully. Commune with the spirit of the place. You might want to create a circle at a certain spot, but does that place, its history, and its present community of plants and creatures want you and your circle there? Who legally owns the property and what are the terms for using it as a circle site? Are there stones nearby? If privacy is desired, how secluded is the site? What is the incline and texture of the terrain? Flat and grassy or sandy are best if the circle is to be used for reclining meditations. If dancing and drumming are to be activities, the terrain should be flat and the design should include a fire ring as a focal point for dancing, as well as a place to tune drums. If the circle is for a group, it should be big enough for the present membership, and have room to allow for guests and future growth. Once the site is selected and legal arrangements made, the circle design should take form in such a way that it harmonizes with the natural features and the environment.

3. *Consecration & Construction.* Before construction, do an honoring of the spirit of place and ask its cooperation in the work. This can be done through meditation and/or divination. Do this at the center of the circle-to-be. Once contact is made and spiritual collaboration has been established, proceed with the consecration. Remove debris from the area, and then, using incense, a crystal wand, sound or other ceremonial tools, clear away any unwanted spiritual energies, thoughtforms, and emotional influences resulting from past human activities and other factors. Begin at the center and spiral clockwise and outward around the entire area as you do the consecration. Once this is done, begin construction. You may wish to work clockwise around the circle, beginning in the North. Construction may be done in phases, or all at one time.

4. *Alignment & Activation.* After the stones have been placed, align them with each other and with the spirit of place through ritual. Use toning in combination with imagery, such as visualizing a ring of light circling clockwise, connecting the stones with each other. Then dedicate and honor the circle as a sacred site and begin working with it. If the circle has cairns or standing stones as quarter markers, these can be worked with as portals for connecting with the sacred directions. Many circles also have a centrally located altar stone which is linked to the circle as a whole, as well as to the Divine.

5. *Development & Upkeep.* Once the circle is created, it develops in power each time it is used. Treat the stone circle with respect and encourage others who visit and work with it to do the same. Keep the site free of physical and psychic debris. Adjust the positions of the stones as necessary. When stones are added or repositioned, do another alignment. If the circle is no longer to be used as a sacred site, do a thanksgiving, and then deactivate it. Stones may be left in place unless the circle is to be relocated, but any portals that have been in use should be closed and sealed. Let the circle sleep until the time it is to be reactivated. For stone circles that are intended to be long-lasting, knowledge of circle protocols, history, and traditions should be passed on to succeeding generations. Today's new circles may become the old circles of centuries yet to come.

MINILITHIC CIRCLES

This classification includes common garden-variety circles made from stones that one or two people can easily maneuver into place; stones of, say, 20 to 200 pounds in weight, 12 to 30 inches in height. An example that you've already seen in chapter 6 is the little "druid circle" in Julie's back garden near Bristol. Margaret Curtis also made a small stone circle in her garden at Callanish, less than a mile from the famous one. These circles each make use of special stones that their builders collected over the years. They each have symbolic and actual alignments incorporated into their designs. Likewise, there is a delightful little "faerie circle" near the reception building at Columcille.

Our "afterthought" ring of stones at the Earthwood circle would stand alone quite nicely as a minilithic circle. It would not have to surround a fire pit; it could simply be featured in the center of a flower or herb garden. Because it does surround a fire pit, the Potsdam sandstone miniliths are subjected to very high temperatures. Two of them have broken after particularly hot bonfires, and had to be replaced. I keep spare stones in stock for just such an eventuality.

Safety

Minilithic circles are fairly safe to build, and the larger miniliths can be good practice for developing techniques that you might use on a full-size circle. However, one potentially dangerous situation could occur while building a minilithic circle that is not likely to occur with a megalithic circle. With miniliths, the builder is inclined to simply lift the stone and place it in its socket, but these stones are just of the size that can cause a real back injury if you don't lift correctly. Keep the stone close to the body and bend from the legs, not the back, when lifting. Unless you do regular weight training, don't

A minilithic "faerie circle" at Columcille.

The "afterthought" ring of stones surrounding the firepit at Earthwood. The removable pipe at the center is used gain accurate alignments.

try to lift anything off the ground that is more than a third of your body weight. If you're really out of shape or have pre-existing back trouble, don't lift stones at all. Roll them into place and tilt them into upright positions, using a lever and fulcrum. Get some help; let a friend in on the fun. The most dangerous stones are the ones just at or beyond your limit. The trouble is that you don't really know what that limit is, which is why you have to err on the side of caution. And this is why, in this single instance, megalithic stonebuilding is safer; everyone knows better than to try to lift a 500-pound stone. In that vein, be particularly careful about moving and standing stones in that range between minilithic and megalithic, where you might be tempted to

"go it alone," when, in fact, you either need a larger work crew or some heavy equipment.

I'm not trying to sound like an old fuddy-duddy about safety. As someone who has spent a few weeks with virtually incapacitating back pain, I simply do not wish the experience on others. Enough said . . . for now.

Well, there you have it; the smallest chapter for the smallest circles. But, while a circle might be small in stature, it can radiate all the beauty and magic of the larger variety. All the design features discussed in the previous chapter can be incorporated. Give me a carefully designed and aesthetically constructed small circle over a large, ill-conceived lumpy monstrosity any day.

CHAPTER 13

Building Stone Circles with Heavy Equipment

You can't get much more physically manifest than a stone circle. You can't say it's half there. It's bloody well there and it's completely solid. And I love that. I really do. But, also, you have to be able to tap into some of the most rarified energies or atmospheres. It's a complete spectrum.

—IVAN McBETH

Before working with Cliff Osenton and Ivan McBeth, I used to say that there were two ways to build a stone circle—back*hoe* or back*ache*—and that I had experience with both. There is a third way, a better way, which is to use knowledge and brain power. While I have no pretensions about writing the definitive megalithic heavy-lift engineering manual, I will share with the reader some of the techniques that stone circle builders in the 20th century have used successfully.

CHOOSING A METHOD

How do you want to build your stone circle? And *why?* The answers may be connected to the purpose and use of the circle. If you belong to one of the neo-pagan religions, you may have strong spiritual reasons for erecting the stones without benefit of heavy modern equipment. A Christian, Buddhist, or atheist may make the same decision, for no other reason than answering the challenge of building like the ancients. Chapter 14 is for you.

We'll kick off this chapter with the construction techniques that have been most commonly used by modern-day stone circle builders, methods involving heavy equipment, or, as they say in Britain, "plant hire."

The use of heavy equipment has made the stone builder's task a lot easier, albeit more expensive and (perhaps) not quite as personally satisfying. Still, if the end result is more important than the process, this is the route to take. There are basically four choices in powered heavy equipment: crane, backhoe, loader, and bulldozer. We'll presume that, one way or another, the stones have all been delivered to site and that they have been dropped off somewhere near the spots where they have to be erected.

With any kind of equipment hire, you are going to have two costing choices; you can hire the equipment by the job or by the hour. Contractors themselves tell me that it is better to hire by the hour, even for common work, such as excavations and septic system installation. When the job is to build a megalithic stone circle—a *what?*—the contractor is even more likely to add on contingency buffers to his cost estimate. He doesn't know if he'll be standing around while the owner decides what to do next. He has to consider how fussy the owner might be about placement, depth, and verticality of the stones.

If you are qualified to operate the equipment yourself, of course, you can save a lot of money. Maybe you have your own backhoe or have a friend with one. A friend of mine with a backhoe helped me with quite a bit of the preliminary work at the Earthwood circle. At Stone Mountain, you may recall, Stephen Larsen rented a very small backhoe, which he and Russ could operate themselves. It was rather undersized for the job, however.

Warning!

If you think I was being an old fuddy-duddy on safety in the last chapter, listen to this: *Building a megalithic stone circle is dangerous, full stop*. I know, I know. *Everything* is dangerous, to a degree: building a house, driving to town, crossing the street. But consider for a moment what we are actually doing with, say, a six-ton megalith. We are transporting it, and perhaps dumping it out of a large loader or dump truck. It can land and bounce like an American football. Next, we want to stand it up unnaturally. It might decide to return to a more stable horizontal position and not warn you ahead of time. Consider, too, that there are so many dangerous places in which to get fingers, hands, arms, and other things jammed while you are working. Stones are heavy and hard. If they have a consciousness, which many people believe, are we really sure it's benign? Maybe the stone doesn't want to be disturbed, thumped, jostled, and hung by its neck before being half buried in the ground. Okay, maybe I'm over-dramatizing, but you get the idea: *Work with heavy stones at your own risk*. Two more points, while we're at it:

Small children should not be on site, no matter how you decide to move the stones.

Also, hard hats should be used when working with megaliths. Similarly, boots are a better choice of footwear than, say, running shoes, and steel-toed boots are best of all. All I can do is recommend these things; I can't make you wear them.

There, I had to say all that stuff, and it's true. If you choose to continue despite my warning, so be it, but *be careful*. I bring up other safety issues in the text that follows, but their inclusion implies neither sanction of any stone-building project, nor that the discussion exhausts all the concerns that may arise.

CRANE

Crane hire is the most expensive heavy equipment option, about £500 per day for a 50-tonne crane in Cornwall in England and $95 per hour for a 35-ton crane ($175 for an 85-ton crane) in northern New York, plus transport to and from the yard. During my research, I was advised by our local supplier that capacity ratings cannot be taken at face value, because the ways that ratings are determined are nothing like what a layperson would expect. A crane rated at 85 tons, for example, might only be useful for stones of up to 25 tons. Check with the contractor.

As with any piece of equipment, a crane is only as good as its operator, who needs to know exactly what it is that you want to accomplish, and, therefore, what additional equipment might be necessary, such as chains or nylon slings, or a backhoe. A site visit by the contractor is a good idea, and assures that the crane operator can actually find and access the site. It is usual to be charged for crane transportation both ways, so you don't want to add any more to the journey. With other equipment, one-way transportation charges are normal.

Cost is really a function of productivity. If the socket holes are all ready to receive the stones, and the stones are all within a boom's swing of where they belong, a good crane operator can plunk those stones in their holes awfully fast, probably two or three per hour, if there is a ground crew available to do some fast packing.

Because of the high hourly cost, the builder needs to be sure that everything is planned and ready for the crane's arrival. Socket holes should be dug, so that the crane isn't idle. The holes can be dug accurately by hand, if soil conditions are favorable, such as at the site in Surrey; or they can be dug with a backhoe, even a small one. The important thing is that the holes and dirt piles are not in the way of the crane's ability to get to all the necessary positions for reaching, lifting, and setting the stones. Plan this out. Sketch the dirt placement on a site plan, so that the crane is not denied important access points. Find out from the contractor how far the crane's boom can reach and still lift the heaviest stone. Most cranes do have a full 360° swing.

Some of the photos in part 2 show cranes doing megalithic work at Columcille, the Seven Sisters, and the Swan Circle.

Lifting the Stones

Normally the crane operator will arrive on site and find a number of stones lying around on the ground. His or her job is to stand the stones up in holes. Once a nylon sling or chain is properly fastened around the stone, this is easy. The problem, the potential delay, is getting one end of the stone up off the ground enough so that a chain or sling can be passed under the stone. The crane has no way to lift one end of an unchained stone, in the way that a backhoe can.

If the side of the stone resting on the ground has a concave depression in it, you might be able to push a chain through the depression to the other side. Or maybe the ground is soft enough that you can quickly and easily dig out some earth and pass the chain through. Neither of these techniques may be available, however, and one end of the stone may have to be lifted to get a chain or a sling around the stone. Manual lifting is too slow when you're paying high hourly rates for idle machinery, so the reasonable lifting choices are a backhoe, loader, or even a farm tractor with a loader bucket or fork lift. Remember that the metal teeth of a bucket can slip off the stone all too easily, and fastening equipment can break. Therefore, never, *under any circumstances,* should you have your arm, leg, or any other part of your body under a stone being temporarily suspended off the ground. Pass the chain under the stone by pushing it through with a long pole.

Digging socket holes and lifting stones for chaining are just two of many reasons to have a backhoe on site while the crane is working. Ivan McBeth, Ed Prynn, and Bill Cohea have all found that backhoes are the perfect complement to a crane. Sometimes the crane cannot stand the stone straight in its socket, because of the way the stone is hitched, or because of some vagaries in the shape of the stone or the socket. The boom of a backhoe can straighten the stone, and/or hold it in place while some packing stones can be thrown in, or the soil tamped. The cost of the backhoe might add 25 percent to 40 percent to your hourly expenditure for the crane, but it will earn the money back in time saved on the aggravating little things that the crane can't do. Ideally, you want two operators who are used to working with each other, a situation that greatly improves efficiency.

Vertical Lifting

The key to quick and easy stone installation is to lift the stone vertically. To do this, the lifting chain, cable, or nylon sling has to be over the center of gravity of the stone along the vertical axis of the stone, not off to the side (see photos opposite). With a chain, use the "megalithic hitch," described below. A crane operator used to working with heavy nylon slings may have a similar trick. A potential problem may be supporting the stone while the best lifting hitch is being tied on, another good reason for having a backhoe on site.

This stone, lifted vertically with a megalithic hitch, will be easy to place in its socket . . .

. . . while this stone, hanging with an inferior hitch, is much more difficult to stand upright.

The Megalithic Hitch

You need a good long chain to make the megalithic hitch, and the chain's load rating must be equal to the task. Because there are so many qualities of chain available on the market, the rating of chains (or of cables or nylon straps) is a complicated and technical subject, and beyond the scope of this book. Get working load limits and advice from your local supplier, or rely on the experience of your operator. Whether you use chain, cable, or nylon straps, inspect the material for damage. Remember the adage about a chain being no stronger than its weakest link? Well, frayed sections of cable or strapping are like weak links.

Half the battle with the megalithic hitch is getting the stone to a position where you can tie the hitch on, particularly if the crane is unaccompanied by a second piece of equipment, such as a backhoe or front-end loader. With a backhoe or loader bucket, the stone can be temporarily stood up while the hitch is tied. Support the stone with a block of wood for safety during the tying procedure. Always be ready to move quickly whenever working near heavy stones, and know your escape route.

The ideal chain to use for, say, a stone like our South Stone, would be about twenty feet long, with a chain link hook at each end, the kind of hook that can clamp onto a link at any point along the chain's length. See the diagram on the following page. To fasten the megalithic hitch, follow these steps:

(1) Hold hook A against one side of the stone, about a quarter of the stone's length from the top.
(2) Wrap the chain laterally around the stone once and fasten hook A to one of the links of the chain, creating the *lower loop*. Make this connection as tight as possible.
(3) Keeping tension on the chain at all times, fasten the second hook, B, to a link at a point diametrically opposite the first hook, creating what we will call the *upper loop*. Keep tension upward on the chain.

(4) The crane or backhoe lifts the stone by the upper loop. As tension is increased, the lower loop tightens like the well-known Chinese finger puzzle. If the hitch doesn't "grab," you'll know right away as you try to lift. Even when it grabs really well, stay clear of the lift. Chains can slip or break, and stones can bounce in strange ways.

If there is too much chain, it is okay to make two or more loops around the stone instead of just one. Sometimes, this helps the chain to grab the stone, or makes it easier to maintain tension until the lift takes place.

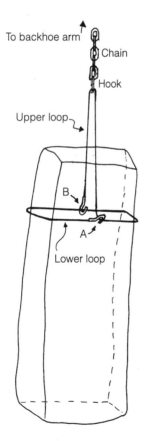

Tying the megalithic hitch. See text. To simplify the drawing, the monolith is transparent, like a block of clear ice. Also, to clearly show its path, the hitch is drawn with a cable instead of a chain.

A megalithic hitch made with a chain will lift this monolith straight up without slipping.

Nylon straps or slings are often used to lift stones instead of chains or cable. Personally, I have no experience with nylon straps, but many Club Meggers prefer them over chains, because they do not mar the stones in any way. Nylon straps can be very strong. Use one that is rated to handle the load you intend to lift. Do not use a worn or damaged strap or sling. Never stand underneath or, for that matter, anywhere *near* a megalith that is floating up in the air. The overriding law of nature on this planet is gravity, and the stone's natural predilection is to make its way toward the center of the planet, and at its earliest opportunity.

Sockets

An advantage of vertical lifting of a long stone is that the socket need not be very much bigger than the base of the stone itself. The individual sockets can be tailored for size and shape. A stone's long point, for example, can be accommodated with a little bar and shovel work. Less grass will be damaged and less topsoil disturbed. The whole job will be neater and tidier. We may as well discuss other socket considerations while we're on the subject, as they are common to any of the construction methods chosen.

The depth of the socket is normally between a fifth and a third of the stone's length, depending on the shape of the stone. If a heavy end is planted in the ground and the rest of the stone is comparatively light in weight, like our Sunrise backsight stone at Earthwood, then 20 percent might be plenty. If the stone has a long narrow point to go into the ground, and the top of the stone is heavy, like the East Stone at Stone Mountain (chapter 3), then a third of it should be in the ground, even a little more. If the site is particularly prone to frost heaving, the socket should be at least as deep as the extreme frost depth for your area, a figure available from the local building department. None of my socketed stones are planted as deep as forty-eight inches, the design depth for frostwalls in northern New York, and none of those stones have heaved in over ten years. The only ones I've had trouble with are the unsocketed stones. My unproven theory is that the stones socketed two feet in the ground simply move uniformly up and down with any frost heaving that might occur, with no long-term impact. Poorly drained soils might disprove this theory real fast.

If local maximum frost depth is too deep for your liking, leaving insufficient stone above ground for the desired effect, plant a large squat bulky stone down to frost level. Pour concrete over your foundation stone and set the standing stone in the concrete. Use strong mortar (1 part cement to 2½ parts sand) instead of concrete if only a shallow layer is required. The concrete (or mortar) should be continuous from the top of the foundation stone to the bottom end of the standing stone. For the best result, clean both stones of loose earth with a wet brush, and apply Acryl-60 bonding agent (Thoro Corporation) or the equivalent, to both the top of the foundation stone and to the underside of the standing stone. We set the Bunny Stone this way and it has stayed put through years of dicey spring frost conditions. The reason for the large stone in the hole, instead of lots of concrete, is to save money. Also, if you fill the entire hole with wet concrete and try to set a six-ton stone in it, the stone may sink quite deeply, displacing most of your concrete. If money is not a problem, you can pour the concrete footings one day, and set the stone in a little fresh concrete at least three days later. See the sidebar on page 277 to see how the Church of Four Quarters handled a similar situation.

Stones with wide flat bases can be set without a socket, but they will be more subject to frost heaving. For more on setting unsocketed stones, see chapter 2 and chapter 3. Unsocketed stones, obviously, are not as stable. Bill Cohea, Ed Heath, and others have stood stones up with little or no socket, but added frost protection, stability, and safety by mounding the stones. Less of the stone is showing than without the mound, but the full height is still there with respect to surrounding grade. In November of 1998, we set up our new nine-ton summer sunrise stone this way. For additional frost protection, I like to bury a piece of 6-mil black plastic about six inches beneath the surface to shed water away from the base of the stone. An alternative to the plastic layer is to finish the top of the mound with three inches of clay, or more. The clay, like the plastic, will shed water away from the base of the stone, decreasing the likelihood of frost heaving. I've used both methods successfully on various stones at Earthwood and Stone Mountain.

While I'm on the subject of foundations and frost depths, I'd like to share with you an interesting methodology developed by Orren Whiddon and the Stone People at Four Quarters Farm, even though the work was done mostly by hand, the subject of the next chapter. The sidebar on the opposite page describes their use of concrete receiving notches to aid in erecting stones.

BACKHOE

Technically a "backhoe loader," the machine in question is more commonly known in America as, simply, a backhoe. In Britain, a backhoe is commonly called a "JCB," after one of the leading manufacturers, just as a cola drink is called a coke. A backhoe has a smaller articulated bucket (the hoe) on the back end, and a larger loader bucket in front of the engine.

For setting stones in the 500-pound to 2-ton range, a small backhoe is probably the most versatile piece of equipment. Socket holes are easily dug with a small backhoe and any operator worth his or her salt can maneuver the stones into the sockets with one end of the machine or the other. In the United States, the John Deere 310E, with 4,420 pounds of "dipperstick lift," or the 410E (5,655 pounds), would be good choices, or a Caterpillar 426C with 5,650 pounds of lifting capacity. (In Britain, backhoes tend to be smaller and you really need to consider a tracked excavator for any serious lifting.) The manufacturers listed are only representative. Ask locally for advice on backhoes or any equipment.

If your stones are in the 2- to 4-ton range, you will want a bigger backhoe than those mentioned above, perhaps a John Deere 710D, a Caterpillar 446B, or the equivalent, or you may need to move up to an excavator, described below. While it's true that a skilled operator, like my friend Ed Garrow from chapter 2, might be able to use his experience to coerce a 3-ton stone with the smaller machine, the project will go more easily and faster with the larger backhoe. The machinery will be grateful, too.

Circles made of small stones can be built with loaders and backhoes attached to farm tractors.

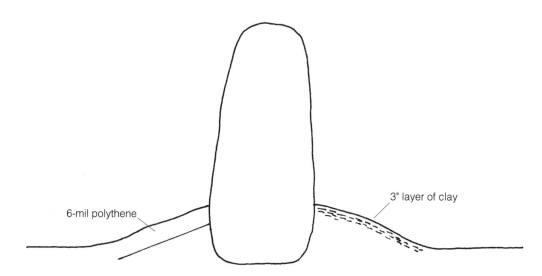

Cross-section of a mounded stone. The left side of the drawing shows 6-mil black polythene buried 6 inches below the surface of the mound, a good method where clay is not readily available. The right side shows the mound finished with 3 inches or more of clay. Either method helps to keep water away from the base of the stone.

Erecting Stones at the Four Quarters Stone Ring

In Pennsylvania, the Church of Four Quarters is building a large oval ring that will have as many as 120 stones by its completion date sometime in the 2020s. The builders like to leave as much of their beautiful sandstone slabs above grade as possible. They have developed a method that has worked well for them over the years. Each June, the Church hosts a large stone-raising ceremony, which about three hundred people attended in 1998.

Prior to the stone raising, Orren Whiddon, president of the Church, and his fellow "Stone People," prepare the requisite number of foundations. They dig down to frost level for their area (about two feet) and fill the holes with concrete. While the concrete is still plastic, they cast a receiving notch in it for each individual stone. This notch, also called a *keyway* or *turning groove,* serves as a very shallow socket. The stones are positioned over these notches, and manually levered upward to about a 25° angle, so that the leading edge of the stone will be caught by the vertical edge of this custom-made keyway. The stones are very long and narrow, like the Stones of Stenness in Orkney, so they are not very heavy compared with their heights.

Orren's receiving notch technique is very much like one used by the ancient Egyptians while erecting their gigantic stone obelisks. Unlike megalithic standing stones, obelisks were not socketed. For stability, the Egyptians relied on large, sturdy, flat stone foundations, without even pins to hold the obelisks in place. Great weight, precision construction, and relatively aerodynamic shape were the keys to their standing, in some cases, for thousands of years. The danger when standing an obelisk was that its lower end might "kick out" as the top end was raised. To prevent that, the ancient Egyptians carved a turning groove into the rock base, very much like the notch cast into the concrete foundations at Four Quarters. The leading bottom edge of the stone was footed by this groove, which prevented the bottom of the obelisk from sliding as its top was elevated.

After the stone had been taken to its 25° angle, it was propped by stout poles. This work was performed before the ceremony and a yellow plastic safety banner around the area kept celebrants from getting too close to the leaning stones. Only experienced Stone People were allowed to cross this safety line.

Aided by a block and tackle, celebrants would pull the stones upright. Once the stone was made vertical, it was trussed with cables tied off to metal stakes which had been driven firmly into the ground. Each of the four supporting cables had a ratcheted winch attached, so that the cable could be pulled to a very high tension. (The ratcheted winch is colloquially known as a *come-along* in the United States, and is similar to the Tirfor winch in Britain.) Next, the Stone People furiously mixed concrete in a portable mixer and wheelbarrowed the concrete to the stone being worked upon. The concrete was dumped, screeded, and troweled with practiced efficiency. In forty-five minutes, the four stones went from their leaning positions to vertical and were concreted into place. The cables needed to be tensioned so that the stones would be held absolutely rigid for at least two days while the concrete cured. These techniques safely accomplished the goal of obtaining the maximum height of each stone above ground.

A notch or turning groove *is cast in the concrete foundation in order to foot the tall playing card stones as they are stood up.*

Orren Whiddon directs teams of twenty helpers to pull the stones into place. A block and tackle, tied up to a tree, gives mechanical advantage.

The stones are tightly trussed into place with winches tied off to metal stakes driven in the ground. They must remain rigid while fresh concrete sets around their bases.

Maybe the neighbor down the road would help you out on a barter or labor-exchange basis. It's worth asking. I'm all for keeping costs down. A stone circle should be *Mortgage-Free!*

EXCAVATOR

If your stones are getting up over four tons, a rubber-tired machine is probably not going to do the job for you. You need more lifting power.

Think of a really big articulated backhoe bucket on bulldozer-like tracks. This machine is known in the United States as a hydraulic excavator, in Canada as a "shovel," "digger," or even a "big digger" (Ed Heath), and in the U.K. as a tracked excavator. As an example, the John Deere 200LC or the Caterpillar 322BL each have the capacity to lift an 11-ton stone off the ground, and to lower a 7- to 8-ton stone into its socket from above, and even larger excavators are available, such as the Caterpillar 375L, with roughly three times the capacity of the 322BL.

On an excavator, the entire operating deck can swivel 360°, which is a great advantage, but can also be very dangerous to the unwary. Operators of this kind of equipment don't want anybody in their swing path, including you. Tell them what you want done, then stand back. They'll tell you when they need help, such as chaining a stone, or whatever. The "big digger" allows Ed Heath to create the amazing megalithic pieces featured in chapter 9, but it's his close relationship with the operator that makes it happen so well. The techniques are essentially the same as with a backhoe, but the scale of things is an order of magnitude greater.

FRONT-END LOADER

The front-end loader (or *wheel loader*) is just what the name implies. The business end, located in front of the driver, is a large loading bucket, which can be lifted high in the air for loading material into a dump truck. In addition to being an excellent load-ing tool, the machine can even be used to flatten a site (if there aren't any large boulders in the subsoil) and for gathering up stones and bringing them to site.

To give an idea of capacity, the Caterpillar 914G could lift a stone of almost six tons so that the top of the stone would be ten feet off the ground. The huge CAT 980G would increase these figures to nearly twenty tons, and twelve feet.

The front-end loader is not as nimble as the backhoe, nor as flexible. It is hard for the loader to pull a stone out of the ground or to extract one stone from a pile of several. However, an experienced operator could build a stone circle with a front-end loader, particularly if the stones are scattered around the site with working space between them. Stones can be chained to the big bucket, but enough of the stone needs to be exposed beneath the lower lip of the bucket to set the stone in its socket. Some loaders have a very high reach, and it might be possible to put the stone in the hole by using a megalithic hitch. As with backhoes, the lift capacity of the loader should be greater than the weight of the largest stone.

If I had to choose between a loader and a backhoe, I'd generally go with the backhoe. After all, the backhoe has the loader feature as well as the backhoe arm and bucket. The backhoe gives more lifting options, can dig socket holes, and can sort stones out better than the front-end loader. However, if you have access to a front-end loader at a bargain price, you should be aware that it can, in fact, do the job (except for socket holes, which, perhaps, could be done ahead of time by hand).

Bulldozer

A *bulldozer?* Yes, it's been done. The key ingredient is an operator who knows a trick or two with chains. Again, as with the loader, the stone must be pushed to an upright position. Next, the blade is raised high enough so that when the stone is chained to it, the operator can lower the stone into its socket. The

A large bulldozer was one of several different pieces of equipment used to set megaliths at the Australian Standing Stones. (John Tregurtha, photo.)

photo above shows a large bulldozer installing a major mehhir at the Australian Standing Stones, one of the featured stone circles in chapter 15. Again, a bulldozer wouldn't be my first choice, but if it's available for cheap, don't discard the possibility *too* quickly.

Most of the stone raising I've done at Earthwood and at Stone Mountain has been with heavy equipment. More and more, though, I have become caught up in the pleasures of moving large stones by hand, the subject of the next chapter. Still, I try not to be dogmatic, and will probably take advantage of a backhoe once in a while in the future, when there are compelling reasons to do so.

If asked what single piece of advice I would give to someone who wants to build with heavy equipment, I would say this: Find—and *keep*—the right operator. The right operator is one with experience in moving stones, but, even more importantly, the right operator is one with whom you can communicate.

CHAPTER 14

Building Stone Circles
by Hand

*Sometimes you just have to stand back and ask "Am I sure?" before
going on to the next move.*

—Cliff Osenton

We all draw lines on the question of technique. No modern person or group, to my knowledge, has yet built a stone circle as "a complete axe job," to borrow Cliff's phrase. The axe job requires going into the woods with an axe, and building a megalithic monument. It means making all the required tools and equipment (the axe, too, if you really want to be Thoreau), extracting the needed timbers and stones from the land, transporting the stones and timbers, and erecting the monument. That would make quite a book, all by itself. Imagine the variety of skills that would have to be learned: toolmaking, ropemaking, logging, woodworking.

The most "authentic" stone circle-building effort, so far, has been the Dragon Circle project in Surrey, described in chapter 10, and updated in this one. Even there, though, compromises had to be made. Ivan chose the stones at a quarry, and the stones were delivered to the site on large lorries. Once the stones were on site, however, the project moved forward without benefit of machinery. Ivan and the others involved in this project have catapulted themselves to the forefront of megalithic

heavy-lift engineering with the knowledge and experience that they've gained at Surrey. The camp itself has also given us a look at how a prehistoric society might have dealt with housing, sanitation, food preparation, and other support systems.

If you choose to go the manual route, great. You have my respect and I wish you every success. If all goes well, you will be rewarded with a special kind of satisfaction. But the chances are that you will have to make some modern compromises. Don't let this bother you. Unbending dogmatism is rarely a virtue, and it can sure make life difficult, even intolerable, for yourself and perhaps for others. So draw a philosophical line, if you like, but don't become a slave to it. Be gentle on yourself. Be happy.

The first question to ask yourself is whether or not your state of physical conditioning is up to the task. No, you don't have to be a big strong Ivan McBeth–type of character. Even Ivan moves most of a stone's weight with his brain, not his brawn. What you *do* need is a basic level of conditioning and muscle tone, not so that you can lift or push *heavier* loads, but so that you don't strain yourself moving the *same* load. Even couch potatoes can tone muscles a great deal with just a month's weight training. Start any such program gradually and work your way slowly to heavier weights. Strength is not the goal, although it will probably increase by 20 percent or more, particularly if you have not been very physically active for quite a while. Muscle tone and overall conditioning is the goal. Before embarking on a program, get advice from a physical trainer or consult books, videos, or magazines that deal with basic fitness. See a doctor if you have a pre-existing condition that might throw a monkeywrench into a training program . . . or into the megalithic work itself.

Stretching exercises before physical activity are also very effective in reducing the chance of muscle strain and damage to other connective tissue. You should never have to actually *lift* heavy things in this business. Roll, walk, tilt, or lever things into place. In fact, a large lever itself might be the heaviest lift you have, and even that can be done one end at a time to great advantage. But when you do lift anything, please, remember to keep your back vertical.

And if you have a bad back already, supervise! (Take my advice, I'm not using it.)

Raising stones manually requires certain equipment, albeit of the simpler, gentler, preindustrial kind. We'll start with the most basic tool in your kit.

THE LEVER

We'll get to physical properties real soon, but, first, it is good to know a little about lever theory.

Levers come in three classes. A *class one* lever has the fulcrum between the *load* and the *effort,* as in the normal use of a crowbar. The early pages of chapter 7 discuss how a class one lever works, how to figure its mechanical advantage, and how to obtain compound lifting advantage by Cliff's method of working close to the axis of the stone. I need not repeat those discussions here.

A *class two* lever, however, places the load between the fulcum and the effort, as in the diagram opposite. A wheelbarrow is a typical example of a class two lever. Sometimes this type of lever is useful, to move the end of a recumbent stone, for example, or to pry a smaller stone away from a larger one. Be careful, though, as a class two lever involves genuine lifting, which can put quite a strain on the vertabrae, whereas class one levering usually involves nothing more strenuous than leaning on the effort arm.

A *class three* lever places the effort between the fulcrum and the load. The classic example is a baseball pitcher's arm: His elbow is the fulcrum, the ball in his hand is the load, and the effort is supplied by the muscles and tendons between the elbow and the hand. A good effort can project the small ball at high speed. A class three lever is not much help to a megalithic engineer, who is looking for slow and easy movements, but it is very useful if you happen to need a catapult.

At Surrey, the intent was to employ methods that could have been used by the ancients. This precluded the use of iron levers, although one or two were used by Cliff one day to quickly maneuver little Titch along the megalithic railway. If the reader has no qualms about using an iron bar, then more power to you, literally. A six- to eight-foot iron bar can provide incredible mechanical advantage (m.a.), because the effort arm can be many times greater than the load arm. Let's use as an example a 6-foot (72-inch) bar set up as a class one lever, with the fulcrum placed just 4 inches from the edge of a stone to be lifted. The effort arm in this case is 68 inches, giving an m.a. of 17 (68/4 = 17). If the load arm is as low as 3 inches, an advantage of 23 can be realized. With a 2-inch load arm, the m.a. is a whopping 35. Do the math. Even without compound lifting, a 200-pound man could lift 7,000 pounds! (Ever so slightly.) With wooden poles, it is virtually impossible to get this degree of m.a.. Not only is the thick wooden lever difficult to position at the close tolerances required, but, with m.a.'s above 12

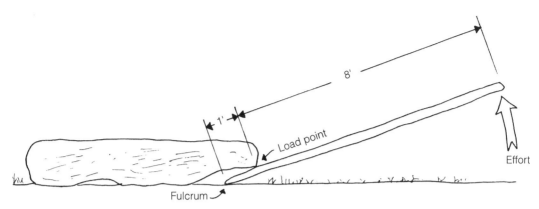

With a class two lever, the mechanical advantage is determined by dividing the distance from the effort to the load point by the distance from the load point to the fulcrum. In this example, the m.a. is 8.

or so the lever could easily suffer shear failure (discussed below) just above the fulcrum.

I like a bar with a chisel-type point, so that it can be worked between a rock and a hard place. Once the stone is lifted a little, it can be held with wedges while the lever and fulcrum are repositioned. Always have plenty of good hardwood wedges on hand. Twenty is plenty. Two bars are often useful, with one leap-frogging the other to gain a better purchase or advantage. More than two levers should not be necessary.

Six-foot-long chisel-point iron bars can be bought new for about $30 to $35, and you might get a good bar cheaper at a farm auction. Some of the best ones are quite old. Dave Brandau of Standing Stones Perennial Farm, whose work graces the next chapter, inherited a great 8-foot bar from his grandfather, and it has greatly eased the way for Dave to erect the wonderful menhirs at his gardens.

Wooden levers should be made of a hard wood having good bending strength as well as good shear strength. *Bending failure* is almost self-descriptive. When a lever bends so much that the tensile strength of the wood fibers is exceeded, the timber snaps. With a class one lever, this will usually occur between the fulcrum and the effort, although bending failure could occur if the "short arm" of the lever—the one between the fulcrum and the load—

is too long. *Shear failure* is the term used to describe all or most of the wood shearing perpendicular to the length of the lever. Usually, this occurs very close to the fulcrum.

Oak is a good choice for a lever, having excellent hardness, shear strength, and bending strength. Other hardwoods are also good, such as ironwood, beech, and ash. I recently made a number of levers of different sizes out of straight hard maples, the tops of which were broken off during an ice storm. They have done the job, so far. Pines vary a great deal in hardness from species to species. Some of the pressure treated pine fence posts have excellent lever characteristics, and they will last a long time. Use a mask and eye goggles when cutting pressure-treated material, as the sawdust is particularly nasty. Hemlock, by the way, is a poor choice for a lever, as it has low shear strength. Very soft woods, such as basswood and poplar (quaking aspen), should not be used. Wood with obvious deterioration already present, such as surface rot, should be rejected out of hand.

Wooden levers can also benefit from a chisel point, because it is easier to fit such a shaped end under a stone.

No discussion of levers is complete without a word about the fulcrum, the point upon which a class one lever pivots. A fulcrum is called upon to take tremendous concentrated loads, hundreds of pounds per square inch in some cases, so the primary characteristic we're looking for is hardness, which can also be thought of as resistance against crushing. Again, oak and ironwood are good choices, but a metal or stone fulcrum can also be used. If stone, use hard stones like granite, gneiss, basalt, and dense sandstone. Avoid schists, soft sandstones, and the like, which will crush.

In cross-section, the ideal fulcrum should have the shape of an upper case letter D lying on is side. A cylindrical log sawn almost down the middle will yield two good fulcrums. A little trick is to cut or split the log an inch off center, so that the two resulting fulcrums have different thicknesses. Some-

The chisel point makes the wooden lever easier to manuever in tight spaces.

Under the direction of Ivan McBeth, wooden levers are used to elevate a large stone a few more inches so that rollers can be installed. Two levers are being used, side by side, but the near one hides our view of the one next to it.

times one will work when the other does not. If the log is split with an axe, make sure that the split is clean and straight. A twisted split will make for wobbly fulcrums. A fulcrum with a good base will distribute the stone's point load onto the surface below, which might be the ground itself, or other timbers. These other timbers can be softwood, as they are not subjected to great concentrated loads. If the ground is soft, timbers will be needed to distribute the load over several square feet; otherwise, the lever will simply drive the fulcrum down instead of lift the stone up.

Practice your levering technique with smaller stones before going after Big Bertha.

ROLLERS AND STONEBOATS

If you've decided to build by hand, you'll need to be able to transport stones, if only fifty feet or so across the site.

In chapter 2 I describe how we moved the two-ton South Stone at Earthwood. Eight of us struggled to move the stone sixty feet in an hour. As I learned at Surrey, and at Four Quarters, we would have been more successful with better rollers and better ground preparation. In Surrey, using large rollers of about twelve inches in diameter and eight feet long, we were able to roll Center, weighing almost seven tons, to the lip of its socket in just a few minutes. Granted, we had thirty-three people pulling on ropes, but this was probably more than necessary. The real keys were the level ground and the large rollers.

The *stone boat,* also called a *stone sledge,* was a common piece of equipment in Britain and in North America up until recent times. Many different designs have been used over the centuries, and a stone boat can be as simple as two short planks, such as we used in Surrey to move Titch, all the way to a well-made sledge that a farmer will use over the years to transport large stones to the edge of his field. With an axe, Cliff Osenton built a simple stone sledge of tree trunks for his Preseli Mountains stone-

moving project in 1998.

While Cliff was in Wales, a documentary was being made for a British TV network, showcasing various efforts to figure out how the bluestones might have made their way along the "Stonehenge trail." Cliff's work was featured in the program, as well as the work of others, including the interesting if modernistic method shown on the next page.

The stone sledge takes the irregularities out of the stone, making it easier to slide or roll along the ground. If your stones have been delivered to site and are close to their sockets, it's probably not worth making a sledge. But if you need to transport awkwardly shaped stones some distance, either by hand, or, for example, across a field with a tractor, the stone boat is worth constructing. Again, you must decide on the degree of "authenticity" that you wish to employ on the project.

At Surrey, we rolled the 1.75-ton Titch into place very rapidly on Cliff's Megalithic Railroad, described in chapter 10. In this case, laying wooden tracks ahead of the work, placing Titch on a simple stone boat made of two 3 x 10 planks, and using truly cylindrical rollers, caused the stone to fairly fly to its destination. The downhill slope didn't hurt any, either.

Providing that the trackway is wide enough, long rollers are preferable to short rollers. Just as with larger diameter rollers, longer rollers lessen the concentrated load on the ground, making them easier to roll. Also, the longer rollers are less affected by slight undulations in the ground, although they will still be vulnerable to blockage by a hard bump on the surface, such as a rock or tree stump.

I am aware that there is a slight controversy among experts about whether or not rollers were used in neolithic times. My personal view is that it would be extraordinary if these intelligent people did *not* discover the benefit of rollers quite early in the game. Some experts say that rollers would crush under the weight of very large loads, such as Stonehenge-type megaliths. With a large stone boat

Cliff made this large wooden sledge "as an axe job, right back to the basic techniques." Later, the three-ton stone was dropped almost perfectly into its socket right off the elevated sledge. (Cliff Osenton, photo.)

This large and very regular concrete block was rolled along the Welsh countryside by a number of strong lads who unwound the ropes binding the "stone" to carefully selected logs. The picture is included for interest's sake. It is hard to imagine ancient ropes—or even modern ones—lasting very long subjected to this punishment. (Cliff Osenton, photo.)

to distribute and "soften" the load, and wooden tracks to "harden" soft ground, I am convinced that oak rollers would be up to the task.

OTHER HAND-POWERED EQUIPMENT

Ron and Margaret Curtis, you may recall, took down an improperly placed stone at Callanish VIII, and re-erected it in its ancient socket a few feet away. They accomplished their task with help from just two men and the kind of seat-of-the-pants engineering that Cliff speaks of in chapter 7. You do the job with the equipment you have available. From long experience, Ron has developed a sixth sense for this sort of work.

In Scotland, the Curtises used ropes and a pair of ratcheted Tirfor winches, similar to the American come-along. To save us all the proverbial thousand words, please turn again to the photo on page 182, which shows how they made use of the most rudimentary unmotorized modern equipment.

Another great tool, older than the come-along, is the *block and tackle,* a rope looped several times through two wooden pulley blocks. Each block has a strong hook on it for fastening a load or for hooking it to a fixed point. The mechanical advantage of the block and tackle is equal to the number of lengths of rope connecting one block to the other, minus friction. My block and tackle, shown below, has a mechanical advantage of almost six. As an example, if my Winter Sunset Stone is leaning away from the other stone in the picture, I can fasten my block and tackle as shown and pull the top of the Sunset Stone straight with just one-sixth of the effort required by a rope alone.

The author is able to straighten the Sunset Stone thanks to the mechanical advantage gained by the block and tackle. The Sunrise Backsight Stone is on the right.

The Ancestor

In November of 1998, using manual techniques, we moved and erected a spectacularly beautiful but awkwardly shaped 1,700-pound stone that we call the Ancestor. His history is a matter of speculation, but he appears to have derived from lava that rose up from the netherworld through a fissure in the anorthosite bedrock, an unusual geologic formation called a *dike infusion*. In fact, the geology class at nearby Plattsburgh State University visits Murtagh Hill each spring to study our dike infusions. The one that produced the Ancestor was composed of basalt, one of the densest of stones at 190 pounds per cubic foot, and was probably blown up with dynamite about seventy-five years ago when the quarry was in operation. The Ancestor has part of a blasting hole still intact near his belly.

The stone was brought up the hill on a lowboy, and deposited a hundred feet from its intended location. It would be the primary stone of a sculptural group forty feet south of the stone circle. An Irishman, Brendan Kelleher, was living on the Hill at the time, and took great pleasure in helping me with projects megalithic. Together, we built a stone boat composed of long 4 × 8 runners and a plywood deck. We used two iron bars to lever the Ancestor onto the boat and used wedges to stop him from wandering around the deck in case of rough seas. We made rollers from 6-inch-diameter poplar logs, each about 40 inches long. Then, on a cold, gray, but dry Saturday, we assembled a team of hearty souls to move the Ancestor and stand him up. Two of the volunteers were friends from downstate, Scott and John, who had come up that day on cordwood masonry business, but got caught up in the excitement of stone building.

First, the team rolled the Ancestor on tracks made of long 2 × 6 material. Then, as a test, we tried moving the stone without tracks. Ground conditions were firm, even when we rolled across the front lawn, and

We made a custom stone boat for the awkwardly shaped Ancestor. But, with firm ground underfoot, it made little difference if we rolled the stone on tracks or straight along the grass.

The Ancestor is walked up the ramp by little steps, first one side, then the other, a bit like a duck.

the tracks were not missed in the slightest. In fact, it actually seemed easier to control and steer the stone boat without them.

We used both iron and wooden levers to remove the Ancestor from the stone boat and stand him up on his good flat feet, still twenty feet from his one-ton anorthosite base stone, placed by machine several days earlier. Once the stone was standing, we lashed a 2 x 6 to his midsection so that we could "walk" him, as we had done with the Sunset Stone at Stone Mountain two years earlier. We turned the stout stone boat over and used it as a ramp to march the Ancestor right into position. Scott and John took one end of the bar, and Jaki and our neighbor Pete handled the other end. Brendan's job was to resist the stone's natural inclination to fall over backward, while I took pictures and chanted instructions. "Now, up with Pete, down with Scott! Good! Now, up with Scott, down with Pete!" Somebody has to do it.

I applied Acryl-60 bonding agent to both the foundation stone and the Ancestor's robe-covered feet, and mixed up a half bag of Sakrete concrete mix. The team rocked the Ancestor up and down on wedges while I applied concrete between the menhir and his anorthosite foundation stone, and the ancient one stood proudly without assistance. The wedges were knocked out with a hammer and I smoothed the half-inch-thick concrete joint with a mason's pointing knife.

We were all amazed at how easily the project had gone, but I knew that fifteen years earlier, I would have struggled with a similar task. There is something to be said for experience after all.

John and Scott had come north with heads full of cordwood, and on the return journey, I am told, spoke of nothing but finding and setting up their own standing stones. Megalithic work will do that to you.

And, as Ed Prynn would say, "It were a pleasurable afternoon!"

The block and tackle can be useful for moving a stone short distances and for bringing it to a standing position. A necessary ingredient, however, is an immovable object to tie one hook of the block and tackle to, such as the base of a large tree or the base of a well-planted standing stone. Be careful, though. The limiting factor is still the strength of the loop tied to the hooks themselves. For that reason, use a chain or, better, a strong nylon strap (not a rope) around the stones and back to the hooks. Also, a nylon strap will protect a tree much better than a chain, cable, or rope.

STANDING STONES UPRIGHT WITH LEVERS

Experience has shown that levers have certain limitations when it comes to standing stones upright. It is a relatively easy matter to roll stones into the correct position to dump them into their socket holes. The problem has been getting them to land in the socket at an angle great enough so that they can be easily pushed or pulled to the vertical position. At the Dragon Circle, the first fourteen stones consistently ran into the opposite side of the socket before the bottom of the stone hit the bottom of the hole. Invariably, the stone would get hung up on the far side of the socket at an awkward angle of about 35°, a very difficult angle from which to maneuver it upright. The primary difficulty with levering from this position is finding something solid enough to lever against. Many drawings have appeared in books over the years speculating on how stones *might* have been stood up. Almost all of the theories speculate on the creation of a wooden platform next to the socket, for the purpose of creating an elevated leverage station for workers to stand on and for the placement of a fulcrum. Margaret Curtis's drawing from *New Light on the Stones of Callanish* illustrates this theory quite well, and is reproduced here with permission. But her book reminds us: "There is no archaeological evidence at Callanish to

support this hypothetical description which is based on research at Stonehenge and elsewhere" (p. 23).

The stone in Margaret's drawing has the so-called playing card shape: rather narrow and, therefore, relatively light in weight for its height. The method illustrated might have worked with this type of stone. I have doubts about levering really heavy stones from such a wooden platform. The quantity of hewn timbers would be extraordinary: a mile of hewn six-by-sixes, for example, to build a sixteen-foot-square platform ten feet high. Small platforms, as we have discovered in the field, tend to "kick out" away from the stone when leverage is applied to a stone at an angle of 30° or greater. At the Dragon Circle, the 3,300-pound Titch provided a great backstop for levering for a while, but as Chieftain went higher, the additional blocking between Titch and Chieftain became unstable. Other methods were employed to bring Chieftain to a standing position, described below.

Lifting a stone horizontally, as discussed in chapter 7, is actually much easier than tilting it upright. The earth itself makes a good solid place to stand and to lever against. Even as small wooden stacks are constructed to go higher, the reactionary thrust on the fulcrum is vertical, so the stack remains stable. With standing up a stone, the reactionary thrust is at an angle, downward and outward, and the stack tends to collapse. At Surrey, we discussed the idea of having a much larger support stone than Titch, something blocky in shape and weighing six tons or more, which could be rolled from stone to stone and used as a massive portable leverage platform. Once a large stone is on rollers, it is relatively easy to move around the outside of the circle, particularly in combination with a wooden track system. Later, it could be the last stone to be erected, perhaps in the middle, or it could be used as a recumbent stone. In fact, some of the Club Meggers speculated that the recumbent stones in the Grampian Region of Scotland could have been used for exactly that purpose. They certainly have the right shape.

A socket hole has been dug and partially lined with short stakes. The stone has been brought into position.

The sledge and rollers have been removed. Workers begin levering at the top of the stone.

As the stone is raised, so the leverage platform is built up with logs. The stakes prevent damage to the edge of the hole.

When the leverage platform is high enough, ropes are attached to the top of the stone. Ropes are used to prevent the stone falling sideways as well as to help pull it erect.

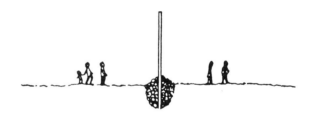

The stone is held in position with ropes while the hole is filled with packing stones and lumps of clay.

The stone is now capable of withstanding Lewis gales for four thousand years, despite the fact that little of it is beneath the surface.

Margaret Curtis drawing.

Consider an unsocketed stone for a moment. Thankfully, lifting one end of a 10-ton stone does not require a 10-ton lift. In the diagram on the next page (top), the end of the stone is just beginning to be lifted. At this point, only half the weight is being lifted by the lever, because the other half is being supported by the ground at the stone's other end. As the stone rises, the actual lift (and, therefore, the effort) becomes less and less. The shaded area of the stone (center), represents offsetting weight, effectively lightening the stone. More and more of the stone's weight is being supported by the end still on the ground.

Working against us, though, is the awkward levering angle, already discussed. The toughest part of the lift is from about 30° to 70°, and this varies somewhat with the shape of the stone and how its weight is distributed. After 70°, most of the weight is supported by the ground, not the levering platform, and it becomes a very serious concern that the stone does not get away from us and continue right on past center and back onto the ground beyond. A sling around the top of the stone, manned by plenty of people, is probably the best way to prevent overshooting the mark. Or, loop the rope around the base of a stout tree trunk or pre-existing menhir to

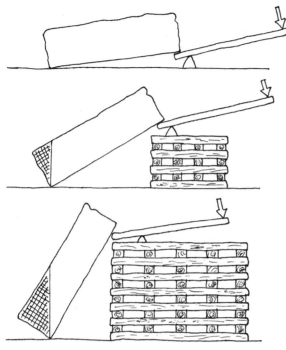

Large, awkward stack

A stone being lifted by its end gets effectively lighter as its angle increases, because the vector of force is being transferred ever more to the ground. Working against us, however, is that levering becomes ever the more awkward as the stone is raised.

aid in slowing the stone's gathering momentum. Put scrap wood or rags around the tree to protect its bark from being pulled off. Yet another possibility is the construction of a heavy timber stop to prevent the stone from going on past center, like the buttresses used at the end of railroad spurs.

At the Dragon Circle, at least through the first fourteen stones erected, it proved to be difficult to get the stones to land into their sockets at an angle greater than about 35°, leaving the team with difficult handwork to maneuver them upright. We'll look at some of the ways they have dealt with the problem of raising a stone from an unfavorable angle. A little later, we'll look at ways to get the stone to land in its socket at a steeper angle.

The Spanish Windlass

I had to return home from Surrey before the awkward Chieftain was fully upright. After conventional levering had taken the stone as far as it would go, Ivan, Cliff, the Curtises, Balin, and the others came up with new ideas to straighten it up. In phone conversations, the people I spoke with referred to the use of a "Spanish windlass," but, in fact, the device they created really works more like a turnbuckle. The following photos show what they did.

A nylon sling was passed around the top of Chieftain and fastened to the base of Center by two strong ropes, made as tight as possible. An 8-foot-long pole was inserted along the doubled rope length, midway between the megaliths. As the pole was turned by the crew, the rope tightened and tried to pull Chieftain up vertically toward Center. Center, of course, was firmly socketed four feet into the ground. At the same time, Titch was repositioned closer to Chieftain, and Cliff set up a vertical lever with a rope attached near the top. Some of the crew pulled on this lever while others turned the "Spanish windlass." It was discovered that bouncing on the twisted rope also encouraged Chieftain upward, ever so slowly.

Two problems occurred with the windlass method. First, after a few stones were straightened in this way, Center became quite loose in its socket, like a giant wiggling tooth. Second, as the windlass is tightened, a great deal of pent-up energy is created. Once, the large turnbuckle arm suddenly slipped, injuring one of the helpers. Greater caution was taken after this accident, and the windlass method proved to be a valuable tool throughout the long period during which the stones were landing at awkward angles in their sockets. However, it is my opinion that the "windlass method" was probably not used by the ancients, at a time when rope was no doubt even more valuable than it is today, and probably not as resilient as our modern varieties. It is more likely that the neolithic builders learned quite early on how to drop the stones in straight.

Chieftain was large, awkward of shape, and "seemed to be stuck in a groove" according to Ron Curtis. Levering at this angle is extremely difficult. (R. & M. Curtis, photo.)

The windlass consisted of a long, strong doubled rope tied from the base of Center to the top of Chieftain. The rope was tensioned by turning the 8-foot-long pole located halfway along, which, in combination with continued levering, managed to pull Chieftain straight . . . almost. (R. & M. Curtis, photo.)

A Giant Lever

The windlass, in combination with levering, had pulled Chieftain vertical in the direction that had been giving all the trouble (toward Center), but its awkward shape had thrown another problem at the team. During the windlass and levering operation, the stone had taken a decided lean to the side, and levering alone wouldn't bring it back. The brain trust came up with a new idea, using the principle of the class two lever.

A very large stone laying about fifteen feet from Chieftain served as an excellent fulcrum for the giant lever that the crew constructed. (See below.) The lever itself was one of the 36-foot-long poles that we'd cut in the forest. The load, of course, was Chieftain, and to gain the greatest mechanical advantage, a sling was tied near the top of the stone, and fastened to a point on the lever about six feet from the fulcrum point at the large recumbent stone. This left about thirty feet of lever for people to push against. The mechanical advantage of the lever, therefore, was about five to one (30/6 = 5). By tying the sling near the top of the stone, Chieftain himself became another class two lever with a mechanical advantage much more difficult to determine. The compound advantage, however, was such that a team of eight or ten could easily straighten the top of the awkward Chieftain.

THE PROBLEM

Instead of having to wrestle with the stones from an awkward 35° angle, wouldn't it be nice if the stones would land in their sockets at a steeper angle? The Dragon Circle team tried dropping the stone off of rollers and letting it pivot on the edge of the socket. In both cases, when the stone wanted to move, its horizontal momentum was so great that it hit the far side of the socket, causing what I call "the Problem."

In my search for a solution, I gained a clue from an experiment done in England in 1995 as part of

A compound class two lever was constructed from available materials on site. (Margaret Curtis, photo.)

a TV series called *Secrets of Lost Empires,* a NOVA Production by the WGBH Science Unit and BBC-TV. In a program about the building of Stonehenge, structural engineer Mark Whitby was called upon to devise methods of construction that could have been used by the ancients. Archaeologist Julian Richards was assigned the task of assuring that the methods chosen would have been available to people four thousand years ago. A book and a video about this project are listed in the bibliography.

The specific project chosen was to demonstrate ways by which the largest stone at Stonehenge, Stone 56 of the great trilithon, could have been transported and erected. An accurate 40-tonne (45-ton) concrete replica of the stone was made. Whitby anticipated the problem of the 29-foot-long monolith getting hung up at an awkward angle, and came up with a plan to try to prevent that from happening. First, he elevated the approach to the socket by constructing a long gradual ramp. Pictures in the video's companion book indicate that the end of the ramp next to the socket was between four and five feet high. (The pictures also show that the end of the ramp was constructed of a massive tilt block of concrete. I'm not quite sure how Richards, the archaeologist, allowed this bit of poetic license.) Second, Whitby assured a rapid drop into the socket by a rather ingenious technique, one that many people—including Cliff Osenton—have criticized as being a 20th-century solution to a 4,000-year-old problem. Still, it is worth relating. Instead of lifting the top of the stone with levers, Whitby's idea was to increase the effective weight of the stone as it was perched over the socket, so that it would fall quickly into the hole. To accomplish this, he put six tons of concrete slabs on a wooden sledge on top of the great monolith. (Hearsay and visual evidence from the book and video suggest that this was done with a crane.) Workers could pull this heavy sledge along greased wooden tracks, a method of stone transport that Whitby had already tested successfully. The replica was positioned with its center of gravity just

behind the edge of the hole. As volunteers pulled the sledge toward the replica's bottom, the center of gravity shifted, and rapidly. The replica quickly pivoted on the fancy concrete end of the ramp, and the stone landed brilliantly at about 70°.

While I agree with critics who say that the test applied 20th-century engineering to an ancient problem and could only work with a very flat stone like their replica, it is impressive to watch the event on video. The lesson that I came away with was the importance of elevating the stone above the socket, and of rotating it quickly on its balance point, and with an *accelerated* lift, discussed below.

A Solution?

To test my idea, I experimented with a 440-pound stone, a perfect little model of a typical Dragon Circle menhir. First, I simulated the basic method that had been used at Surrey: rolling or sliding the stone to the edge of the socket, and then lifting the top edge with levers until the stone moved forward. As the effect of scale is important in such experiments, I wanted to first find out if I could duplicate "the Problem". I dug a socket 13.5 inches deep, equal to one-third of the 40.5-inch height of the stone, and, aided by Jaki, Darin, and surveyor friend George Barber, rolled the stone so that it was overhanging the socket. Gradually, we levered the top of the stone with an iron bar, until the stone started to shift forward. The stone got caught on the back of the socket at a 50° angle, a very awkward angle from which to pull it to the vertical.

Using a tripod, George's come-along, and a megalithic hitch, we extracted the stone from the socket. (Note: Tripods, or quadrapods, in combination with ropes, pulleys, or a chainfall, are quite useful in raising even large stones manually.

For our main test, we elevated the same stone onto a platform, using levering techniques we'd learned from Cliff. The platform height, including the 1-inch rollers, was 8.5 inches above ground level. Next, we slowly and carefully rolled the stone for-

On the control test, the stone landed at this awkward angle, typical of "the Problem." We extracted the stone with a tripod, a come-along, and the megalithic hitch.

your hand accelerates to the ground. George understood the importance of this acceleration on the lift: The stone would never get the chance to even think about moving laterally; it would simply pivot on the last roller at the edge of the hole. This was our theory, and, when all was ready, George began his lift, a slow lift that constantly accelerated, like a large rocket leaving the launch pad. The top photo on the next page shows the stone just after George released it. Note that its motion is predominantly vertical, not horizontal. The stone landed at about 80°, and actually straightened to vertical, before settling back to 75°. Had helpers been ready with pieces of jamming timber, the stone could have been held at the vertical position with ease. As it was, it took virtually no effort to stand the stone upright, where we jammed it in as a memorial to our experiment.

We celebrated with glee! Our experiments with the 440-pound stone had exceeded expectations. Everything was kept in scale with the Dragon stones.

ward along the platform until its balance point was on a roller right at the edge of the platform next to the hole. Little wooden shims in front of the rollers prevented the stone from going too far too soon, but, in case it did, we had built a safety platform over the hole, as shown in the diagram on the next page. This safety block, another good idea from Cliff, actually saved us once, when the stone prematurely started its descent. When the stone was finally perched over the hole at very close to its balance point, a child could tip it in, but we chose seventy-year-old George to do the honors, because he had a clear idea of *how* to perform the lift to prevent "the Problem."

I'd watched the NOVA video over and over again, at regular speed and in slow motion, and noticed that the effect of the counterweight, sliding along the Stone 56 replica, actually was to *accelerate* the drop into the socket, just as a rock dropped from

We positioned the test stone on a platform with a height about five-eighths as deep as the socket hole. Fairly uniform rollers allowed us to roll the stone easily along the platform.

To duplicate the experiment with a 12-foot stone and a 4-foot socket, the platform would have to be 2.5 feet high, and about as long and as wide as the stone. Rollers would have to be regular in size, at least 5 inches in diameter, like the rollers Cliff used to roll Titch into place. Instead of a man standing on the platform and lifting it by himself, I surmised that a long pole could be passed under the top end of the stone, so that two or three people could lift each end of the pole. Assuming that the stone was positioned so that only a 200-pound effort could lift it, it would be an easy matter, I thought, for four to six people to apply an accelerated lift on the bar and tip the stone almost vertically into its hole. Other helpers could be ready with large timbers to jam the stone in a truly upright position. A safety should be erected across the socket in case the balance point is exceeded during set-up.

I was so excited by our success that I started to call some of the Club Meggers. I reached Ed Prynn first, who listened very carefully to my report. Then he said, "You know, Rob, that's exactly how the granite fence posts were stood up. Some of 'em was 15 hundredweight (.75 tonne). The stone was loaded

George flips the 440-pound stone into its socket with the greatest of ease.

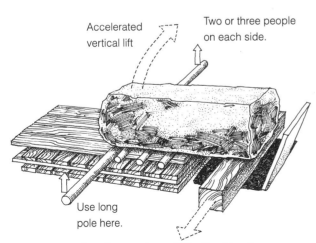

Accelerated vertical lift

Two or three people on each side.

Use long pole here.

Practice accelerated lift without load!

Roll a stone to near the balance point. A safety block spans the hole in case of misjudgment. With the block removed, the top end of the stone can be raised with an accelerated lift (Michael Middleton, drawing).

Success! After landing almost vertically, the stone rests at a favorable 75° angle, from which it is easy to maneuver.

onto a farm cart and the horse would back him right up to the hole, and the men would lift him right in, straight up, just like you said." Ed's boyhood memories confirmed that the technique would work with stones four times larger than my model. I called Ivan in Surrey, who was still raising a stone each month. I reached him too late for the November raising, but he assured me that he would try my idea at the next opportunity. I faxed him a description and the illustrations that appear in this chapter. In February of 1999, I spoke with Ivan, Alan Connett, and others to find out how the full-scale test had gone.

The news was disheartening. The stone had been elevated, but not nearly enough, and the lower surface of the stone chosen was irregular, making it difficult to roll the stone toward the socket and to the critical balance point. In the event, six men had all they could do to lift the stone; they could not accelerate it. The stone rose slowly and, at the critical moment. lunged forward and got caught up again. Another failure. Or was it? Some of the team were able to see the problems and felt that they could make the system work. What's more, Ivan, the chief, was willing to have another go. But work slowed during the winter, as weather conditions deteriorated. The long Easter weekend of early April would be the next opportunity.

I got back to my writing and watched the calendar.

THE A-FRAME

Another interesting idea to came out of the NOVA television series was the use of an A-frame to help straighten the stone to a truly vertical position. The replica of Stone 56 was very tall, about twenty-one feet out of the ground, so even though Mark Whitby and his team did well to get the massive monolith to fall into the socket at 70°, it was still quite a task to straighten it up. To assist in the straightening process, Whitby built a large timber A-frame to increase the leverage of his pulling team. The diagram below shows the process schematically. In essence, the A-frame served as a vertical class two lever, increasing the m.a. to perhaps four or more. The system worked. In the video, it appears that about ninety volunteers were able to pull the stone straight. Other volunteers packed behind the replica as it went up.

LINTELS

The only modern stone circles that I know of with lintels joining the tops of two uprights have been built with heavy equipment. Individual monuments such as the Thor's Gate trilithon at Columcille and Ed Prynn's Angel's Runway dolmen in Cornwall were built with cranes. Ed Heath in Quebec has

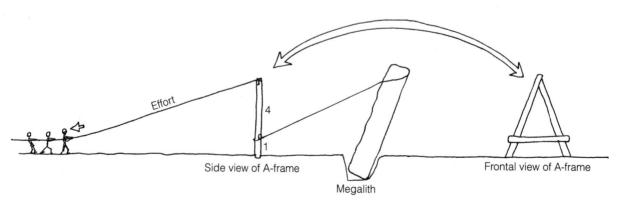

The A-frame works as a class two lever to increase the team's pulling power on the stone to be straightened.

built several trilithons and dolmens using a large excavator. The only lintel that I am aware of that was installed "by hand" was the concrete replica of the Great Trilithon built by Mark Whitby and others for the NOVA television series *Secrets of Lost Empires*. Using old railway ties (or *sleepers*) for stacking material, the team experimented with the timber crib method popularized by Richard Atkinson in his book *Stonehenge,* and something like Margaret's drawing on the next page. This method worked quite well in lifting the nine-tonne (ten-ton) lintel off the ground a few feet, but the team ran out of railway sleepers, so they were not able to find out about the kinds of problems that occur twenty feet off the ground. Author Cynthia Page, writing in the companion volume to the series, also called *Secrets of Lost Empires,* describes what happened next:

> Despite the fact that Julian Richards regards the timber crib as the more likely method, Mark felt it would be more dramatic to drag the lintel up a ramp. Again, budget and time considerations meant that this ramp could not be built using earth, so modern scaffolding tubes were used as a substitute. It was a huge ramp and it looked dreadful. And it was made even worse because all sorts of handrails were needed for safety reasons. It certainly did not have a prehistoric look about it. But the basic idea was the same, and Mark was not, after all, trying to prove that ramps could have been built by the people of Stonehenge. (p. 40)

Curiously, the narrator in the video says that British safety officers insisted on the steel scaffolding.

I relate this story for interest's sake. The NOVA production had a big budget, expert engineers, and lots of volunteer help. You are not advised to try this at home. My strong suggestion is that if your stone circle design calls for heavy lintels perched on uprights, do the work with a crane or a large excavator, and stand well clear of the action.

FINAL THOUGHTS ON CONSTRUCTION

The choice of using modern equipment or not is up to the individual, but, either way, safety should be the paramount consideration. Until they are solidly jammed in their sockets, these stones must be considered dangerous. Unsocketed stones are also dangerous, and should be limited to stones having large flat horizontal bases.

The order of erecting stones in a circle is a personal decision. You can build clockwise around the circle or erect stones here and there based on convenience. Builders with a mystical outlook strongly counsel against building "widdershins," or counterclockwise. Ivan and I have both built stone circles by beginning with the four directional "corner" stones, and we've also built them in a more random way. The most important consideration here is to not "paint yourself into a corner." Make sure that the order of events never closes off important lines of access.

Part 3 does not stand alone in its commentary on building stone circles. Many important building techniques have been discussed parts 1 and 2, so if, in your impatience to get building, you've started right in at this part, you are advised to go back and catch up on all you've missed. All the chapters are a part of the whole picture, but, in particular, there is lots of building detail to be gained from chapters 2, 3, 7, and 10. In addition, certain "traditional" theories of megalithic building are discussed in books listed in the bibliography. And there are even some design and building ideas in the next chapter, "A Guide to Modern Stone Circles."

Remember, too, that many old-timer farmers and experienced contractors know as much about moving heavy stones as "the experts." Seek their counsel and help.

Above all, have fun. If building a stone circle becomes a chore, you're doing something wrong. Take a break and read part 2 again.

SPRINGTIME IN SURREY:
AN EPILOGUE TO PART THREE

The work on this book was winding down and I was past deadline, but it bothered me that we had not yet solved "the Problem" on a large stone, and I couldn't do anything about it in the frozen north of New York. But Surrey, now there's a temperate climate . . .

I'd known for weeks that a long Dragon Circle camp weekend was planned for Easter in early April of 1999. Apprehensively, I called the barn at 5 P.M. on Monday, April 5, and spoke with John, one of Ivan's assistants. After learning that Ivan was out for the evening, I popped the question.

"How'd the stone raising go, John?"

"Great. Just perfect. We stood a huge stone up within a few degrees of vertical! Bloody marvelous!"

John told me some of the details and Ivan phoned me two hours later with the rest.

The stone chosen for the new test was Whale, 10'7" long and marked as 8.31 tonnes, although Ivan had reason to believe that the actual weight was a tonne greater, or about 10 U.S. tons in all. The stone team felt that they could get Whale to drop into the planned 3'6"-deep socket if they elevated the stone high enough and found a way to get the accelerated lift. They came up with several new ideas to add to the mix, and the combination was spectacular.

First, the two 36-foot-long pine poles were set as rails on either side of the socket location, although the hole was never dug until after the stone was perched in its final position! With no hole in the way,

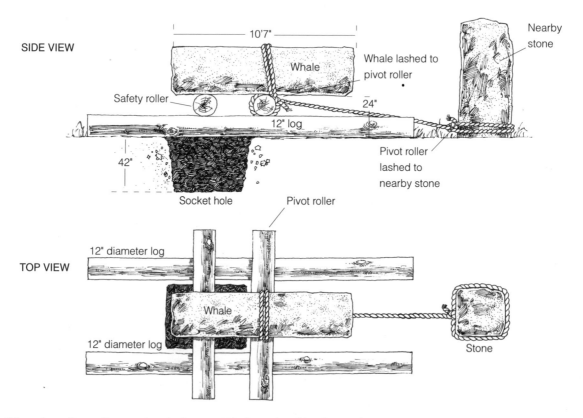

When the safety roller was knocked away, Whale, assisted by the accelerating effect of gravity, dived straight into the socket (Michael Middleton, drawing).

the team was able to maneuver Whale up onto two transverse tree trunk rollers that spanned the long rails. The thicknesses of both the rails and the rollers were similar at about a foot, which meant that the stone was elevated 24 inches above the planned 42-inch-deep socket, just about right. The team marked the socket on the ground and positioned Whale on the two rollers in such a way that just over half of its weight was beyond the balance point of the pivot roller. The other roller was used as a substantial safety spanning the socket area. With the stone firmly in place, and well supported by wedges, four people dug the socket hole. Two could work from the sides, but two had to be under the stone. The conditions were cramped, but not unsafe, thanks to the huge pine timbers overhead.

The socket was dug quickly and the team made one final important improvement to the plan. With ropes, they lashed Whale down tightly to the pivot roller! Plus, the pivot roller itself was wedged and firmly tied to another stone already in place, so that it was now an immovable beam.

When all was ready, the team began to knock out the leading (safety) roller with sledgehammers. I can only imagine their anticipation and wish I could have been there. When the roller finally escaped from Whale's pressure, the great stone began to pivot on the other roller, and its fall was accelerated by gravity according to Newton's best-known

law. Whale did not lunge forward like all his mates before; he pivoted on the cross-beam and dove straight into the hole, landing "within a few degrees of vertical." About forty people were present at the event and when Whale went in (I was told) there was an almost anti-climactic feeling, a sense of, "Well, of course. It had to land straight up. What have we been doing these last fourteen stones?" But a day later, I could still hear the joy in Ivan's voice. The big breakthrough had been made, and it had come after much testing and experimentation, little successes, and learning from failures, building on the work of many others who went before. No longer would the team have to struggle with straightening stones in their sockets.

In fact, at the May 1999 meeting, Ivan's team, using the same method, set two more stones perfectly into their sockets. The "breakthrough" is now routine procedure.★

★ Stop the press! September 24, 1999. I just returned from the inauguration of the Dragon Circle in Surrey, held Saturday evening, September 18, 1999. Ivan McBeth presided over the ceremony in his white robe and shamanic headdress, festooned with deer horns. Ella Anstruther, captivating in her gold mask and gold gown, welcomed all present and assisted Ivan in bringing the Dragon Circle "on line." The ceremony was similar to the druid ceremony I'd witnessed in Julie's Garden exactly two years earlier, with all the great spirits called upon to bless the circle. This time, though, there were over one hundred thirty people present, most of whom had played a part in the construction. Ron and Margaret Curtis, Ed Prynn, and Cliff Osenton were all there. I cannot adequately express the powerful emotion I felt as I witnessed the inauguration of such a spectacular circle. At the turn of the third millennium A.D., we were transported back to the third millennium B.C.

THE MEGALITHIC
REVIVAL

— CHAPTER 15 —

A Guide to Modern
Stone Circles

This is fun. I can't think of anything I'd rather be doing.

—ORREN WHIDDON,
CHURCH OF FOUR QUARTERS

There's a certain kind of brawny excitement about erecting large standing stones, which may explain why most are built by men. But women are getting involved too, women like Reva Seybolt, who has built a perfect stone circle on a rounded hilltop in northeastern Connecticut; and astronomer Judy Young at the University of Massachusetts, who is building a stone circle at the university as an astronomical learning tool. And Selena Fox has been creating stone circles all over the map for twenty-five years.

Megalithic construction dates back at least seven thousand years, as evidenced by the recent discovery of stone circles and standing stone alignments at Nabta Playa in Egypt. The golden age of megaliths in Europe was from about 4000 B.C. to 1500 B.C., and building stone circles in the British Isles was in vogue from about 3000 B.C. to 1500 B.C., during which time approximately 1,500 circles were built, an average rate of about one a year. Curiously, during the last quarter of the 20th century, humankind has been matching this rate again, if not exceeding it.

After the end of Europe's golden age, the megalithic mantle passed to other parts of the world, places like India, Japan, Korea, Africa, the South Seas, and even North America, as we'll see in the final chapter. William Danby built his "Druid's Temple" in the early 19th century and megalithic construction has certainly persisted well into the 20th century in India's Assam state. Noted Cambridge archaeologist Glyn Daniel, in his article "Megalithic Monuments" (*Scientific American,* July, 1980), writes: "We can gain some information about [prehistoric megalithic builders'] probable techniques by studying the methods used today in areas such as Assam, where megalithic monuments are still being built." Lack of funds and current unrest in Assam prevent me from conducting this research myself, as much as I'd like to.

We are in the midst of what Glynis Kent has aptly called a "megalithic revival," which really got underway around 1980, although a few small circles were built in the 1970s by Selena Fox, Chuck Pettis, myself, and others. In 1981, Bill Cohea built his stone circle at Columcille, and a year later Ed Prynn began the Seven Sisters Temple in Cornwall. I didn't build my first full-scale standing stone circle until 1987. The Australian Standing Stones were built in 1991 and Ivan McBeth hatched his Swan Circle in 1992. In that same year, on the Isle of Lewis, Margaret Curtis struck a blow for the ancients by discovering a "new" prehistoric circle. Meanwhile, in Quebec, sculptor Ed Heath was entering his megalithic period in the early 1990s with commissioned work. Many of us (myself included) thought that we were building the first stone circle in four thousand years! Partly as a result of the serendipitous circumstances described in part 2, we finally began to become aware of each other's work, and Club Meg was born. Now, it seems, I hear of a new stone circle every month. In one day, in fact, while surfing the Internet, I learned of five new stone circles, while another came through the mail in the form of a letter and pictures from Cliff Osenton.

Stone circles are in no danger of dying out, and our learning curve for discovering possible ancient building techniques is very positive, although we've got a long way to go to match the work of the people who built Stonehenge, Avebury and the 350-ton Grand Menhir Brisé.

This chapter is a guide to 30 modern stone circles that can be visited (although sticklers for accuracy will observe that one of the "stone" circles is actually made of concrete, while another is built out of old cars!). I've divided these circles into three categories: public monuments (12 examples), circles at businesses or private institutes (7), and private circles (11). Within each category, American sites are listed first, followed by Canadian, British, and other international sites. Statistically, half of the entries are in the United States, two in Canada, eleven in the United Kingdom, and one each in Australia and Costa Rica.

I am aware of several other stone circles that, for one reason or another, are not included in this chapter; and probability dictates that there are dozens more out there that I don't know about. I welcome correspondence and photos from anyone who has

already built a stone circle and from anyone who builds one as a result of reading this book. It is my intention to keep a database on modern stone circles, and this chapter is just a beginning.

CATEGORY ONE: PUBLIC MONUMENTS

This first section lists several circles and other modern megaliths that are public monuments, or practically so. Just show up and enjoy, at reasonable hours. The first three entries are American Stonehenge replicas.

Stonehenge at Maryhill

Location: Maryhill, Washington.

History: This full-scale replica of Stonehenge was built by Sam Hill, a wealthy entrepreneur, as a memorial to the soldiers of Klickitat County, Washington, who gave their lives in World War One. It was dedicated on July 4, 1918, and completed eleven years later. A second dedication was held on Memorial Day, 1930.

Description: Although the original plan called for stone, local stone quality was found to be inferior, and the monument was built of roughly 1,650 tons of reinforced concrete. The use of wrinkled metal molds gives an interesting if not quite accurate texture to the "stones." Unlike the ruined original, the replica is complete: 30 standing stones and lintels in a circle, the 5 great trilithons, the Heel Stone and Altar Stone, even 45 smaller "bluestones," although these are probably not so accurately placed.

An astronomer helped Hill to select the site and lay out the summer sunrise alignments; however, tests by Ernest W. Pinni and friends on June 21, 1978, showed the monument to be off by a full 3°. The error was probably due to the designers not taking proper account of a hill in the way, which critically changes the horizon. In his book, *America's Stonehenge,* Pinni shows all of the alignments at the replica, and how they compare with the original Stonehenge. He also shows how the monument can still be used for astronomical purposes, despite the 3° error.

Rohan Roy visited Maryhill in 1997 and tells me that the site itself is spectacular, overlooking the Columbia River Valley, and with views to Mt. Hood, Oregon's highest point, fifty miles away. But Rohan, and others who have visited, report that the monument itself, while impressive in scale, is not well presented to the public. The "stones" are still in good condition, but the place could benefit from some thoughtful landscaping. Still, it is the only place I know to get a sense of what Stonehenge might have been like completed.

Stonehenge at UMR

Location: University of Missouri at Rolla, at 14th Street and Bishop Avenue.

Web site: www.umr.edu/~stonehen

History: Built by the university's Engineering Department, this Stonehenge model was dedicated on the summer solstice of 1984.

Description: Unlike the Maryhill Stonehenge, this one is made of stone: 160 tons of granite. The diameter of the circle (50 feet) and the height of the trilithons (13 feet) tell me that the monument is intended to be a one–half-scale replica of the original, a very impressive project. Like Maryhill,

America's Stonehenge at Maryhill, Washington. (Rohan Roy, photo.)

the Rolla Stonehenge is complete, or nearly so, having 29½ standing stones in the outer ring, a "heelstone," and five complete trilithons arranged in the familiar horseshoe shape. The outer ring, however, is composed of granite blocks of sitting height, and they are not joined by lintels.

In addition to the familiar solstice alignments of the original, the Rolla model has several additional astronomical features, including the equinox alignment that makes use of the east and west "compass stones," and a true north alignment (the other two compass stones). One of the more interesting features, called an *analemma,* was incorporated into the "south-facing trilithon." On any clear day of the year, the noon sun shines through an aperture in the lintel and casts its image on the "horizontal and vertical stones at the base of the trilithon." A figure eight has been carved into the stone, and, precisely at noon, the analemma can be used to determine the date. In addition, the north-facing trilithon is

equipped with an opening through which Polaris, the North Star, can be viewed. To assist visitors, tables for the rising and setting of the sun and moon are posted monthly at the site.

The granite used in the Rolla Stonehenge was cut to the proper dimensions by the university's waterjet equipment. Two waterjets at a pressure of 15,000 pounds per square inch traversed the surface of the granite, cutting like a saw. The waterjet holes, which rotate at 180 rpm, are about the size of the wire in a paper clip. Moving at a speed of about 10 feet per minute, the jets could cut between one-quarter and one-half inch deep on each pass.

Thanks to Patricia Robertson and Dr. David A. Summers for maintaining UM-Rolla's Web sites on the model. Ms. Robertson tells us that "every year the National Society of Professional Engineers makes up to ten awards for outstanding engineering accomplishment. These awards are given for

Half-scale model of Stonehenge, University of Missouri at Rolla. (Photo courtesy of David Summers.)

Moon rising over Carhenge. Provided by Friends of Carhenge.

completed projects and typically go for such items as the space telescope. In 1984 the University of Missouri-Rolla received one of these Outstanding Engineering Achievement Awards for its Stonehenge model."

Carhenge

Location: U.S. Route 385, 2½ miles north of Alliance, Nebraska.

Contact for group tours: Paul E. Phaneuf, President, Friends of Carhenge, P.O. Box 464, Alliance, NE 69301. Tel: 308-762-4954.

History: Jim Reinders, an engineer, developed a keen interest in Stonehenge while living in England, and dreamed of one day building a replica of the ancient monument on the family farm in Alliance, Nebraska. When Jim's father died in 1982, the large Reinders family gathered at the farm and agreed to meet five years later to build a memorial in the form of a Stonehenge. But Nebraska was blessed with far more junk cars than stones, and, as the family believed in the use of indigenous materials, Carhenge was born.

The Reinders clan, about thirty-five strong, drove and dragged the first twenty-eight cars of 1950s and 1960s vintage to the ten-acre site. Family members dug socket holes and the upright cars were tipped into place. A forklift was used to set the lintel cars on their uprights, where they are welded in place. Friends of Carhenge later added ten "stones," and the memorial is now a fairly accurate replica of Stonehenge in its present-day ruined state, and even has cars that correspond to the Slaughter Stone, the

Heel Stone, and the two remaining Station Stones. The monument is oriented the same as Stonehenge in England, which perhaps wasn't the best choice, as the summer solstice sunrise alignment is some 7° off, due to the different latitude in Alliance, Nebraska. Nonetheless, the memorial was dedicated on the summer solstice, 1987. One of the builders quipped, "While Stonehenge may have taken centuries to complete, we built Carhenge in just seven days using blood, sweat, and beers."

At first, the monument was viewed disparagingly by some of the local powers-that-be, and even considered by some as an eyesore that should be removed or screened. But Carhenge is now painted a uniform stone gray and is preserved by the Friends of Carhenge as one of Nebraska's most popular tourist attractions, with over 80,000 visitors annually. There is no admission charge.

The Sunwheel Project

Location: University of Massachusetts, Amherst, Massachusetts. The Sunwheel stone circle is at the south end of the campus, "near the maze," between Stadium Drive and Rocky Hill Road.

Contact: Professor Judith S. Young, Department of Physics and Astronomy, University of Massachusetts, Amherst, MA 01003. Sunwheel Web site: www.umass.edu/sunwheel.

History: The Sunwheel was conceived by Astronomy Professor Judy Young. It serves as an astronomy teaching tool, with special emphasis on the movement of the sun during a full year. Solstice and equinox risings and settings are incorporated into the design, and, when completed, the declination of the sun will be easy to observe as a function

Marker stones at the UMass Sunwheel. (Judy Young, photo.)

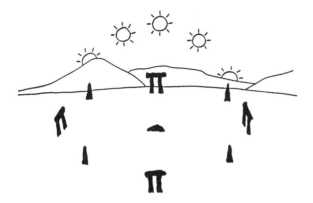

Plan of the UMass Sunwheel, by Judy Young. The view faces south on the shortest day.

of the shadows of the 8-foot-high stones. The project was given official approval in October of 1995 and Judy completed a full year of observations at the site in March of 1997, with the significant alignments marked with wooden stakes.

Description: On May 13, 1997, the first preliminary circle of marker stones was completed, using 12 "small field boulders" of 500 to 1,000 pounds each and projecting about two feet above ground. These stones are all 50 feet from the center and they are already useful as instructional aids. Judy has taken many groups to the site for equinox and solstice observations, and has shown the circle to hundreds of primary school kids who come to learn about the seasons. Gatherings are open to the general public at the beginning of the four seasons. Write Judy Young for exact dates and times.

The next phase of the project involves erecting 12 rough-cut granite standing stones in a circle of 120 feet in diameter. These stones will be about 1.5 feet square in cross-section and about 11 feet in length, which would make them about 2 tons each in weight. Four of the stones, the single monoliths, will mark the rising and setting of the sun on the summer and winter solstices. The other 8 stones will form four gates at the cardinal compass points,

and, because the site has a relatively good horizon, the east and west gates will also serve as equinox sunrise and sunset markers. Finally, the four gates will be capped with granite lintels, as seen in Judy's drawing.

The cost of the project has been estimated at $50,000, about a third of which has been raised as of April 16, 1999. The figure seemed high to me, but Judy explained that $20,000 is needed for site work because of poor drainage, and another $7,000 is needed for wheelchair accessibility. Donations are accepted at the address above and you can purchase excellent T-shirts, sweatshirts, etc., to help support the project. Like many other stone circles, the Sunwheel is a work in progress.

Georgia Guidestones

Location: 7.2 miles north of Elberton, Georgia, on Highway 77.

History: Late on a Friday afternoon in June, 1979, a well-dressed and articulate stranger identifying himself only as "Mr. Christian" walked into the office of Elberton Granite Finishing Company president Joe H. Fendley Sr. and asked about the cost of building a large granite monument. Fendley told the stranger that he was a wholesaler and didn't handle individual orders. But Christian—he admitted it was an assumed name—persisted, and described a monument with stones of Stonehenge proportions, a monument larger than anything ever made in Elberton, which promotes itself as the "granite capital of the world."

Fendley's banker, Wyatt C. Martin, of Granite City Bank, became the intermediary for the project, which suddenly got on the fast track when Martin advised Fendley that the stranger had opened an escrow account with a deposit in the upper five-figure range, an account intended to finance what would become known as the Georgia Guidestones.

Christian said that he represented a group that had been pondering the fate of humankind for the past twenty years, and had come to the conclusion that a great holocaust would occur in the world. The purpose of the Georgia Guidestones would be to help humankind to make a new start. To further this end, the group had come up with a list of ten "guides" pertaining to ecological issues such as population control and maintaining a balance with nature, as well as advice on good governmental practice, such as "Avoid petty laws and useless officials," and "Balance personal rights with civil duties." The guides were to be inscribed on the monument in eight major languages, two lists on each of the four largest stones.

Martin, the mediator of the project, promised Christian that upon completion of the project he would deliver his file on the affair to the anonymous sponsors. Martin was as good as his word, and the secret benefactors of the Georgia Guidestones were never made public.

Description: Jaki, Darin, and I visited the Guidestones late in the afternoon of February 26, 1999. The monument can be seen from quite a distance, because it occupies the tallest point in Elbert County. We parked in a little lay-by made for the purpose and were glad that we had the place to ourselves. Even Darin was impressed with the power of the monument. The stones were huge, hard, nicely proportioned, and beautifully inscribed. From my research, I knew that this project had stretched modern granite quarrying, shaping, inscribing, and monument erection to the limits of the available local resources. The four largest stones are 16'4" tall, 6'6" wide, and 1'7" thick. The wide flat faces are polished and contain the ten messages in the eight different modern languages. The narrow edges of the stones are left rough. At the center of the monument is the Gnomon Stone, of the same dimensions as the four large corner posts, except only half as wide. The Gnomon has an eye-level hole cut through its 3'3" width, which aims precisely at the North Star. I was hoping to see Polaris through this

narrow hole, even though the sun was still up, but had no such luck. Another half hour, perhaps, but we couldn't wait, as we had an appointment in North Carolina that same evening.

Another slot is cut through the narrower 1'7" thickness. This slot forms a "window" that can be used to observe the rising sun at the summer and winter solstices, as well as at the equinoxes, depending on whether the observer sites along the right side of the slot, the left side, or directly through the center. We were months away from any solstice, but, through the slot, we could see clearly the four points on the distant horizon marking the sun rise and set for the longest and shortest days. "Mr. Christian" had certainly selected a spectacular site for the stones.

The five standing stones are capped by a large lintel measuring 9'8" long, 6'6" wide and 1'7" thick, or about 9 tons. This capstone brings the height of the monument to 18 feet. A hole is drilled through the capstone in such a way that the sun's lightbeam at noon of any clear day shines on the south face of the Gnomon, telling the day of the year, a similar

The Georgia Guidestones, Elberton, Georgia.

Knowledge, Innocence, and Growth, *a megalithic sculptural group by Jon Isherwood, Plattsburgh, New York.*

feature to the analemma at the Rolla Stonehenge. Clearly, great care was taken by the designers to get these alignments correct, and I knew that all calculations had been checked by an astronomer.

Each of the standing stones is supported by a large flat granite base. These base stones, in turn, are set into a large, heavily reinforced concrete slab extending well below grade. The five standing stones are pinned to their individual granite bases with stainless steel pins. Years earlier, I had seen large sculptured granite stones installed exactly the same way at a stone circle in Plattsburgh, New York, just twelve miles from Earthwood . . .

Knowledge, Innocence, and Growth

Location: Plattsburgh State University college campus, where Rugar Street meets Broad Street, Plattsburgh, New York.

History: In 1994, while a visiting professor at Plattsburgh State, sculptor Jon Isherwood created a beautiful group of 7 shaped granite megaliths, a permanent sculptural piece.

Description: Three of the 7 stones are true menhirs, the tallest standing twelve feet high, while the other stones are recumbent. The stones were shaped by Isherwood, who also directed their placement by crane. The standing stones are founded by large concrete footings, extending at least four feet below grade. The bottom of each standing stone was cut smooth and flat to rest on the flat foundations. Each standing stone is pinned to the foundation by two stainless steel pins, which appeared to me (on stone-raising day) to be about an inch in diameter and sticking about five inches out of the foundation. The use of two pins prevents any possibility of the stones rotating on the footing. An expensive project made possible by donations, *Knowledge, Innocence and Growth* was built for the ages. If undisturbed by future development, it will give pleasure to people for millennia. The Earthwood circle is just twelve miles away.

313

The Druid's Temple

Location: On Forestry Commission lands near Ilton by Masham, North Yorkshire, England. Ask directions at Masham and make good notes, as the site is not signposted. The site is open to the public and is a good spot for a picnic. The Druid's Temple is only about twenty miles northwest of the ancient Devil's Arrows, described at the end of chapter 7.

History and Description: We visited the Druid's Temple en route from our chapter 7 adventures to meeting with the Curtises in chapter 8. I was particularly interested in the site because it is neither prehistoric nor truly modern; it links the very old with the very new. The Temple was built about 1820 by a Mr. William Danby, who, unfortunately, left no record of his intentions for the monument, although local folk memory is of a kindly landlord who thought up the project to provide employment to men during hard times. But what he created was nothing short of spectacular, and were the site to be dated to 3000 B.C. instead of A.D. 1820, it would be one of Britain's best-known and most visited attractions. But, luckily, it is not heavily visited, so you can usually enjoy it in peace and quiet. It is one of my favorite circles, ancient or modern, and well worth the difficulty of finding it.

The site is actually spread over several acres, with at least a half dozen beautiful dolmens, a tall cheese-ring-type stack of stones surrounded by a small stone circle, and other assorted monuments, all of which look like they could be *very* old, instead of simply

Jaki and Darin play at pubbing at the huge stone table at one end of the oval. The multi-ton slab, unfortunately, cracked into a large and a small piece, fairly recently, by appearances.

old. But the main attraction of the site is the temple itself, a fantastic oval of large *millstone grit* megaliths—the same stone as at the Devil's Arrows—and featuring trilithons with six-ton capstones, huge stone tables and benches, grottoes, and a ring of almost continuous standing stone orthostats.

We were fortunate to have the site to ourselves for quite a while before a small busload of tourists arrived. Still, it was good to see a number of people milling about inside the monument, giving an idea of scale and use. And it was fun to hear their varied comments and theories. It was also good to see them leave again, so that we once more had this magical place to ourselves.

A walk along the nearby dirt road took us to several dolmens, which Cliff would have appreci-

ated for the building skills in evidence, and done by country folk without modern equipment. All in all, megalithomaniacs will find that combining the Druid's Temple and the ancient Devil's Arrows (chapter 7) makes a great morning or afternoon outing. See the color section for another photo.

Diana's Stone Circle

Location: Powys County Observatory, Llanshay Lane, Knighton, Powys, Wales LD7 1LW.

Contact: Brian Williams or Cheryl Power. Telephone: 01547 520247

History: John Martineau is a stone circle designer, as well as a writer and researcher in the field of sacred geometry. He has designed and/or built three

Diana's Circle, Powys County Observatory, Wales. (Photo courtesy of Powys County Observatory.)

stone circles, including this one and the next entry. The third is in Slovenia, and not listed.

Description: Built in 1997, John Martineau's original design forms the core of this 100-foot-diameter stone circle, but Brian Williams and Cheryl Power added a few stones to mark the solar and—interestingly—stellar alignments, an important feature at an observatory. The rising and setting of Aldebaran, Cygnus X1, Altair, Rigel, and Sirius are all marked with stones. An inscription on one of the 24 standing stones dedicates the circle to Diana, Princess of Wales. The ring is used as an instructional aid when school groups visit. An "observation stone" at the center helps interpret the alignments, including one odd pair of alignments that only an astronomer would think of; according to the interpretive chart, the same two stones that mark the most southerly moonrise and moonset also mark the spots on the horizon where the galactic center rises and sets on any clear night. Curious. Visitors are well received by Brian and Cheryl, who are happy to share this kind of lore.

Greenhenge

Location: Near Warminster, Wiltshire, England. Access by footpath. Go three miles south of Warminster on the A350 to the village of Longbridge Deverill. Go west on the B3095 to Brixton Deverill (just over a mile) and turn left immediately after the bridge. Go past the church to the end of the lane and walk up the steep grassy track for about a half mile. When the track ends at the double gates, keep on going to the ancient beech clump, and the circle is just beyond.

Contact (to use the circle for ceremonies or for disabled people who would like vehicle access): Mark Houghton Brown. Tel: 01747 820205. E-mail: yjp51@dial.pipex.com

History: This circle was designed by John Martineau for Mark Houghton Brown, who was hosting the Green Gathering at Pertwood Farm in 1995.

Description: The circle is about 150 feet in diameter, a good size for a gathering, and is composed

Greenhenge, Brixton Deverill, Wiltshire, U.K. (Mark Houghton Brown, photo.)

of 21 stones, each roughly four feet high. The stones were selected at a quarry in Worth Matravers (near Swanage in Dorset), brought to site on a lorry, and unloaded near their respective positions with a large forklift. After that, all the work was done by hand.

The circle was designed and built with regard to the various leys that criss-cross this magical part of Britain, including the well-known Stonehenge-Glastonbury ley, which also includes the Swan Circle described in chapter 6. Martineau incorporated all sorts of sacred geometry considerations into the design, too complex to list here. John is listed in the sources as a contact person for information on sacred geometry.

The circle continues to be used for the annual Green Gathering, a convention for people pursuing alternative patterns of living, including (among others) druids, New Agers, neo-pagans, astrologers, dowsers, friends of the ecological movement, and anyone interested in sacred spaces, numbers, or geometry. Music is a big part of the Gathering, which drew 10,000 people to the stone circle in 1998. When not used for the Gathering, the circle does quieter but no less important duty as a focus for healing, and for fertility rites.

Touchstone Maze

Location: Strathpeffer, Ross and Cromarty, Scotland. Accessible by footpath. Ask directions at the Information Center at the village square.

History: Known primarily as a maze or labyrinth, this wonderful monument is also a series of five concentric and definitely megalithic stone circles. The maze/circle was designed and built by Helen Rowson on behalf of Forest Enterprise and Touchstone, a charity in the Highlands and Islands Region of Scotland. Stones were donated by individuals, quarries, and county councils throughout Scotland, including stones from the islands of Skye, Lewis, Orkney, and Shetland. The construction was gifted by R. J. McLeod

Contractors, Ltd., who completed the maze in the early 1990s.

Description: A labyrinthine pathway winds and turns among 81 large stones arranged in five concentric circles. The stones are also arranged like 17 spokes of a wheel, with at least nine of the rows aligned with such events as the solstice and equinox sunrises and sunsets, and the sunrises and sunsets for the four Celtic cross-quarter days. Two outliers near the entrance cast their shadows to the center of the circle at midsummer sunrise and at the major northerly moonrise. On any day of the year, an observer at the center of the maze can look back at these two outliers and see where both the sun and the moon have their most northerly rising.

We always enjoy our visits to this wonderful monument, and marvel at the color, shapes, and varieties of the stones from all over Scotland: Lewisian gneiss, Skye marble, Grampian granite, Orkney sandstone, and so many more. I am trans-

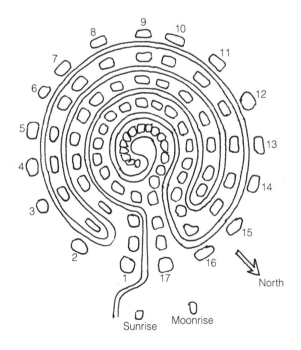

Touchstone Maze alignments.

The Touchstone Maze, Strathpeffer, Scotland.

ported to many of my favorite ancient stone circles by seeing their doppelgangers at Touchstone. The circles serve as lessons in both geology and astronomy, with a flavor of ancient history thrown in for good measure. But Helen Rowson's greatest accomplishment at Touchstone, in my view, is that she has created a sculptural work of genuine art, and in a beautiful setting. Children, particularly, love the maze.

After viewing the maze, continue east along a footpath, which begins to rise up Knockfarrel, a distinctive and prominent hill, also known locally as the Cat's Back. A climb to the top will be rewarded by spectacular views in every direction, and will place you in the center of an ancient hill fort, one of many situated on coastal promontories over twenty-five hundred years ago. Of particular interest is the way that the stones have been fused together under great heat, a process known as *vitrification*. Opinions differ as to how and why many of these hill forts became vitrified.

The Australian Standing Stones

Location: Glen Innes, New South Wales, Australia.

Contact: John "Trigger" Tregurtha, Glen Innes Standing Stones Committee, P.O. Box 121, Glen Innes, NSW 2370, Australia. Phone (02) 6732 3066.

History: The Celtic Council of Australia wished to build a stone monument in Australia to commemorate those early settlers of Celtic origin who helped to build the nation. Cities and towns all over Australia were invited to submit proposals. Glen Innes submitted a 46-page package prepared by tourist officer Lex Ritchie, complete with photographs of the proposed site, site plans and surveys, and an enormous amount of supporting data. The hard work paid off and the town's proposal was selected.

The chairman of the committee charged with building the stone circle monument was local pharmacist John Tregurtha, and he joined forces with

Ritchie to find suitable stones. The plan called for stones extending twelve feet out of the ground, so they needed to find monoliths nearly eighteen feet in length, a tall order indeed. The two spent months "scouring the bush," but, in the end, could only find three stones of suitable height to be used in their natural state. All the others had to be split from larger rocks, a task accomplished by drilling rows of holes and filling them with a substance called *expandite*. When the drillers returned the next day, the stones had split. No explosives were used.

Transporting these huge stones was itself a monumental task, but eventually they all made it to site. With great pomp and ceremony, and in bright spring sunshine, the first stone was hauled upright on September 7, 1991. The hole was dug with machinery but the stone was stood upright using people power only. Using a "quadrapod" made of four long stout poles, and a six-to-one (m.a.) block and tackle, the local Volunteer Rescue Squad and an Australian national champion tug-of-war team pulled the stone upright. Like all of the 37 remaining stones in the array, the socket hole of the ceremonial first stone was packed with concrete.

After the first stone—one of the smaller ones at 15 tonnes!—the remaining stones were completed with heavy equipment, including a backhoe, a large bulldozer, and a crane. The array of 38 stones, including 24 in a true circle, was completed in just three months, on December 13, 1991.

The array had been laid out by local surveyor Ian Macdonald, and the plan called for alignments for the summer and winter solstice sunrises and sunsets. Macdonald was so worried about "getting it right" that he consulted his old tutor every inch of the way. In a personal correspondence, Trigger Tregurtha takes up the story:

At dawn on the summer solstice, just a few days after the completion of the array, a huge number of Celts gathered at the stones, including Ian Macdonald, who was observed to be skulk-

ing away behind some nearby trees. Others were observed to be sipping away at an amber beverage being poured from flasks that fitted neatly into hip pockets! We were waiting for the sun to arise, and arise it duly did, causing the marker stone in the southeast quadrant of the circle to cast a shadow on the northernmost of the three central stones, just as intended, thus giving great cause for rejoicing! At this time, said Macdonald was further observed to be strutting throughout the gathering with chest thrust forward as though he personally owned the whole show!

Description: The Australian Standing Stones consist of 24 stones in a true 50-foot-diameter circle, with a group of three "guidestones" in the center. The northern guidestone is the Gaelic Stone, for the Scots, Irish, and Manx peoples who spoke Gaelic. The southern guidestone is the Brythonic Stone for the Welsh, Cornish, and Bretons, whose languages derived from the Brythonic. The central stone, a golden-colored unhewn ground rock, is called Australis, for all the Australian people. There are four outliers marking the true compass directions, and seven other stones within the circle that help form a processional avenue while marking the sunrises and sunsets at the solstices. According to Trigger, all of the stones are between 15 and 30 tonnes, and stand between 12 and 15 feet out of the ground, making the array one of the most impressive of all modern stone circles.

The monument shares design characteristics with both Callanish and the Ring of Brodgar, but has its own unique features. For example, the four directional outliers, in combination with the summer solstice sunrise stone, make a very accurate star chart of the Southern Cross, just as it appears on the Australian flag.

Readers fortunate enough to make the journey to Australia should not miss this wonderful stone circle. The best time to visit is during the annual Celtic Festival held each year over the first weekend in May, which Trigger assures me "is rapidly becoming one of *the* annual events not to be missed in Australia."

So much more could be said about the Australian Standing Stones, but I'll let the pictures do the talking. Thanks to Trigger and the Australian Standing Stones Committee for supplying all of the pictures, in chapter 13, in the color section, and on the front cover of this book.

Rebuilt Stone Circle, Costa Rica

Location: In the main square of the southern town of San Isidro, Costa Rica.

History and Description: This little circle of seven standing stones was a pleasant surprise to Jaki, Darin, and I when we visited Costa Rica in 1995. The circle is roughly 25 feet in diameter and the tallest stones are about 4'6" inches high. The stones inspired the creation of my little stone circle on the Pacific beach at Dominical, a few miles to the west. In 1998, I called a Costa Rican friend in San Isidro to see if he could find out the history of the stones. He spoke with an archaeologist and was able to learn that the stone circle is new, built in the early 1990s, but that the stones are very old. A local teacher had collected these ancient stones, which had been used, apparently, as border markers in various old countryside cemeteries.

On the other side of the square is the Cathedral of San Isidro, whose main entrance is flanked by two of the ancient and mysterious perfectly round granite balls that are found in the southern part of

Stone circle on the main square, San Isidro, Costa Rica.

Costa Rica. The ones at San Isidro are about three feet in diameter, but we saw larger ones in Sierpe and at the national museum in San José. The largest known ball was eight feet in diameter and weighed about thirty-two tons, definitely megalithic. It is said that these large balls originated in what is now Panama, and were delivered great distances to southern Costa Rica, where they were used as funerary monuments to the chiefs.

CIRCLES CONNECTED WITH BUSINESSES OR INSTITUTES

The stone rings in this section are all associated with businesses or institutes. Generally, visitors are welcome, but please remember that the places listed in this section have other functions besides showing people around their stone circles. Advance contact is strongly advised, and if you can patronize the businesses, that helps, too. Some, like Columcille and the Seven Sisters Temple, depend upon donations to stay open to the public.

Earthwood Stone Circle

Location: 366 Murtagh Hill Road, West Chazy, NY 12992.

Contact: Rob or Jaki Roy, 518-493-7744. Web site: www.interlog.com/~ewood

History and description: See chapter 2.

Nature of business: Earthwood is a building school specializing in cordwood masonry, underground housing, and stone circle building. Books and videos are available on site or by mail.

Columcille, Inc.

Location: 2155 Fox Gap Road, Bangor, PA 18103 (next to the Kirkridge retreat center).

Contact: William H. (Bill) Cohea Jr., 610-588-1174. E-mail: colum@epix.net. Web site: www.columcille.org

The new megalithic gate at Columcille is, itself, a work of art.

History and description: See chapter 4.

Columcille is "a sacred playground for the odyssey of the human spirit, a sacred earth space and place of myth and mystery, and a salon where everything is sacred and nothing is sacred in the exploration of ideas, forms and paradigms." The eight ancient days of Celtic celebration are observed each year at Columcille. Donations help keep the gate open.

Stone Mountain Stone Circle

Location: Center for Symbolic Studies, Stone Mountain Farm, 475 River Road Extension, New Paltz, NY 12561.

Contact: Dr. Robin Larsen or others at the Center, 914-658-8540. E-mail: css@ulster.net. Web site: www.mythmind.com

Description: See chapter 3. New stones to be added periodically.

Nature of business: Again, see chapter 3, the second paragraph.

Standing Stone Perennial Farm, Royalton, Vermont.

Standing Stone Perennial Farm

Location: Back River Road, Royalton, VT.

Contact: David Brandau or Lynne Hall, 802-763-8243.

History: David and Lynne raise all sorts of perennial plants for sale at their farm, and in 1986 started raising stones as well. Their farmstead, and the surrounding area in general, is a stone circle-builder's dream; long and narrow, the beautiful schist shards are found in old stone walls, as well as in the ground itself, appearing when the earth is tilled.

At first, David stood the stones up as an aesthetic alternative to hauling them to the stone walls, but, after a while, he and Lynne could see that the stones graced the garden paths in a magical way. Visitors liked them, too. Their business became Standing Stone Perennial Farm.

Description: Most of the standing stones, sixty or seventy in number, are individual mini-menhirs, and beautiful in their own right. One forms the backdrop to a Buddha in a peaceful Eastern garden setting. Across the road from the farm itself, near the customer parking lot, Dave's latest standing stone project is underway, and it is more closely related to the stone circles, in that outliers radiate from a center stone, with various alignments incorporated into the design. The center is a leaning stone, and points exactly to the North Star. David raised a new equinox stone in the fall of 1998.

Most of the stones are long and narrow, and, therefore, not particularly heavy, the largest probably weighing about half a ton. The tallest stone is about David's own height, 6'3", with another couple of feet in the ground. In a wall on the edge of the road are two stones that appear to have been originally joined as one. They are an order of magnitude larger than the stones that David has raised previously, over 15 feet in length and probably between 4 and 5 tons each. They will stand someday, and I'd like to be there the first day they do.

There is an interesting aside to David's involvement with large stones. About thirty years ago, while visiting India, David was a participant in a most incredible event. Under the direction of a small elderly Hindu ascetic, David and five other men performed an extraordinary feat of mental or psychic energy. The six men surrounded a large megalith, which, by David's description, had to have weighed at least three tons and probably more. After adhering to the proper preparation techniques as prescribed by their mentor, which included chanting, the six men proceeded to lift the huge stone clear of the ground to chest height, each using just one finger! On average, each man was in some way responsible for at least a half ton. David, as level-headed an individual as you'll ever meet, related this experience exactly as he remembered it. A few months later, I witnessed a similar kind of "levitation" in

the person of Ivan McBeth in Surrey (chapter 10), albeit at a much lesser scale. I told David that if I had experienced what he experienced, it would change my outlook on, well, on *everything*. David assured me that his experience, as well as the whole spiritual scene of his trip to India, certainly had a profound effect on him.

The Farm is worth a visit. Please remember that it is a business where David and Lynne earn their living. You might come away with a flowering plant or a shrub rose, and maybe even a stone. Last time I visited, David had a pile of beauties for sale.

The Crystal Sanctuary at Mines Cristal Kébec

Location: 430, Rang 11, P.O. Box 130, Bonsecours, Quebec, Canada, J0E 1H0

Contact: Gaudry or Pauline Normand.

History: The Crystal Sanctuary, part of Mines Cristal Kébec, has been owned and operated by the

The nineteen stones of the circle at the Crystal Sanctuary are all white crystal quartz, Bonsecours, Quebec.

Normands since 1989. The site comprises a working quartz crystal mine, an interpretation center, and a research and transformation center, offering quartz crystal vessel concerts and various workshops. Gaudry and Pauline Normand, with others, have created several "sacred spaces" using white quartz stones, including a magical little quartz medicine wheel in a wooded glade, a labyrinth designed by Marty Cain (see sources), and the wonderful stone circle laid out by Gaudry himself.

Gaudry agreed to meet Jaki and me on a fine early October day, two days after the Crystal Sanctuary had closed for the season. After a tour of the museum and crystal shop, Gaudry took us upstairs to the transformation center, where he treated us to a short concert on his 49-piece set of "crystal vessels." These vessels are used in the manufacture of silicon microchips, but they can be played like giant wine goblets, by either striking them or by passing the striker around the edge of the vessel, in the way that one can make a stemmed wine goblet sing by passing a wet finger around the rim of the glass. Gaudry had recently rented the Great Pyramid at Giza for a night and, in the King's Chamber, recorded a compact disk of crystal vessel music, a copy of which he gave us. The sound of these vessels was unlike anything we'd ever heard, ethereal and pure. Gaudry is researching the curative powers of the sound, and is a wholesaler for the vessels themselves.

After the mini-concert, Gaudry led us to the crystal stone circle. As we approached, I was reminded of the Dragon Circle, because of the size of the ring and the bright whiteness of the stones. But as we got closer, I thought of the Merry Maidens ancient circle in Cornwall, which is of similar scale and has the same number of stones as Gaudry's, nineteen. Curiously, the Boscawen-Un circle also has nineteen stones, and one of them, the Beltane stone, is white quartz, just like stones at the Crystal Sanctuary. Adding to the Cornwall connection is the fact that the only other all-quartz circle in the world—to my knowledge—is the ancient Duloe Stone Circle, a 37-foot-diameter ring of 8 large white quartzite stones, the tallest of which is 9 feet tall. (Well worth a visit. Just ask at the village of Duloe, Cornwall, four miles south of Liskeard on the B3254.)

Description: The site is spectacular, a large field with a long view out over the crystal mine area and for dozens of miles over the rolling countryside of Quebec's Eastern Townships. The 19 stones are arranged in a true circle, but Gaudry did not lay out the circle with a rope from the center peg. Instead, he paced the radius of each stone from the center. The circle is 28 paces in diameter. This intuitive approach was in keeping with the intent of the circle, which was to amplify the earth energies—or *vortex*—that Gaudry, Gerard Couture (a well-respected local dowser), and others were able to pick up at the site. Beneath the ground is a huge vein of white quartz and quartz crystals. We did not immediately enter the circle, but walked around the outside of the stones, clockwise of course, until we came to a gate that consisted of two stones slightly closer together.

The stones were boulder shaped, large and heavy—six tons on average, I guessed—and not particularly tall at about four to five feet off the ground. I found Gaudry's method of setting the stones to be of particular interest, as it was so easy. After dowsing the stones in his large pool of candidates, to find out where each one was meant to go, Gaudry used a large excavator to set them in their places. The machine's large digging bucket simply pushed a depression into the topsoil, perhaps a foot deep, and then set the boulder-shaped stone into the depression thus formed, very much like marbles on a Chinese Checkers board. This method left the site practically undisturbed, and is perfect for this shape of stones.

The stones were beautiful. Pure silicon quartz, as these stones are, is extremely hard and dense. Some of the stones actually had crystals in them. One in particular had a cavity like the interior of a geode, with many beautiful quartz crystals inside.

At the center of the circle, we spoke of the meaning and purpose of stone circles and other sacred spaces. Gaudry and Pauline are very involved in studying the ancient Vedic philosophies, although, like so many of the other Stone People, they gather wisdom from a variety of traditions: Native American, Celtic, Hindu, and many others. The conversation came around to the usual question.

"I asked myself why I am compelled to do this," Gaudry said with his charming Quebec accent, "and then I realized that there was something deep within me, almost a genetic thing that I could not stop, that was telling me that I needed to protect this site, this crystal mine, and to do a ritual with nature, to connect what I think of as Mother Earth and Father Sky, the creation and the creator. The circle is like a great *mala*—a giant rosary on the landscape—that enlivens the landscape just by itself. The sun comes up and wakes the stones and makes the mala lively. The sun's energy goes in the ground and contacts the vortex. It's a structure that has life in it, life of very high energy because the material is so pure—the quartz is 99.9 percent pure silicon dioxide—and it has so many unique properties.

"Now I see that stone circles are also an art form, and I think that a person who builds one finds that it mirrors back to his real essence. I think the most potent circles are the ones that are not just mirroring the physical, but are doorways to subtler levels of the self. The circles are lively mirrors with doors and windows—energy doors and windows—that allow you to tap into your fundamental nature.

"We do workshops here. People enter and have the symbolic aspect of the sacred before their eyes, which is usually enough to enliven something deep within. At the workshop, I facilitate that experience of taking them back to their essence, so that they are grounded. Eventually, the people leave, and they have smiles on their faces. They feel at home within. In a sense, the stones serve as acupuncture points, not only to the earth, but to the inner being."

Powerful stuff indeed. Jaki and I knew that Gaudry had a deep sense of the transrational. It came across as he described his circle and its effects on people; it came across in his strong interest in the special qualities of the crystals over which he is the guardian; it came across clearly in the way he played the crystal vessels. A visit to the Crystal Sanctuary is an experience in several different dimensions.

The Seven Sisters

Location: Tresallyn Cross, St. Merryn, Padstow, Cornwall, PL28 8JZ, U.K.

Contact: Ed Prynn or Glynis Kent, 01841 521045.

Description: See chapter 5.

Technically, this is a private residence. Ed and Glynis use donations to maintain and improve the Seven Sisters Temple and all the wonderful megaliths contained within it. Ed's books are sold at the Temple, and I highly recommend them for their authentic account of life in Cornwall after the Second World War.

Merlin's Circle at Tintagel

Location: In the garden behind Merlin's Rock Shop in the village of Tintagel, Cornwall, U.K.

Description: Tintagel Castle and Merlin's Cave below it play important parts in the legend of King Arthur. The setting is idyllic, a small but beautiful bay along Cornwall's South West Coast Path. The rock shop in the village, unfortunately, was closed the day we visited, so I was not able to glean any information about the circle, except by observation. Eight light-colored granite standing stones are arranged in a garden circle of about twenty feet in diameter, an excellent example of how a lovely circle can be incorporated into a relatively small space.

Remember that this circle, although it can be seen very well from the road, is part of a private business. It would be nice if you went in the shop and bought a rock.

Merlin's Circle, Tintagel, Cornwall, U.K.

PRIVATE CIRCLES

These circles are built on private land and require permission to view. Please treat such permission as a privilege, not a right, and at the owner's convenience, not yours. By adhering to this protocol, the visit will be pleasant for all, and the likelihood of the sites remaining accessible to those that follow will be enhanced.

Church of Four Quarters

Location: Near Artemis, Pennsylvania, about 15 miles west of Hancock, Maryland.

Contact: Orren Whiddon, President, The Church of Four Quarters, RR 1, Box 62C, Artemis, PA 17211. Tel: 814-784-3075. E-mail: megalith@his.com Web site: www.his.com/~megalith

History: Orren writes, "We began in 1995 with mud, 200 people, and one 8,000-pound stone. In 1996, we finished the East Gate with the Mother Stone and her Consort, almost losing the rigging, the stays, and the Stone People in the process! In 1997, three more stones entered the circle and were raised by over 400 people."

I attended the 1998 stone raising, described in a sidebar in chapter 14, and saw four more stones raised efficiently into place, making 10 so far. The oval will take over twenty more years to complete, but this doesn't worry Orren in the slightest. "This is fun," he told me. "I can't think of anything I'd rather be doing."

Description: The long axis of the oval or ellipse (for the ring is still young enough to develop into either) follows the contours of the landform, a large semi-cleared space in the midst of a mature hardwood forest. A few trees have been left inside the ring, giving a decidedly druidic atmosphere to the site. The menhirs are lovely flat sandstones of various warm colors. Now, with ten adjacent stones in place, the curvature of the ring is apparent for the first time, and with more stones being added each year—five more are scheduled for early September, 1999—the curve will rapidly become more definitive. The stones raised so far vary from about 8 feet to 11 feet in height and, like the Stones of Stenness, are quite light in weight for their height: generally 1½ to 3 tons each. An interesting feature is that the stones are relatively close together, just two or three feet between them, reminiscent of the Rollrights in Oxfordshire and the oval of stones at the west end of the Menec alignments at Carnac, Brittany.

The circle is used for celebrations, ceremonies and specialized camps connected with the Church of Four

Dedication ceremony for the new stones, June, 1998, Church of the Four Quarters, Artemis, Pennsylvania.

Quarters, and applications to visit the property at Four Quarters Farm are handled by the Church. Four Quarters Farm is a haven for people tuned in to the various Earth religions.

Ellis Hollow Stone Circle

Location: Near Ithaca, New York.

Contact: Foundation of Light, 399 Turkey Hill Road, Ithaca, NY 14850.

History and Description: Chuck Pettis helped design the original stone ring in 1977, consisting of 8 small boulders in an ellipse. At the two foci are a monolith and a recumbent stone. Twenty years later, 7 new standing stones were added to the ring, and positioned between the original boulders. Lydia Pettis, in an article on the Geo Group's Web site (see sources), says, "The male energy of the new stones blends with the female energy of the existing stones to create a powerful sense of balance and harmony."

Lydia concludes her article by inviting people to the Ellis Hollow Stone Circle. "There are no rules or prescribed rituals to be followed. Simply come into the circle. You may sit, stand, touch, lean against and hug the stones, be quiet, talk to yourself, listen, and watch. Empty your mind or explore. This is your space as well as our space. Enjoy it and let us know how it feels."

Thanks, Lydia.

While in Ithaca, you might like to check out the Carl Sagan Memorial model of the solar system, done in standing stones. The model is exactly one five-billionth (1/5,000,000,000) scale, so the distance from the Sun Stone, downtown, out to the Pluto Stone at Cornell University, is about 1,200 meters, or ¾ mile. These stones are polished granite, while the other planets, spaced at their right scale distances from the sun, are cast concrete. I suspect the cost of the granite had an influence on the material makeup of the solar system. Taking the walk gives a great sense of the scale of the planetary neighborhood.

Ellis Hollow Stone Circle. (Lydia Pettis, photo.)

Reva's Stone Ring

Location: Woodstock, Connecticut.

Contact: Reva Seybolt, P.O. Box 49, East Woodstock, CT 06244. Telephone: 860-928-0754.

Description: I learned about this stone circle from geomancer Patrick MacManaway (see sources). My regular megalithic companions, Jaki and Darin, joined me to meet with Reva at 3 P.M. on February 12, 1999. Although the snow had recently disappeared, the day was gray and threatened rain. Reva, a friendly lady in her 40s, came out of her large farmhouse to greet us and congratulate us on our punctuality after our six-hour drive. As light was failing, we went straight to the circle . . . and were immediately impressed.

Eight stones form Reva's Ring, and consist of the usual four station stones and the four stones that mark the rising and setting of the sun on the summer and winter solstices. At the center is a squarish stone, upon which rests other small stones of special interest, crystals and the like. The stones average about five feet in height and two or three tons each. They were "ground rocks," stones found at the edge of the farm's several fields. They varied in shape and texture and each had its own personality. Some may have been indigenous to this rocky corner of Connecticut, while others may have been delivered to site by glacial action. Long weathering assured that the stones were compatible with one another. The circle might have been ancient, but we learned that it was built in a day, less than five years before our visit. I paced the diameter at 19 megalithic yards (my pace), which translates to about 52 feet. What made the circle special, however, was its siting at the top of a field with a 360° panoramic view to the distant woods and surrounding countryside. Later, after visiting and walking Reva's

The Sun Stone at the Carl Sagan Memorial, Ithaca, New York. (John Ober, photo.)

nearby labyrinth, we went inside for tea and learned of the circle's creation.

Reva, of course, had identified the stones ahead of time and had some local equipment operators gather them with a backhoe and deliver them to site, where they were placed in a row like soldiers for inspection. "I took everything over five feet high that didn't look like a potato head," Reva told us.

Reva was studying for an apprenticeship with the well-known dowser, author, and geomancer Sig Lonegren. The stone circle was Reva's final project for a class called "Construction and Uses of Sacred Spaces," and the whole class gathered for two days

at summer solstice, 1994, to help complete the project. A couple of weeks earlier, with a friend, she had "dropped north" from Polaris, and so knew the meridian for alignment purposes.

On building day, the crew spent the morning talking about the project and laying out the alignments. Sig is one of the world's experts on finding correct alignments with geometry, and his clear instructions occupy many pages of his excellent Web site, listed in the sources. The site had an almost perfect horizon, so the east and west station stones also act as equinox sunrise and sunset markers.

"We measured everything," Reva said. "The guys who operated the backhoe and loader thought we were nuts. After the holes were dug, we ran around putting tobacco into the holes and things like that. After a while, the drivers started to come around. They were kind of macho–type guys so they didn't quite say it like this, but you could tell that they thought it was cool."

We'd heard this type of story before. Stonework has a way of gentling people down.

"Why the tobacco?" asked Jaki.

"Well, one of our our group said that the Native Americans would bless with tobacco before they would do something with the earth, so we figured we'd do the same."

"And why the loader?" I asked.

"The backhoe was too small for the stones," Reva explained. "We lowered the stones with the loader bucket, using nylon straps. At the end, we had this square stone left over, so Sig says, 'Let's put it in the center for an altar.' We'd had no plan for doing that, but I said 'Cool!' By this time, I'd lost it. I was on overload. But we put it in the center. The next morning, at sunrise, we had a ceremony to welcome the stone circle, inaugurate it, and to celebrate the summer solstice."

I looked at her inquiringly.

"It was in the right place," Reva smiled. "They all came out right in terms of sunrise and sunset, as I learned during the next year."

Reva's Ring, North Woodstock, Connecticut.

I was impressed. I'd never actually met Sig Lonegren, although we had many mutual friends, but this condition was rectified at Earthwood exactly three weeks after our visit with Reva.

Reva showed us her construction pictures and taught Darin how to dowse with a little plumb bob, which she had given him. Then we had to leave to rendezvous with my brother and his wife, who were expecting us for dinner sixty miles away. We were on our way south for a fortnight's vacation, and to visit a few stone circles, including the Georgia Guidestones and the Starbridge Circle (next entry). We thanked Reva for her wonderful hospitality and knew that our megalithic journey had found us a new friend and kindred spirit.

The Starbridge Circle

Location: Near Dahlonega, northeastern Georgia.

Contact: Brian or Donna Sullivan, 396 Star-bridge Road, Murrayville, GA 30564. Or call, 706-864-0882.

Description: Like the Four Quarters stone ring, this stone circle is in its early stage of construction. We visited on February 26, 1999, and only the East and South Stones were erected, as well as several outliers. We were met by Brian Sullivan, one of the permanent residents at the Starbridge Community, a group of families interested in all things Celtic, including music, myth, and ancient religion. After introductions, Brian, an overgrown leprechaun with the trademark twinkling eyes, took us to the circle by way of the Grandfather, a tall handsome tanned stone who greets people on their way to the circle site.

"The Grandfather Stone represents the chieftain diety who provided food and fertility to the tribe. He is the hunter, the provider, the responsible father. That's what we tried to set up as an image to aspire to."

We walked up the hill, through the woods, and came to the circle site, and I saw instantly that we were in a large earth bowl. Was it natural?

"I was influenced by the great stone works in England and in Scotland," said Brian. "Here, the cultivated land meets the wild land. In order to control the erosion, we converted this natural valley into a terraced amphitheater. There are three main levels, created by our gentlemen's group, the Starbridge Hunt Club, who used the trees in the valley as retaining timbers to stop the erosion."

"This is the first time I've seen a stone circle in combination with an amphitheater," I said.

In reply, Brian told us that the Starbridge Circle, although in its infancy, already serves as a venue for community meetings, bardic competitions, weddings, and religious ceremonies. His Celtic band, Emerald Rose, performs concerts there.

"The ancient stone circles were related to the religion of the time," Brian explained, "so in an effort to take that spirit and bring it into the future, we have incorporated the stone raisings as part of our seasonal celebrations. All our friends come for a day of festivity and feasting and getting to know each other. At night, we build a fire here in the center and do our ceremonies. The whole thing culminates in a huge energy-raising spiral dance, as was performed in ancient traditional British cultures. Imagine 150 people doing a spiral dance here. We really get going, and dance on all the levels, and past the stones and around the stone's hole. When everyone is in a great fevered pitch, we bring in the ropes."

"Ropes?"

"Yes, we use four ropes with maybe twenty-five people on each. Five or six druids, big strong fellows, man the wooden levers and the 4 × 4 wooden brakes which stop the stone from overturning. Then I instruct the people on the ropes in the raising of the stones."

We moved over to the lovely south stone, a triangular piece of gneiss about seven feet tall and probably close to three tons.

"When you get that many people excited and singing and with that divine power flowing through them, things fly up. We roll the stone to the edge of the socket hole, and the druids put levers down and place the 4 × 4 brakes on the back of the hole. Then the pullers, positioned on the downhill side, begin to haul. The druids lever the stone up to get it started and the pullers just pull it up, the brakes stopping it from moving forward. If the hole is big enough, it'll go right in. If not, they hit the brakes and begin to tilt up. As soon as it tilts enough, it drops in."

I did a quick mental calculation: 6,000 pounds divided by 100 people equals 60 pounds apiece. Yeah, that sounded quite doable.

"Would this be more difficult on a flat site?" I inquired.

"Not really. The secret is to get people that excited, that joyous, that uplifted, and, I'm tellin' ya, these things just leap up!"

Brian Sullivan shows us the impressive supply of recumbent stones at Starbridge Community, just waiting to be stood up in the circle.

Brian told us about their future plans, which include the four station stones and sunrise and sunset stones, much like Reva's Ring, which we'd visited two weeks earlier. The difference here was that the circle was in a depression, which meant that the East Stone did not mark the equinox sunrise. "We'll need another stone there, just to the right of the East Stone," Brian said.

"Are all the stones from the land?" Jaki asked.

"Yes, there's plenty of stones here. A lot of them that we use come from a pile that was pushed up many years ago by the people who put in the pond you saw. It's like a standing stone graveyard. We push them to site with a dozer, then do everything by hand, including digging the holes."

Brian showed us the collection of wonderful stones from which they could pick and choose. Sometimes, he and his friends would stand a particularly lovely stone right where they found it, and it would become a pleasant friend to visit during walks in the woods. We also visited a stone sweat lodge that the men of the tribe had built, a smaller stone ring of eight or ten standing stones, and a labyrinth created by the Starbridge women marked in stones on the forest floor. Brian had to excuse himself, as Emerald Rose had gathered for a practice session. Later, we met the group back at the house, traded a book for their excellent CD of Celtic music, and thanked Brian for the lovely visit. I look forward to seeing the circle in another two or three years' time.

The Stone Circle at Circle Sanctuary Nature Preserve

Location: Thirteen miles west of Mount Horeb in southwestern Wisconsin.

Contact: Selena Fox, Circle Sanctuary, P.O. Box 219, Mt. Horeb, WI 53572. Tel: 608-924-2216. E-mail: circle@mhtc.net. Web site: www.circlesanctuary.org

Author's note: During my first conversation with Selena Fox, I asked how many stones were in her circle, expecting a number like twelve or nineteen in reply. She laughed and said, "Thousands!" I soon learned that even a large circle (39 feet in diameter) can be composed of small stones. The Equinox Stone is the tallest, about 2½ feet above the ground. The article and picture that follow will tell and show how this unique stone circle is built, one stone at a time.

History and Description, by Selena Fox. (Note: This article is a condensed version of one that first appeared in *Circle Network News,* summer 1998, at the contact address above. It is used here with permission.)

The Stone Circle is the best known and one of the most frequented of the ritual sites at Circle Sanctuary Nature Preserve, and is positioned atop a naturally occurring sacred mound known today as Ritual Mound.

I first journeyed to the site during the summer of 1983 as a part of my explorations of the property prior to purchase. Sensing an energy vortex, I climbed the mound and found a grassy space at the center of a grove of oak and birch trees. I knew immediately that this would be a great site for the Stone Circle we had planned to create on Circle land.

The first Shamanic Wiccan ritual at the site occurred on Halloween night, 1983, just prior to Circle's signing of purchase papers. I did a Samhain rite there with Jim Alan, my spiritual partner at the time. During this rite we discovered that this circle area was a gateway to the Spirit world and to the realm of Faery. As we held up the offerings we had brought, a radiant white ball of light appeared in the darkness to the North. In our meditation that followed, we connected strongly to the Spirit of the land. We knew it was right to proceed with the land purchase and to create a Stone Circle at this place.

Construction of the Stone Circle began at Yule, 1983. Eight of us carried rocks and trekked to the circle area through a foot of snow. In the center of

The Stone Circle at Circle Sanctuary Nature Preserve, near Mt. Horeb, Wisconsin. (Selena Fox, photo.)

the circle, we placed our large altar stone, which had been at Circle's first stone circle near Sun Prairie, Wisconsin. Then we placed smaller stones next to the altar to form a fire ring. We lit a fire to celebrate the solstice and this group ritual space.

At Spring Equinox, 1984, a much larger group joined together for ritual at the Stone Circle. We brought and placed stones of various shapes and sizes to form cairns at each of the four sacred directions. At the Fall Equinox, we moved the cairns several feet outward to accommodate the growing number of people taking part in rituals. We also began forming the stone ring connecting these quarters and did a realignment ritual. Just prior to Yule, 1985, a group of us dedicated the Stone Circle to Mother Earth as a planetary healing place.

Over the years, numerous rituals have been held at the Stone Circle. In addition to Sabbat and Moon ceremonies, there have been rites of passage, weddings, baby blessings, and memorial services. My husband, Dennis Carpenter, and I had our legal handfasting there on June 7, 1986.

The Stone Circle is open for visitation by participants in festivals and other events at Circle Sanctuary land. Except for our Earth Day and Fall Equinox ceremonies, most rituals and spiritual activities are closed to the media. However, the Stone Circle has been featured in several documentary films, the print media, and on television. University professors and Sunday School teachers have arranged field trips here and brought classes to the Stone Circle as part of multicultural and interfaith learning experiences.

Our Stone Circle has continued to grow and now comprises several thousand rocks, pebbles, boulders and crystals from all over the world, some coming from contemporary and ancient sacred sites.

Our ring also includes a variety of small objects, such as amulets, coins, shells, fossils, and beads left as offerings, as well as unique items like a piece of the Berlin Wall and a pottery shard from ancient Crete. Each stone or offering is placed with a prayer or wish for well-being for the planet.

Contributions of stones and crystals are welcome. Mail stones to us at the contact address above or place them there yourself at one of our events.

Ed Heath's Megalithic Sculpture Park

Location: Sutton, Quebec.

Contact: Ed or Barbara Heath, 226 Elie Road, Sutton, Quebec J0E 2K0. Tel: 450-538-6692.

Description: Stone circles, dolmens, inuksuit, megalithic sculptural groups. See chapter 9.

The Swan Circle

Location: Pilton, Somerset, U.K. (about 5 miles east of Glastonbury).

Contact: Michael Eavis, Worthy Farm, Pilton, Somerset, U.K. BA4 4BY. Tel: 01749 89047. Web site: www.glastonburyfestival.co.uk

Description: See chapter 6. This lovely stone ring, designed and built by Ivan McBeth, is best visited during the Glastonbury Festival, the last weekend of June each year, when it is open to the public as a part of the festival.

The Dragon Circle

Location: Hascombe, Surrey.

Contact: Ella Anstruther, Lodge Farm, Dunsfold Road, Loxhill, Surrey GU8 4BL. Telephone: 01483 208626.

Description: See chapter 10. This circle by Ivan McBeth stands on a beautiful hill in the middle of a working farm. It is not a local tourist attraction. Permission is required to visit the circle.

Welsh Farm Circle

Location: Trevine, Dyfed, Wales. The circle is on the Pembrokeshire Coastal Path, a public right of way, so permission to visit is not strictly required. There is access to the path from the road between Llanrian and Trevine. The access is marked by a sign at a set of stone steps leading up into the field, on your left if coming from Llanrian, on the right from Trevine. Follow the path into the field and you'll see the stone circle. The owners of the farm, Mr. and Mrs. Phillip Matthews, live in a house near the path.

Contact: Ken Marpole, 8 Brynhedydd, Mathry, Haverfordwest, Dyfed, Wales SA62 5EX. Tel: 01348 831124.

History: Ken Marpole and Terry Gethin built this stone circle in 1996, "just for the fun of it." Ken is amused by how many people seem to take an interest in the circle and enjoy it. Sometimes, Ken and Terry enjoy hearing people theorize on how ancient the circle is and speculate about its purpose.

Description: The circle is composed of eleven or twelve stones in a ring of about fifty feet in diameter, with a stone in the center. Some of the work was done with a JCB backhoe, and some by hand. The largest stones are over four feet high and probably weigh about a tonne. The setting for this circle is spectacular, with a clifftop view over the Irish Sea. For fun, compare this circle with the similar but ancient Gors Fawr circle near Mynachlogddu, about twenty-four miles to the east. You will see why the casual visitor might be confused.

The Dutch Circle

Location: Near St. Austell, Cornwall.

Contact: Roy Dutch, Roy Dutch Camera Corner, 2 East Hill, St. Austell, Cornwall PL25 4TW. Telephone 01726 72960.

History: Roy Dutch was impressed by Ed Prynn's stone circle and decided that he'd like to build his own. He was fortunate in finding the perfect standing stones at a farm auction: a consignment of old granite fence posts, each eight to ten feet long. Some had metal gate-hanging pins still in them.

Description: The site is cozy and friendly: about an acre of field enclosed by hedgerow and hardwood trees, just off a quiet country lane. Ten of the stones are arranged in a circle about forty feet in diameter, and an eleventh stone is placed at the center. The tallest is 6 feet high. We visited the circle with Ed and Glynis, and Ed couldn't resist dancing around among the stones. I sensed that Ed thought of this circle as one of his children.

We met Roy at his camera shop in town and found him to be amiable, even taking time to help me with a camera problem. Later, he supplied me with a construction photo that shows the large fence posts being unloaded from a flatbed lorry with a long hydraulic boom and grappler, fastened just to the rear of the deck. The grappler had no trouble with the fence posts, which probably averaged about half a ton each.

The Welsh Farm Circle overlooks the Irish Sea, Trevine, Dyfed, Wales. (Cliff Osenton, photo.)

The Dutch Circle, St. Austell, Cornwall, U.K.

The Stones at High Canfold, near Cranleigh, Surrey, U.K.

The Stones at High Canfold

Location: Cranleigh, Surrey, U.K.

Contact: Dominic and Abigail Ropner, "Spring-fold," Barhatch Lane, Cranleigh, Surrey GU6 7NH. Tel: 01483 274190.

History: This delightful little circle was built by Dom and Abigail and their friends over several weekends in 1996 and 1997. After Dom's sister, Carey, died, he felt compelled to build the circle in her honor.

Description: A few of the Club Meggers (Ron, Margaret, Ed Prynn, and I) took time out from work on the nearby Dragon Circle to visit the Ropners. "The Stones" are perched on the slope of a lovely grassy tree-lined field, which adjoins an ancient trackway running through the property. I paced the circle's diameter at about 16.5 megalithic yards, or roughly 45 feet. Nine stones form the circle, including two quite close together that make up the South Gate. Constructed of Horsham sandstone, the tallest of the mini-menhirs stands about 4'6" high. A grouping of four smaller stones stands near the center, including a recumbent that falls along the north/south axis. From the center, looking south through the "gate," one's eyes are drawn to a spectacular fully crowned yew tree some 500 feet in the distance. This ancient sentinel has a trunk over five feet in diameter and it would be difficult to estimate the population of the rabbit warren underneath it. Roughly northeast of center is a stone marking the summer sunrise.

This chapter brings the state of stone circle construction up to the end of the 20th century. The final chapter brings us full circle back to the past, and sends tendrils out into the future.

— CHAPTER 16 —

Backward and Onward

If you build it, they will come.

—FIELD OF DREAMS

The megalithic journey is long, it is deep, and it is wide. It's beginnings are lost somewhere in the deepest crypts of our racial memory, with tantalizing clues about its beginnings being discovered even as we traverse the millennial gate. Archaeological evidence is showing up in unlikely places: a lifeless desert in southern Egypt; the tidal waters of Holme next the Sea in Norfolk, England; even at an excavation for a highrise development in Miami Beach, where a curious "inverted" stone circle was discovered in the limestone bedrock in late 1998.

My personal journey has led me to the inescapable conclusion that the ancient sites were closely connected with a deep spiritual or religious outlook. There is nothing original, profound, or very controversial about this view, shared by many if not most experts in fields as diverse as archaeology, archaeo-astronomy, and anthropolgy. I see the stone circles as forerunners to other religious temples and even the Christian churches of Europe's Middle Ages. In Britain, at least, recent evidence suggests that the great megalithic sites were preceded by even earlier round buildings constructed of massive oak

tree trunks, buildings that were probably roofed with a radial rafter system and, perhaps, thatch.

That many of the modern circles described in the previous chapter are built for similar purposes—rituals and observances by people following Earth-based spiritual paths—gives another clue, I think, to the original uses of the circles, even though there is no link of hard evidence. I do not have as many intuitive insights as I would like to have, but one that I feel quite sure about is that people have not changed a great deal in six thousand years. What may have been of interest in 4000 B.C. may be equally of interest to people today, particularly those that unplug the TV and step outside to look at a clear winter night's sky.

When I started out on my megalithic journey in October of 1966, I knew only of the great Stonehenge temple, but I was drawn to it like a moth to light. By osmosis, I soon became aware of other great stone circles, such as Avebury, Callanish, and Brodgar; megalithic tombs such as Maes Howe and Newgrange; mind-boggling rows of menhirs like Carnac; the great dolmens with 100-ton capstones;

and stupendous standing stones such as Kerloas, Dol, and the Grand Menhir Brisé. I visited other ancient sites that exhibited a much more advanced approach to what was still, essentially, Stone Age technology: the Egyptian pyramids, the temple at Baalbeck in Lebanon, and the Acropolis in Athens.

We discover the wonders in our own backyard last. North America wasn't (and still isn't) known for its megalithic monuments, but they are there, particularly in my native New England and in nearby southern New York. I grew up in Massachusetts, oblivious of nearby megalithic sites, and didn't get on their trail until the mid-1980s. There are a number of books dealing with pre-Columbian contact between North America and other parts of the world, such as Europe, the Middle East, China, and Japan. *America B.C.* by Barry Fell remains one of the popular titles, and introduced me to many enigmatic megalithic structures that I did not know existed. Meeting Enrique Noguera in 1987—the reader met him briefly in chapter 3—accelerated my knowledge and interest in these seemingly pre-Columbian sites. Enrique was executive director of the now-dormant Early Sites Preservation Association, and leads people to ancient sites in southern New York, through "Megalithic Journeys." (He may have been the first to coin the phrase.) I went on a journey with him in 1987, and visited structures that seemed to have no business in North America, structures remarkably close in design to ancient counterparts I'd seen in Britain. On my first journey with Enrique, we visited the Eye Chamber (discussed below) and several other corbelled arch stone slab chambers, as well as a most enigmatic dolmen in North Salem, New York.

The Eye Chamber

There are over three hundred corbelled arch stone slab chambers in New England and southern New York. They are found along roadsides, near colonial farm settlements and in the deep woods. They exist singly, in pairs, and in groups of three or more. The basic structure is quite similar from one to the next: Drystone walls rise up from the ground straight for a ways, and then begin to corbel inward, stone courses cantilevered over previous courses, very much like Maes Howe, described way back in chapter 1. When the builders thought the corbelled walls were close enough together, they somehow spanned the gap with giant stone slabs, sometimes weighing two tons or more.

The first such chamber I visited—in America, at least—is known as the Eye Chamber for its uncanny resemblance to a human eye, or with its vertical aperture, perhaps a cat's eye. The chamber is twenty-eight feet long and is capped by about seven large stone slabs of two to four tons each, spanning the six-foot gap between the tops of the corbelled sidewalls. The whole structure is covered with earth, forming a cozy earth-sheltered bump in the landscape.

The Eye Chamber does not stand alone. Downslope, and within a couple of hundred yards, are two more similar stone slab chambers, one in not such good condition. Also, just fifty yards away, a curious earth-covered stone slab bridge reaches over a stream, a stream that seems to emanate from a spring at the back of the bridge. I was reminded of the ancient clapper bridges on Dartmoor in Devon, bridges made of giant stone slabs spanning from buttress to buttress.

Some local historical societies are adamant that these structures are "colonial root cellars," although there seems to be little if any direct evidence to support the theory. Because of structural similarities, I was already open to a connection between these chambers and similar chambers of known antiquity in Cornwall, the Orkney Islands, and France. But Enrique told me something that was to strongly reinforce my willingness to accept the connection, something that seemed to put "colonial root cellar" theories to an almost impossible test.

"The openings of the structures in this complex, even the penetration under the stone slabs from which

The Eye Chamber near Carmel, Putnam County, New York.

the stream emanates . . . all but one point to the rising sun on the shortest day of the year. And the other chamber has an equinox sunrise orientation."

"Are you sure about the winter sunrise alignment?" I asked Enrique.

"Oh, yes, quite sure. Last December 22, we came here with video and still cameras and recorded the event. The sun rises and hits the back wall of the Eye Chamber about the same time that it penetrates into the spring under the stone slab bridge. It is as if the intent was to purify the spring, once a year, at the winter solstice."

"I'd like to see that. I witnessed a similar event at the Maes Howe burial chamber in Orkney, although it was the winter sunset, not the sunrise."

"Well, you can join us this year. We need to do some further documentation, so we'll be coming back to this site at the winter solstice. The site is being threatened by a major housing development, and we need to do anything and everything we can to save it. Come the night before and stay at my house."

A few months later, I made the four-hour trip to Enrique's place near Kingston, accompanied by Rohan, then twelve years old. We had a fine meal and an early night, as we needed to be up at 4 A.M. and on the road by 4:30. We picked up architect and art historian John Friedman en route and arrived at the Eye Chamber, near Carmel in northern Putnam County, when the sky was just beginning to brighten in the southeast. It was cold, but we didn't care, because stars shining told us that it was also clear. And the chamber itself provided a warm womb in which to await the sunrise. John Friedman, who knew exactly what to expect, took a seat at the very rear of the chamber, at the center. He turned his body at right angles to the opening, so that the first rays of the sun would hit his left profile. He sat still like Queequeg waiting to die in *Moby Dick,* except that John, I felt sure, was awaiting something more akin to a rebirthing. Enrique, Rohan, and I waited just outside the chamber, where we could scan the "horizon." Here, the place where the land meets the sky is nothing at all like the flat distant treeless horizon of Orkney Mainland. Here, there was a hill a half mile or so to the southeast, and lots of large hardwood trees, leafless, of course, at this time of year.

At about 7:30 A.M., the sun rose in the southeast. Conditions could not have been better; the clear cold air admitted the first solar rays gloriously. Enrique smiled. Rohan and I marveled at the precision of the alignment. John's profile, twenty-eight feet back from the chamber's entrance, was lit perfectly by the first rays.

John Friedman meditates at the back of the Eye Chamber, bathed by the first purifying rays of the rising sun at the time of the ancient new year.

After a few minutes, Enrique suggested that we catch the rays at the opening where the stream came out of the hillside, beneath seven large horizontal stone slabs, set one beside the other, their ends supported by large stones $1\frac{1}{2}$ to 2-feet high. Sure enough, the first rays gave a clear view of the stream sparkling and gurgling over a small waterfall, ten feet back under the slabs. It was magic, and we knew it. Later, I was to learn from Enrique that he'd traced the stream back to a spring emanating near an ancient tree about thirty feet back from the opening, traveling under slabs all the way.

"Now, we can catch one more sunrise event," said Enrique.

Rohan and I were not sure how this would be possible, as the sun seemed to be rising rapidly now, but we followed Enrique down the slope in front of the Eye Chamber and soon walked into the shadow of the distant hill, thereby witnessing an apparent sunset. In a few minutes, we arrived at another chamber, one I'd seen on a previous visit, and waited. About forty minutes after the spectacular sunrise event at the upper chamber, we witnessed a virtually identical repeat performance at the lower chamber. This punctuated the magic of the complex. One chamber could be just coincidence, even the spring and stream alignment thrown in, but this lower chamber could not have the same orientation as the other two, and I expressed as much to Enrique.

"You're right. We figure that there is about an 11° difference between this chamber and the one above, but they're both perfectly aligned with the sunrise."

"Colonial root cellars?" I asked.

Enrique just smiled.

The North Salem Dolmen

After taking a well-appreciated breakfast at a nearby eatery, we drove just a few miles southeast so that Rohan could see the well-known and spectacular North Salem Dolmen—or Balanced Rock or

Perched Stone, depending on your point of view. The structure is right on the side of State Route 116 in the village of North Salem, New York, very close to the Connecticut state line.

Rohan enjoyed this one. The dolmen consists of a ninety-ton boulder of pink granite resembling the head of a turtle, and four to six standing stones beneath it, depending on how you make your count. Some of the stones have sheared into two halves under the great weight; do you count them as one or two? Enrique had told me that probing with rebar reveals that the pegstones supporting the boulder extend at least four feet into the ground. If true, it would mean that these stones are long and narrow menhir-type stones, not simply squat gumdrops.

In 1988, the local historical society posted a sign next to the structure, calling it a "glacial erratic."

We laughed and marveled at the skill and intelligence of the glacier that would erect four standing stones and plunk a ninety-ton stone on top. Fell's book has a photo of a very similar dolmen, which he describes as being of the "cromlech type," at Trelleborg, Sweden, while Michael Balfour, in *Megalithic Mysteries,* includes a beautiful color plate of the De Steenbarg *hunebed,* or chambered tomb, also remarkably similar, near Noordlaren in Holland. Like North Salem, the De Steenbarg capstone "looks preposterously large and must surely be the largest natural boulder ever to be heaved on to supporting stones for funerary purposes in The Netherlands." Curiously—and, I am sure, coincidentally—the De Steenbarg capstone also resembles a turtle's head.

In common with both Trelleborg and De Steenbarg, there is little doubt that the pink granite

The Dolmen at North Salem, Westchester County, New York.

capstone at North Salem *was* brought to the site by glacial ice. But, like its European counterparts, I am equally certain it was placed on its supports by the hand of man. Close proximity to the many stone slab chambers in the area suggests a connection with them, but does not prove it. Still, it is remarkable that two distinct styles of known megalithic building are in close provenance to each other in the New World, and three thousand miles from their European cousins. Another origin scenario for the megalithic structures in New York and New England has them erected by Native Americans, without European contact. While the various tribes were certainly capable of building with large stones, there is no evidence that they actually did so.

Whatever its origin, the North Salem dolmen fascinates. On the American segment of our own megalithic journey in February, 1999, I pulled off Interstate 84 at North Salem to show Jaki and Darin the dolmen. They were not disappointed.

America's Stonehenge

For many years, I'd been aware of the site known as "America's Stonehenge" (formally "Mystery Hill") in North Salem, New Hampshire, a complex of seemingly anomalous megalithic remains. I knew vaguely that there were chambers and a so-called sacrificial table, and also that there had been great destruction of the site over the years, particularly when local builders used it as a quarry during the 19th century. It wasn't until August of 1998 that I finally visited America's Stonehenge, accompanied by Jaki, Darin, and Stephen and Robin Larsen. The Larsens were particularly interested to see the sacrificial stone and compare it to their own anomaly at the center of the Stone Mountain Stone Circle. I arrived at the site with low expectations, worried that this might be a place where truth was stretched to help separate tourists from their dollars. My spirits and expectations rose as we entered the tasteful lodge, which served as an entrance house, museum, and shop. Immediately, we were caught up in sev-

eral interesting exhibits, which included artifacts, inscribed stones, original lab reports of carbon dating, a carefully constructed scale model of the site, and much more. It was some time before we made our way to the ticket counter, paid our entrance fee, and were directed to a small theater where a short introductory video is shown. Without hype or outrageous claims, our interest had been nicely tweaked. After our orientation, we walked up the hill to the site itself.

Space does not permit a full description of America's Stonehenge, but one of the most impressive features is the long corbelled "oracle chamber" reminiscent of both the Eye Chamber and Cornish fogous in its construction. The defining features of this chamber are the "secret bed" and the "speaking tube." The "bed" is a stone-built niche 6'4" long and 22 inches wide that a person could slip into and remain unseen. At the business end of the niche is the inner end of the 6-foot-long speaking tube, a 4-inch by 6-inch shaft extending transversely through the thick stone walls of the chamber. The outer end of the shaft comes out beneath the huge "sacrificial table," a 4.5 ton slab with a grooved rectilinear runnel carved onto its face. This runnel, in combination with a carved outlet at one end of the stone, was obviously intended to deliver an unknown liquid to a collecting vessel. One line of thought is that an oracle's voice would emanate from beneath the stone during the ritual sacrifice of small animals on the table, and that blood was the desired liquid. There is even a small stone enclosure nearby that might have served as a holding pen for unlucky goats or sheep. Another view, based upon similarity to a stone in Old Sturbridge Village in Massachusetts, is that the table is a cider press. Still another interpretation is that the sacrificial stone is nothing more than a "garden-variety lye stone." But the physical evidence at America's Stonehenge, while inconclusive, certainly makes sacrifice a legitimate contender as the stone's purpose. But whatever its purpose, the Larsens and I were immediately struck with the similarities to the Magic Stone at the center of their circle.

We'd heard about the sacrificial stone, but we were totally unprepared for what we were to see next. The central part of the site, now marked with a modern viewing platform, acts as an effective backsight to a number of important astronomical alignments, the foresights for which are several triangular-shaped standing stones. America's Stonehenge, in fact, has much in common with ancient British stone circles. At varying distances of about 100 to 200 yards from the center, menhirs mark the winter solstice sunset, the Imbolc sunset, the summer solstice sunrise and sunset, and true north. Fallen menhirs, which will one day be raised again in their original sockets, mark the equinox sunrise and sunset, Samhain sunset, and Beltane sunrise. In a summary of a report to the owners from B. V. Pearson Associates (consulting engineers and land surveyors) of Chester, New Hampshire, senior partner Chas Pearson states, "... it seems surprising that anyone with even a basic background in field evidence evaluation, historical research, archaeology or astronomy would reach any conclusion other than the fact that [America's Stonehenge] was constructed as, and most likely used as, a calendar site ..."

Although the site is wooded now, it is thought that at the time of construction this rocky hill would have been quite barren of trees. The present owners, the Stone family (honestly!), maintain clearings behind the menhirs so that the sunrise and sunset events, as well as the North Star, can be observed

The author is impressed with the similarities between the "sacrificial table" at America's Stonehenge, in North Salem, New Hampshire, and the magical centerpiece at Stone Mountain Stone Circle.

The Winter Sunset Stone at America's Stonehenge, looking back toward the central observation platform.

FULL CIRCLE

My megalithic journey had come full circle, beginning at ancient Stonehenge in 1967 and, in 1998, returning to the distant past at the enigmatic "America's Stonehenge." Along the way, I'd met the cast of extraordinary characters whom you met in part 2, plus several other stone builders whose work appears in chapter 15.

As my two-year labor of love winds down—I refer to the book in your hand—I ponder the future, my future and what part the stones will play in it, and wonder what the other Stone People have in mind. It is April 22, 1999, hopefully my last day of major writing, just a few paragraphs to go. I've had occasion to speak to most of the Club Meggers in the past couple of weeks, and can report on their doings. And I'll phone the others today for an update.

Stephen and Robin Larsen continue to lead the busiest—and most interesting—lives of anyone I know. They always seem to be on some quest for myth in some fascinating part of the planet: India, Ireland, the Australian Outback, the far reaches of the Amazon. They have recently visited Callanish, where they were shown the sites by Margaret Curtis. And when they're not out in the world, the world comes to them, at the Center for Symbolic Studies. Stonewise, they plan on setting the winter sunrise and sunset stones at their circle very soon, and the plan is to do it by hand this time.

Bill Cohea finds that retirement doesn't make life any the less busy. He still oversees Columcille, Inc., which continues to grow and to thrive. Bill conducts an occasional workshop, such as one on Sacred Placemaking this year; he goes back to Iona whenever he can; and, yes, he continues to set stones at the Megalithic Park. At the moment, Bill and I are planning to erect a nine-footer by hand, the first large stone to be set at Columcille without heavy equipment.

Ed Prynn and Glynis are getting ready for the big solar eclipse in Cornwall, which happens on

from the center observation point. With good long alignments and nice pointy stones, America's Stonehenge is one of the best ancient sites for astronomical observation that I have seen. And how old is the site? Carbon dating and astronomical evidence point to use as early as 1500 to 2000 B.C. One of the pieces of astronomical evidence is particularly interesting. On either side of the North Stone are smaller pointy stones that mark the right and left extremes of the orbit of the star Thuban, which, in 2000 B.C., was the best star we had to gives us north. Bisecting the extremes of Thuban's orbit give us the true meridian, and all of this evidence exists in the stones.

August 11th, 1999, at 11:11 A.M. A week of festivities is planned, including setting up a new Eclipse Stone. "We'll all have a damned good time," says Ed. And that's what stone circles are all about.

Ivan McBeth is in the process of finishing up the Dragon Circle as I write, and it should be completed about the time of publication.

Cliff Osenton continues his research. Presently, his attention is focused on forest management and learning the woodworking skills that the ancients would have had to master to build the megaliths. They would have needed levers, rollers, stone boats, good timber stacking material . . . and the main tool they had to create all this was a stone axe. Cliff now feels that he can trace megalithic building back to an axe, and hopes to demonstrate the entire process very soon; that is, walk into a forest with an axe and extract, move, and erect a large standing stone, or even a dolmen. I'd like to be there.

Ron and Margaret Curtis continue to act as guardians over the Callanish complex of monuments. They have successfully saved an ancient cairn at the extreme north end of the Callanish I stone circle from further destruction by roadworks, and have rebuilt it on council land, just a hundred meters or so from Olcote, their home. They have another stone to reset at the Bernera Bridge site (Callanish VIII), one of the distant foresight stones across the water. And Margaret, as resident archaeologist, continues to guide visitors at Callanish. See sources.

Ed Heath seems to be finding it difficult to leave the megalithic phase of his sculptural career, with new commissions coming in all the time, the latest being a stone circle for a small town in Quebec.

Stone circles are here to stay. The next twenty-five years may even bring about a new golden age in their history. At the least, I expect that we will learn a great deal about how the ancients might have built their masterpieces, if not why.

And myself? Well, I can't imagine life without the stones any more than I can imagine life without my family. We'll continue to do megalithic work, most of it by hand, at workshops here at Earthwood Building School. Tomorrow, in fact, some young people are coming over to help me finish the Ancestor Group in front of the stone circle. Jaki would like a dolmen someday, and who am I to say no? It makes more sense than a diamond ring. And I expect that, like cordwood masonry and underground housing, stone circles will continue to take us around the world.

And that leaves just one more person that we must speak of, the person who may turn out to play a major part in this ongoing story, and take it well into the Third Millennium. And that person, of course, is you, dear reader. By now, you know me better than I know you, and you know that I can run off at the mouth now and again. So, caught up in the sadness that is a part of finishing this labor of love, I hope you will permit me one final bit of personal pedantry.

Everybody's different. This remains the one thing I know that might qualify as a universal truth. At the very least, those two words imply tolerance; at least they do to me, and I'm as different as the next guy. And tolerance is important. How important? Well, the two big stories in this morning's news are about *in*tolerance: a horrific shooting in a Colorado high school, perpetrated by young people who hate those different from themselves; and ethnic cleansing in Kosovo, the result of intolerance that dates back hundreds of years. Of course, the real answer to the world's problems will be found beyond tolerance. It will be found in love. Better authorities than I have spoken eloquently about this for thousands of years, and, eventually—we can only hope—the idea of love will reach critical mass. Only then will our species, at last, deserve to be called a civilization. But universal love is a goal, a distant destination, and we have to pass through tolerance to get there. We can't love somebody we can't even tolerate.

What does all this have to do with stone circles? Everything. Two years ago, I thought I'd write a book about stones circles and how to build them.

The Ancestor Group at Earthwood Stone Circle.

It turned out that the book was as much about people as it was about stones. Good people. Maybe even *eccentric* people. But certainly tolerant and, yes, loving people.

Building stone circles has to be good for the human condition. In a time when religion takes on ever greater overtones of intolerance (instead of the love that was supposed to have been the main idea), I have found that the kind of Earth-based spirituality often connected to the building and using of stone circles is inspiring and invigorating, and it extends hope for the human/planetary relationship. Stone circle people are peaceful people, to a man and woman, and they have a deep respect for Gaia.

No, your humble author has not taken on a new religion, a worship of the stones. My spiritual journey may have been rekindled by my megalithic journey, but enlightenment can come from a variety of sources. We need to keep our eyes, ears, and hearts open. Nor do I say that everyone should run out immediately and built a huge circle of megaliths. Aubrey Burl would have a fit. What I *do* say is, take what value you can from this book and use it in your own way, even if this is nothing more than enjoying the read. Build a window box circle, a garden circle, a mini-circle on your next visit to the beach, even a megalithic circle if you're so inclined. I'd love to receive a picture of and a paragraph about your creation. Or visit the work of others, ancient and modern. To continue your studies, start with the sources and bibliography sections that follow.

Time and space grow short. With equal weight, I offer two final pieces of advice . . . and a *koan* to dwell upon.

1. Be careful.
2. Have fun.

Stone circle: never begins, never ends.

West Chazy, New York

347

Sources

Here is a list of sources to lead you further into the stone circle field. Like the bibliography, I found it convenient to separate the entries into six categories, as follows:

Stone Circle Builders
Labyrinths
Dowsing and Geomancy
"Megalithic" Web sites
Magazines and Periodicals
Guide Extraordinaire

To visit individual stone circles, see chapter 15

STONE CIRCLE BUILDERS

Do you have a project that you're not sure you can handle alone? Here's a list with a good geographic spread.

Cliff Osenton, 14 Lower Close, Bodicote, Banbury, OX15 4DZ, U.K. Tel: 01295 273506.

Ivan McBeth, c/o Ella Anstruther, Lodge Farm, Dunsfold Road, Loxhill, Surrey GU8 4BL, U.K.

Tel: 01483 208658 or 01483 208626. *Also conducts stone circle camps.*

Ed Heath, 226 Elie Road, Sutton, Quebec, Canada J0E 2K0. Tel: 450-538-6692.

Rob Roy, 366 Murtagh Hill Road, West Chazy, NY 12992. Tel: 518-493-7744. *Also conducts workshops.*

Jay Markel, Viriditas, 304 Camino Bosque, Boulder, CO 80302. Tel: 303-444-7888.

John Martineau, The Walk Mill, Discoed, Powys, Wales LD8 2NT. Tel: 01547 560251. John has designed and built Greenhenge in England and Diana's Memorial Circle in Wales, both listed in chapter 15, as well as a circle in Slovenia. He'll only build a stone circle with public access, believing that private circles are "against the whole idea of the thing." John integrates sacred geometry and astronomical alignments with careful siting.

Chuck Pettis, The GEO Group. *See entry under "Dowsing and Geomancy" below.*

Dominic Ropner and M. Scott, Monument ("Specialists in Outdoor Sculpture and Stone Monuments"), Springfold, Barhatch Lane, Cranleigh, Surrey, GU6 7NH, U.K. Tel: 01483 274190.

LABYRINTHS

Labyrinths are distant relatives of stone circles, but they have a lot in common. They're great for those wishing to create sacred space, but lacking in large stones. At least four of the sites in chapter 15 have labyrinths in close proximity to stone circles, and Jaki and I are even thinking of building one.

Marty Cain, 55 Park Street, Newport, NH 03773. E-mail: marty@sugar-river.net. Tel: 603-863-7343. Marty is a dowser, artist, geomancer, and labyrinth designer with extensive experience in sacred spacemaking. She has created over 75 labyrinths in North America and Scotland and sells a one-hour video on the subject.

The Labyrinth Society. P.O. Box 144, New Canaan, CT 06840. Toll free in the U.S.: 877-446-4520. Web site: www.geomancy.org/tls/. Formed in 1999, the society's mission is "to support those who create, maintain, and use labyrinths and to serve the global community by providing education, networking, and opportunities to experience transformation."

Sig Lonegren, labyrinth creator, geomancer, author of books on dowsing and labyrinths. See Mid-Atlantic Geomancy below.

DOWSING AND GEOMANCY

Here are some useful contacts if you want to check out the earth energies and other transrational considerations before you site or lay out your circle.

Mid-Atlantic Geomancy. This group gets its name from the locations of its two organizing geomancers, Dr. Patrick MacManaway, a Scotsman living in Vermont, and Sig Lonegren, a Vermonter living in England. The best way to reach them is by e-mail or through the MAG Web site,

one of the most interesting sites I've seen. You'll find: how to align a stone circle with solar events; an introduction to geomancy; how to lay out a labyrinth; and a whole lot more. The 200-entry bibliography, built up over many years, is a valuable resource. Web site: www.geomancy.org. Sig's E-mail: sig@geomancy.org See also the next entry.

Dr. Patrick MacManaway, Whole Earth Geomancy, 2031 Shelburne Road, Suite 6, Shelburne, VT 05482. Analysis and balancing of earth energies and chi flow; feng shui and pre-construction assessment of sites; location and design of sacred space; labyrinths. Tel: 802-985-2266. Web site: www.geomancy.org.

The American Society of Dowsers (ASD), P.O. Box 24, Danville, VT 05828. Tel: 802-684-3417 or 800-711-9530. E-mail: ASD@dowsers.org. Web site: http://dowsers.newhampshire.net. The Society is a great source of information about all sorts of dowsing: finding water, minerals, electromagnetic fields, "geopathic fields," spriritual dowsing, you name it. There are 78 local chapters in the U.S., and over 30 new pending chapters. The ASD Bookstore is at 99 Railroad Street, St. Johnsbury, VT 05819. Tel: 802-748-8565 or, within the U.S., 800-711-9497. It's a one-stop source of books, videos, and dowsing equipment. Subjects of interest include: dowsing, earth mysteries, feng shui, labyrinths, and more. The Web site has an on-line catalog and order form.

British Society of Dowsers, Sycamore Cottage, Tamley Lane, Hastingleigh, Ashford, Kent TN25 5HW, U.K. Tel: 01233 750253. This is the British counterpart to ASD.

The GEO Group, P.O. Box 602, Medina, WA 98039. Tel: 425-637-8777. E-mail: crp3@msn.com. Web site: www.geo.org. The Geo Group is "a nonprofit organization that promotes and facilitates the creation of sacred spaces and environmental art for the purposes of world peace and planetary healing." The Group's president, Chuck

Pettis, was one of the first to build a modern stone circle (Ellis Hollow in chapter 15). He does spiritual dowsing, conducts geopathic surveys, and searches for both underground veins of flowing water and leys, thought by some to follow energy lines.

"MEGALITHIC" WEB SITES

People interested in modern stone circles are almost invariably drawn toward the ancient sites as well. In addition to books listed in the bibliography, the following Web sites are great sources of information. Web addresses change frequently. If an address does not work, use the Internet search function.

The Stone Circle Webring: www.webring.org/cgi-bin/webring?ring=stonecircle;index. Last time I checked, late April of 1999, this amazing Webring provided links to 118 sites about stone circles and related subjects. Even the newly discovered Miami circle was listed. This is a good place to start your surfing. The site is maintained by the thorough Andy Burnham, who is always looking for new input by e-mail: aburnham@easynet.co.uk.

Megalithic Map in Association with Aubrey Burl: www.megalith.ukf.net/bigmap.htm. This incredible map will lead you to about 1,000 different stone circles and megalithic sites in Great Britain and Ireland. Descriptions and locations are given, and, in many cases, pictures and links to Web sites with further information. This is another amazing Andy Burnham creation.

Prehistoric Web Index of Ancient Sites in Europe: http://easyweb.easynet.co.uk/aburnham/database. Track down almost any megalithic site in the U.K. by location or by name in an alphabetized list.

Stone Pages: www.stonepages.com/. The most interesting megalithic sites in the British Isles, France, and Italy: stone circles, dolmens, cairns, standing stones. Click on "Archaeo News" for the latest on recent archaeological meetings, digs, and discoveries. Recent on-line issues had articles on the Miami Circle, the Timber Circle at Holme next the Sea, megaliths in Korea, and the latest about plans to improve the environment around Stonehenge. Click on "Archaeo Links" for nearly 300 frequently updated links to Web sites about stones and related subjects. Stone Pages provides solid information, well organized.

New England Antiquities Research Association (NEARA): www.neara.org. NEARA is a preservation group concerned with saving ancient stone sites in New England. Current events are listed. Lots and lots of photos.

The Gungywamp Society: www.goudsward.com/gungywamp/. Focuses on the megalithic sites of the northeastern United States. Write them at 334 Brook Street, Noank, CT 06340 or e-mail at Noank2@AOL.com.

America's Stonehenge: www.stonehengeusa.com. See chapter 16. Tel: 603-893-8300. Or write at P.O. Box 84, North Salem, NH 03073

MAGAZINES AND PERIODICALS

I subscribe to three mainstream American magazines that all do a good job of reporting new megalithic and related finds within six months of first publication of discovery. Followed by their subscription phone numbers and Web sites, they are:

National Geographic (800-548-9797, www.nationalgeographic.com); *Discover* (800-829-9132 in U.S., 515-247-7569 from outside U.S., www.discover.com); and *Archaeology* (877-275-9732, www.archaeology.org).

The following publications are not as well known:

Antiquity. This is an old and well-respected journal of archaeology. Lots of articles about matters megalithic. Subscriptions are managed by The Company of Biologists, Bidder Building, 140 Cowley Road, Cambridge, CB4 4DL, U.K. Tel: 01223 426164. Web site: http://intarch.ac.uk/antiquity/.

Current Archaeology, 9 Nassington Road, London NW3 2TX, U.K. This one is written for the layperson and deals almost exclusively with British archaeology. A well-illustrated forty-page issue is published bi-monthly and without advertising. A subscription form can be printed off their Web site: www.archaeology.co.uk/.

Third Stone Magazine, P.O. Box 961, Devizes, Wiltshire SN10 2TS, U.K. Tel: 01380 723933. "The magazine for the new antiquarian." Cliff Osenton is among the many contributing writers. The Web site has current subscription info: www.thirdstone.demon.co.uk.

The Ley Hunter Journal, P.O. Box 180, Stroud GL5 1YH, U.K. In Britain, call 01453 768804. In the U.S., subscriptions are handled by B & S (USA), Dept. TLH, Box 940, Beacon, NY 12508. Tel: 914-838-4340. "The journal of geomancy and earth mysteries." Paul Devereux and Aubrey Burl are listed as consulting editors for this journal, which is entering into a new publishing format: one large issue per year. The Web site has a great reading list: www.leyhunter.com/leyhunt/welcome.html

Meyn Mamvro, 51 Carn Bosavern, St. Just, Penzance, Cornwall TR19 7QX, U.K. This small journal, published three times a year, is devoted to the ancient stones and sacred sites in Cornwall. With lots of good stuff in a concise package, *MM* also serves as a newsletter for the Cornish Earth Mys-

teries Group and the Sacred Sites Network Group. Subscriptions are £6 a year in the U.K.

Archaeoastronomy: The Journal of Astronomy in Culture. One volume is published annually in two parts. This is a scholarly journal with juried articles, and a good source for the latest discoveries in the field. A sister publication, *Archaeoastronomy and Ethnoastronomy News,* appears quarterly. Both the journal and the newsletter are published by The Center for Archaeoastronomy, P.O. Box "X," College Park, MD 20741-3022. Tel: 301-864-6637. Visit their Web site at: www.wam.umd.edu/~tlaloc/archastro/.

Stonehenge Viewpoint, 2261 Las Positas Road, Santa Barbara, CA 93105. Tel: 805-687-9350. Strictly speaking, *Stonehenge Viewpoint* is no longer published as a journal, although collections of back issues are available. The focus now is a catalog of books, covering geomancy, pre-Columbian visitors to North America, off-center speculation on megalithic sites, Atlantis, the ancient Earth Goddess religion, crop circles, etc. This is a place to find the uncommon and the controversial; it's fun just browsing through the titles and descriptions.

GUIDE EXTRAORDINAIRE

Margaret Curtis, "Olcote," New Park, Callanish, Isle of Lewis, Scotland, HS2 9DZ, U.K. Tel: 01851 621277. Personalized tours of Scotland's greatest stone circle complex, by the leading authority. She is so good that I created a special category just to list her. Also, with Ron Curtis, Margaret writes, publishes, and sells books about the Callanish group.

Glossary

Alignment. Three or more points placed in a straight line; for example, two standing stones and the midsummer sunrise.

Analemma. A device that can be used to determine the date from the sun's image. A hole in a stone, for example, can serve as an aperture for the sun. At noontime each day, the sun's image can indicate the date on a figure-eight pattern carved into another stone.

Anorthosite. A hard igneous rock found in the Adirondacks of New York, Glen Torridon in Scotland, and on the moon. Consisting largely of soda-lime feldspar, it is like a quartzless granite.

Archaeo-astronomy. The study of ancient astronomy.

Avenue. Two rows of standing stones, often leading to a stone circle.

Azimuth. A compass bearing based upon true north. East, for example, corresponds to an azimuth of 90°.

Backhoe. 1. The articulated hydraulic digging arm of a tractor or excavator. 2. A backhoe loader, a piece of equipment with a backhoe at one end and a large bucket loader at the opposite end. In Britain, the machine is commonly known as a JCB, a popular brand name.

Backsight. A place from which to observe an alignment, often a standing stone. *See also* Foresight.

Barrow. Earth-covered burial chamber.

Blocking stone. The terminal stone of a stone row, turned at right angles to the prevailing orientation.

Bluestones. The smaller two- to five-ton standing stones at Stonehenge, originating from the Preseli Hills of Wales, 135 miles from the monument.

Broch. Circular drystone fortification, once common along the coast of the highlands and islands of Scotland. Also called a dun, as in Dun Carloway.

Cairn. A round or long mound of stones, often covering or enclosing a chamber or tomb.

Capstone. The horizontal flat stone at the top of a dolmen.

Corbelled arch. Describing stone walls whose courses are stepped inward, until flat slabs can span the remaining gap. The Maes Howe burial chamber is a good example, very old chambers in New York and New England also exhibit the style.

Cromlech. Same as dolmen in Wales, but, in France, a stone ring.

Cross-quarter days. The four ancient Celtic holidays of Imbolc (about February 1), Beltane (May 1), Lugnasad (August 1), and Samhain (November 1). These days fall approximately halfway between the equinox and solstice observations.

Declination. 1. The angle that a freely turning compass needle makes with the imaginary line pointing to true north, also called the magnetic declination. 2. In astronomy, the angular distance of any object from the celestial equator, northward or southward.

Dolmen. A megalithic monument consisting of a large usually flat stone supported by three or more upright stones. Literally, "table stone."

Druid. In pre-Roman Europe, the Celtic shamans or priests. Today, a follower of one of several sects of Earth-based religion.

Ellipse. A geometric figure consisting of a ring and two focal points. The sum of the distances from the two foci to any point on the ring is constant. A common shape used with stone rings.

Equinox. One of two days of the year when the day and night are of equal lengths. The vernal or spring equinox occurs about March 21, the autumnal equinox about September 22.

Excavator. A machine, usually on tracks, designed for excavating. The large bucket is very useful in digging socket holes and erecting standing stones.

Foresight. Usually, the second of three objects that are in alignment. At Stonehenge, for example, the Heel Stone is the foresight for the summer solstice sunrise. *See also* Backsight.

Flattened circle. A geometrical stone ring shape identified by Professor Alexander Thom. *See* chapter 11 for design details.

Fogou. An underground chamber in prehistoric settlements, the purpose of which is disputed. In Britain, fogous occur almost exclusively in Cornwall. In France, similar structures are known as *souterrains.*

Gnomon. 1. The upright on a sundial that shows the time of day. 2. A perpendicular column erected on level land used to find the meridian by equal angle of the sun. *See* chapter 11.

Henge. An earthern ring enclosure with an inner ditch and one or more entrances.

Jammer. One of a number of smaller stones used to tighten a standing stone in its socket.

Ley. Sometimes, erroneously, ley line. One of a number of straight lines that can be drawn across the countryside to connect natural or manmade objects. Popularized by Alfred Watkins in *The Old Straight Track,* many ley aficionados believe that the lines indicate the location of natural "earth energies."

Lintel. A horizontal stone supported by two standing stones.

Low-loader (U.K.), **Lowboy** (U.S.). A flatbed lorry or truck with a low-lying bed, used mostly for transporting heavy equipment, but also good for moving megaliths.

Medicine wheel. In Native American culture, medicine wheels were large round stone rings with a number of spokes radiating from the center. The spokes have been found to align with important astronomical events, much like stone circles.

Megalith. Literally, "large stone."

Megalithic hitch. A method of tying a chain around a stone so that it can be lifted on its vertical axis. *See* description in chapter 13.

Megalithic yard (MY). An ancient unit of measure postulated by Professor Alexander Thom, equal to 2.72 feet or 0.829 meters.

Menhir. A vertical stone, usually standing alone (French).

Meridian. The true north alignment.

Metric tonne. 1,000 kilograms, or 2,204.62 pounds.

Midden. Ancient refuse heap. Provides many clues to domestic life of the period.

Monolith. A single standing stone.

Neolithic period. Literally, New Stone Age. A period when farming began to supersede nomadic life, about 4500 to 2200 B.C.

Ogham. Ancient Celtic writing, characterized by long straight lines. Ogham is sometimes found on ancient stones.

Orthostat. An upright; a standing stone.

Outlier. A standing stone found some distance from the main grouping, often used as the foresight or backsite of an alignment.

Packer. *See* Jammer.

Parallax. The apparent movement of an object when viewed from different positions.

Recumbent stone. A stone deliberately laid on its side in a stone circle, particularly in Ireland and the Grampian region of Scotland. The stones are thought to have served astronomical purposes.

Recumbent stone circle. A stone circle containing a recumbent stone.

Refraction. The bending of light rays from a celestial object, greatest when the object is low in the sky, giving the effect that the object appears higher than its true altitude.

Sarsen. An extremely hard and dense sandstone found on the Marlborough Downs of southwest England. Sarsen is the predominant stone at both Avebury and Stonehenge.

Socket. The hole in which a standing stone is set. In combination with packers, the socket provides an effective foundation for the stone.

Solstice. One of two days in the year when the sun is in its extreme northerly (summer) or southerly (winter) rising and setting positions. The summer solstice, or longest day, occurs on or about June 21 in the Northern Hemisphere, while the winter solstice, or shortest day, usually occurs on December 21.

Standstill. The sun appears to duplicate its rising and setting pattern for several days at the time of the solstice, an effect called the standstill. The moon does a similar but more complex dance every 18.61 years. *See* chapters 8 and 11 for details.

Stele or **Stela.** A standing stone engraved with an inscription or design, and used as a monument. Examples are the carved Pictish stones of Scotland, ancient Mayan steles in Central America, and carved religious stones in India and Southeast Asia.

Stone circle. Strictly, a group of stones arranged in a true circle, but more commonly assigned to any ring of stones. *See* Stone ring.

Stone ring. A ring of stones, usually installed vertically in the ground. The ring can be designed to follow an almost infinite number of geometries, or none at all. Common shapes chosen in ancient and modern rings are the circle, ellipse, egg-shape, and flattened circle.

Ton. In this book, an American or "short ton" of 2,000 pounds.

Tonne. In this book, the British ton, also called the "shipping ton" or the "long ton" of 2240 pounds. *See also* Metric tonne.

Trilithon. Two upright stones supporting a lintel; the best-known examples are the five trilithons arranged in a horseshoe shape at Stonehenge.

Vitrification. The fusing of stones by great heat.

Widdershins. Counterclockwise; considered by some to be unlucky or tempting fate.

Annotated Bibliography

This is an eclectic list, ranging from the important and scholarly (Burl, Thom, etc.) to the lighter-weight overviews and picture books. I've found all to be of interest and some have been invaluable in shaping my understanding of the ancient stones and the people who erected them. Several of the books have their own extensive bibliographies (indicated), often more complete than this one, but not necessarily annotated. ISBNs are given when known; some works predate ISBNs. I have only listed books with which I am familiar, most of them in my research library. Videos and periodicals are included in this list, and identified as such. For convenience, the bibliography is organized into six categories, although some entries could happily fall under more than one. The categories are:

  Stonehenge
  Stone circles and megaliths
  Guide books
  Archaeo-astronomy
  Sacred spaces and earth energy
  Other

STONEHENGE

Atkinson, R. J. C. *Stonehenge* (Hamish Hamilton, 1956; revised edition, Pelican, 1979; Penquin Books, 1990). The publishing history alone shows that this is a very important book, the definitive account of the archaeology and historical development of Britain's most renowned ancient monument. For years, Atkinson's account of the transportation of the stones and construction of Stonehenge has been considered to be authoritative. Now, modern stonemovers like Cliff Osenton and Bertrand Poissonier are developing new "ancient" techniques requiring much less manpower. Nevertheless, Atkinson is required reading for anyone who wants to know what archaeology has to say about Stonehenge. As the lead archaeologist for the most important dig, no one was ever more in touch with the monument than Atkinson. ISBN: 0-14-013646-0.

Balfour, Michael. *Stonehenge and its Mysteries* (Charles Scribner's Sons, 1980). Good overview of Stonehenge: history, archaeology, archaeo-astronomy, the mystical side, etc. In particular, I like his chapter on the construction of Stonehenge, including photographs of its reconstruction, and several intriguing drawings of how the stones *might* have been transported, stood up, and topped with lintels. ISBN: 0-684-16406-X.

Chippendale, Christopher. *Stonehenge Complete* (Cornell Paperbacks, 1987). Chippendale's buoyant tome remains my favorite on Stonehenge. The author takes the unusual and effective approach of looking at Stonehenge from the perspective of several different periods of time, and thus creates a fascinating history, entertainingly written. Contains a wealth of information I've never seen elsewhere. ISBN: 0-8014-9451-6. (A revised edition was published in 1994 by Thames and Hudson.)

Fowles, John; photos by Barry Brukoff. *The Enigma of Stonehenge* (Jonathan Cape, 1980). A sometimes brutally honest and personal overview of Stonehenge up until the unfortunate period of the late 1970s. Excellent photography throughout. ISBN: 0-224-016180-0.

Holland, Cecelia. *Pillar of the Sky* (Alfred A. Knopf, 1985). This novel paints a good picture of the kind of motivation—or obsession—that could have brought about the construction of Stonehenge. Good on structural speculation and a possible social structure of neolithic times. Literarily, it reads like a Thomas Hardy tragedy, but I like Thomas Hardy tragedies, especially *Tess,* which reaches its climax at Stonehenge. ISBN: 0-394-53538-3.

Souden, David. *Stonehenge: Mysteries of the Stones and Landscape* (Collins & Brown, in association with English Heritage, 1997). Thorough, accurate, up-to-date, extremely well illustrated. The latest carbon dating calibrations are used. Good, not strong, discussion of transporting and erecting the stones. Good bibliography, recommended. ISBN: 1-85585-291-8.

STONE CIRCLES AND MEGALITHS

Ashmore, Patrick. *Calanais: The Standing Stones* (Urras nan Tursachan, 1995). Well-illustrated mini-guide to Callanish. Ashmore was in charge of the 1980-81 dig. ISBN: 0-86152-161-7.

Burl, Aubrey. *Prehistoric Avebury* (Yale University Press, 1979). This is the definitive book on what many consider to be the greatest of all stone circles. Burl combines his academic precision, his narrative ability, and a fine selection of beautiful and informative illustrations in this epic work.

———. *The Stone Circles of the British Isles* (Yale University Press, 1977). In my view, this monumental work is the bible of stone circle study. The edition cited still holds up well against all comers, and contains Burl's statistical analysis and diagrams based on his own visits to and studies of hundreds of stone circles (not available elsewhere). Burl is presently working on a rewrite for Yale University Press. The edition cited here has ISBN: 0-300-01972-6.

Giot, Pierre-Roland. *Prehistory in Brittany: Menhirs and Dolmens* (Editions D'Art Jos le Doar, 1995). Good introduction to the megaliths of Brittany. Well illustrated in color. Available in France in several languages. ISBN: 2-85543-123-9.

Hadingham, Evan. *Circles and Standing Stones* (Walker and Company, 1975). A good overview of ancient stone circles for the layman. Not nearly as detailed as *The Stone Circles of the British Isles,* but nicely illustrated, entertaining, and well worth a look. ISBN: 0-8027-0463-8.

Michell, John. *Megalithomania* (Thames and Hudson, 1982). Long a personal favorite of mine, Michell's entertaining tome was instrumental in my meeting Ed Prynn and, perhaps, helping the book in your hands to come about. You see, I too suffer from *megalithomania,* although I do not seek a cure. Michell's book is a compendium of humankind's fascination with megaliths for the past three centuries. (For many, *compulsion* would be more accurate.) Backed by hundreds of wonderful old line drawings, many showing sites either destroyed or changed, Michell takes us on a rollicking, journey. ISBN: 0-500-27235-2.

Mohen, Jean-Pierre. *The World of Megaliths* (Facts on File, Inc., 1990). Although hard to find—interlibrary loan is suggested—this out-of-print work is the best of the mainstream books about the history and construction of megaliths, as well as the continued worldwide use of megaliths in Japan, Korea, India, Madagascar, and other parts of the world. A respected archaeologist, Mohen used actual field testing to learn about construction techniques. ISBN: 0-8160-2251-8.

Ponting, Gerald and Margaret. *New Light on the Stones of Callanish* (G & M Ponting, 1984). An excellent history and overview discussion of the primary stone circle at Callanish. Introduces some of the archaeo-astronomical aspects of the site, expanded upon in later works by Margaret and Ron Curtis. ISBN: 0-9505998-4-0. Available from the Curtises, see sources.

Mysteries of the Ancient World (National Geographic, 1979). Done in typical National Geographic style and quality, this is a series of articles on such topics as pyramids, megaliths, and Easter Island. Conservative and a bit out of date, but worth a look. Try used book stores. ISBN: 0-87044-254-6.

GUIDE BOOKS

Ashmore, Patrick. *Maes Howe* (Historic Scotland, 1990). A 16-page color guide to the great neolithic tomb in Orkney. I particularly like this one for the translations of the messages carved in the sandstone slabs by Viking looters almost a thousand years ago. ISBN: 0-11-493473-8.

Balfour, Michael. *Megalithic Mysteries: An Illustrated Guide to Europe's Ancient Sites* (Parkgate Books, 1997). Big and beautifully illustrated in color, this is more than just a coffee table book. While 58 pages deal with 79 megalithic sites in the British Isles, an additional 107 pages take us to another 119 sites in Europe and Africa, accounts that are hard to find in English-language works. I learned a lot from this one, worth owning for its superb photographs. ISBN: 1-85585-355-8.

Burl, Aubrey. *Prehistoric Stone Circles* (Shire, 1997). A good little introduction to ancient stone circles. Burl lists the fifty circles he "has most enjoyed." ISBN: 0-85263-962-7.

———. *Megalithic Brittany* (Thames and Hudson, 1995). This detailed and well-organized guide to over 350 ancient sites in Brittany was invaluable to us during our megalithic journey in 1997. ISBN: 0-500-27460-6.

———. *A Guide to the Stone Circles of Britain, Ireland and Brittany* (Yale University Press, 1995). Simply the best. Almost 400 different stone rings are discussed, with directions to locate every one. And Burl spices the text with plenty of inside information and fascinating anecdotes. This is the one book I would always have in my pack on a megalithic journey. ISBN: 0-300-06331-8.

Cooke, Ian. *Journey to the Stones* (Men-an-Tol Studio, 1987). Cornwall is one of the world's great repositories of amazing stones, natural and manmade, ancient and modern (Ed Prynn's collection). Cooke divides his book into nine detailed walks, so, used as a guide, the megalithomaniac can enjoy many a pleasant hour, day, or week discovering Cornwall's trove of standing stones, holed stones, stone circles, fogous, and dolmens, and learn a wealth of ancient lore at the same time. Good maps and directions; well illustrated with old engravings, modern photographs, and the author's own linocuts. ISBN: 0-9512371-0-1.

Cope, Julian. *The Modern Antiquarian* (Harper Collins, 1998). This huge and colorful work is divided into Book One, a series of essays, including several about the ancient religion of the Earth Mother, and Book Two, "A Gazetteer of over 300 British Sites." Although the book *could* be used as a travel guide—it has all the Ordnance Survey grid references—its 4¼ pounds would be better replaced in the backpack by Burl's *Guide,* above, and 3¼ pounds of food and drink. Cope's view is highly opinionated, unabashed, and often delightful. The notes from his diary give a sense of here and now to the text. Strongly recommended for those with an interest in the transrational. Over 600 photographs and illustrations, even 50 poems. ISBN: 0-7225-3599-6.

Lambert, Joanne Dondero. *America's Stonehenge: An Interpretive Guide* (Sunrise, 1996). This is the definitive work on one of the most enigmatic megalithic sites in North America. Available at the site in Salem, New Hampshire. ISBN: 0-9652630-0-2. See also next entry.

Hanion, Charles. *Puzzles in Stone* (Tape-It! Video Productions, 1993). This 30-minute video is a good overview of America's Stonehenge. P.O. Box 444, Rogersville, TN 37857.

Jackson, Anthony. *The Pictish Trail: A Travellers Guide to the Old Pictish Kingdoms* (The Orkney Press Ltd., 1989). Scotland is home to hundreds of Pictish stones, generally carved standing stones dating from the 6th to 9th centuries A.D. Most were used to commemorate marital unions between clans. Erected three thousand years after the great stone circles, the Pictish stones are well worth visiting. ISBN: 0-907618-18-9.

Pinni, Ernest W. *America's Stonehenge* (Sarsen Press, 1980). The first half of this 30-page self-published booklet describes the full-sized Stonehenge replica built at Maryhill, Washington, from 1918 to 1929. The second half deals with the original Stonehenge. The astronomical alignments at each site are compared. ISBN: 0-937324-06-X. About the only ways to get a copy are to actually visit America's Stonehenge at Maryhill, or (in the U.S.) send $9 to Maryhill Museum, 35 Maryhill Museum Drive, Goldendale, WA 98620.

Pitts, Michael. *Footprints Through Avebury* (Westdale Press, 1992). A thorough 64-page guide to the great circle. As curator of the Keiller Museum at Avebury for five years, Pitts knows his stuff. Great maps and illustrations, old and new. I found the book to be very useful on site. No ISBN number. Available at Avebury.

Ponting, Gerald & Margaret. *The Stones Around Callanish* (G & M Ponting, 1984; revised 1993 by Ron & Margaret Curtis). The best guide available on the outlying sites at Callanish. ISBN: 0-9505998-9-1. Available from the Curtises, who also publish several other mini-guides to sites on Lewis. See my sources section.

Sykes, Homer. *Mysterious Britain* (Weidenfeld and Nicolson, 1993). Excellent book of color photos by the author, showing nearly 100 enigmatic sites, including many stone circles, holed stones, and monoliths. Includes concise thumbnail texts about each site. I learned about the Druid's Temple from this book. ISBN: 0-297-83453-3.

ARCHAEO-ASTRONOMY

Aveni, Anthony F. *Ancient Astronomers* (Smithsonian Books, 1993). Undistracted by TV, the Worldwide Web, and the light pollution of the night sky, the ancients were far more in touch with practical everyday astronomy than modern humankind. Aveni provides a detailed global look at ancient astronomers in his clear text, backed by carefully selected illustrations. Thoroughly referenced. Bibliography. ISBN: 0-89599-037-7.

Burl, Aubrey. *Prehistoric Astronomy and Ritual* (Shire Publications, 1997). A short and tidy introduction to archaeo-astronomy by a universally accepted expert in stone circles and the leading authority on the astronomically enigmatic recumbent circles. "Archaeo-astronomy is no longer regarded as an activity of the lunatic fringe," says Burl. "It has become a respectable study. It now needs to become a respectable discipline." ISBN: 0-85263-621-0.

Curtis, Ron & Margaret. *Callanish: Stones, Moon & Sacred Landscape* (self-published, 1994). A detailed account of the almost magical lunar event that takes place every 18.61 years at Callanish. Available from the authors. See my sources section.

Hawkins, Gerald S. *Stonehenge Decoded* (Dell, 1965). This one caused a sensation when it was published. Hawkins forced the experts to consider a new archaeo-astronomical viewpoint on Stonehenge and other ancient megalithic monuments. While not the first to connect stone circles to celestial events, Hawkins popularized the idea.

Hoyle, Fred. *From Stonehenge to Modern Cosmology* (W. H. Freeman, 1972). These lectures on archaeo-astronomy by an astronomer give a clear account of how and why the sun and moon appear to take their respective paths through the sky. Hoyle concludes that Hawkins (previous entry) was right about Stonehenge being an astronomical observatory. ISBN: 0-7167-0341-6.

Krupp, Dr. E. C. *Echoes of the Ancient Skies: The Astronomy of Lost Civilizations* (Harper & Row, 1983). Astronomical orientation was important for ancient builders throughout the world. The author builds logically from chapter 1, in which he describes with clarity "how the sky works" and takes the story across the eons almost to the present day. He includes the most complete bibliography on the subject I have found, over 500 entries. ISBN: 0-06-015101-3.

Newham, C. A., *The Astronomical Significance of Stonehenge* (Moon Publications, 1972). A short treatise on the solar and lunar alignments. Often cited in other works.

North, John. *Stonehenge: A New Interpretation of Prehistoric Man and the Cosmos* (Simon & Schuster, 1996). This huge and detailed book makes a strong case that Stonehenge was, first and foremost, an astronomical site, and attempts to dovetail the archaeo-astronomical evidence with the religion and culture of the ancients. While scholarly, thoughtful, and professional, the book may be rather imposing for lay readers. Nevertheless, North includes much valuable archaeo-astronomical information and theory under one cover, and his book may eventually be considered an important work. ISBN: 0-684-84512-1.

Ruggles, C. L. N. ed. *Records in Stone: Papers in Memory of Alexander Thom* (Cambridge University Press, 1988). Over 400 pages of papers by friends and colleagues of Thom's, including Ron and Margaret Curtis. This hard-to-find volume will be of interest to the serious archaeo-astronomer. ISBN: 0-521-33381-4.

Thom, Alexander. *Megalithic Sites in Britain* (Oxford University Press, 1967). Thom's epic surveying of megalithic sites, mostly stone rings, revealed the megalithic yard, various stone ring shapes, and new information on archaeo-astronomy. Hard to find, but worth the effort for the very serious student of stone circles. ISBN: 0-19-813148-8.

———. *Megalithic Lunar Observatories* (Oxford University Press, 1971). Detailed accounting of lunar alignments, refraction, parallax, azimuth, and declination. Complex, almost imposing, but the implications are fascinating.

SACRED SPACES AND EARTH ENERGY

Devereux, Paul. *Geomancy: Overview of Earth Mysteries* (a Mystic Fire Video from Trigon Communications, 1988). The prolific and popular author/geomancer Paul Devereux hosts this fascinating video, some of it filmed on location at the Rollright Stones when Devereux's Dragon Project investigated the electromagnetic and ultrasonic characteristics of the site, long a focal point of transrational experiences. Crystals, leys, geomancy, and sacred geometry are covered. For me, the highlight of an engaging video is the lucid commentary of author John Michell.

Lip, Evelyn. *Feng Shui: A Layman's Guide* (Times Editions Pte. Ltd., 1979; 1987 ed., Heian International). A good introduction into the ancient Chinese art of placement. ISBN: 0-89346-286-1. There are lots of other books on *feng shui* (geomancy) available today. Contact the American Society of Dowsers Bookstore in the sources for other books on *feng shui*.

Mann, A. T. *Sacred Architecture* (Element Books, 1993). Sacred geometry, feng shui, earth magic, man-delas, numerology, megaliths . . . a good overview of what has made certain architecture sacred, from prehistoric times to the present. Excellent bibliography. ISBN: 1-85230-391-3.

Michell, John. *The New View Over Atlantis* (Thames and Hudson, 1983). Michell's best-known work lovingly investigates ancient metrology (study of weights and measure), earth energies, leys, sacred space, and the like. This fascinating book is essential for all who want to build their stone circles from a geomantic perspective. Excellent bibliography. ISBN: 0-500-27312-X.

Mystic Places (Time-Life Books, 1987). Wanders off the mainstream path of science to introduce "mysteries" and, sometimes, to suggest solutions. The range is from scientific through pseudo-scientific to borderline nutsy-cuckoo. Still, the book stimulates and entertains, and some of the megalithic photos are gorgeous. Look in used book shops for low-cost copies. ISBN: 0-8094-6312-1.

Swan, James A. *The Power of Place: Sacred Ground in Natural & Human Environments* (Gateway Books, 1993). A compendium of 25 articles on sacred space, including such topics as holy wells, megalithic sites, ancient and modern geomancy, leys, Earth's electromagnetic fields, Native American sacred space, and the like. ISBN: 0-946551-94-4.

Zink, David D. *The Ancient Stones Speak: A Journey to the World's Most Mysterious Megalithic Sites* (E. P. Dutton, 1979). Zink tours 25 "mysterious" sites, including Maes Howe, Callanish, and America's Stonehenge, often accompanied by a "sensitive" or two who gives his or her impressions, a kind of "psychic archaeology" that sends shivers down the spines of "real" archaeologists. Leys, dragon currents, and the planetary grid system all come into the mix. Lots of work and good research here, some, but not all of it, in the transrational realm. ISBN: 0-525-47587-7.

OTHER

Fell, Barry. *America B.C.: Ancient Settlers in the New World* (revised edition: Simon & Schuster, 1989). Columbus was not the first European to make the journey to the Americas, nor were the Vikings. Harvard professor and epigrapher Barry Fell makes a strong case that many others visited the "new world" over the centuries, including Romans and Iberian Celts. Of particular interest is Fell's study of megalithic chambers, standing stones, and dolmens in America's northeast region. ISBN: 0-671-67974-0. [See the Stonehenge Viewpoint entry in the sources for more on Pre-Columbian visits to America. Other titles of interest in this field are *Maps of the Ancient Sea Kings*, by Charles H. Hapgood (Adventures Unlimited Press, 1966; 1996, ISBN: 0-932813-42-9); *The Rediscovery of Lost America*, Arlington Mallery and Mary Roberts Harrison (E. F. Dutton, 1951, 1979, ISBN: 0-525-47545-1); *American Discovery*, Gunnar Thompson (Misty Isles Press, 1994, ISBN: 0-9621990-4-4); and *The Friar's Map of Ancient America*, Gunnar Thompson (Laura Lee Productions, 1996, P.O. Box 3010, Bellevue, WA 98009, ISBN: 0-9621990-8-7).]

Green, Miranda J. *The World of Druids* (Thames and Hudson, 1997). An in-depth look at the Celtic priests or shamans known as druids, from ancient times up until the present day revival in Earth-based religions. Includes a detailed directory of modern druid organizations and a thorough bibliography of writings about druids, ancient and modern. ISBN: 0-500-05083-X.

Hancock, Graham. *Fingerprints of the Gods* (Crown, 1995). Hancock paints a compelling picture of a world whose civilizations were seeded by a much earlier people. This is not visits-from-spacemen stuff, but carefully researched evidence presented in an organized yet entertaining manner. Hancock introduces stupendous feats of megalithic construction by the ancients, including the great obelisks. Highly recommended. Detailed references and bibliography. ISBN: 0-385-25475-X.

————. *Heaven's Mirror: Quest for the Lost Civilization* (Crown, 1998). Hancock continues his search for the original civilization. This book stimulates on both the intellectual plane and the visual, thanks to the spectacular color photography of Hancock's wife and partner, Santha Faiia, including many plates of almost miraculous megalithic engineering feats. ISBN: 0-517-70811-6.

Hedges, John W. *Tomb of the Eagles: A Window on Stone Age Tribal Britain* (John Murray, 1984). Using a recently excavated tomb in Orkney as a focal point, Hedges provides a close look at neolithic life. ISBN: 0-7195-4343-6.

Heyerdahl, Thor. *Aku-Aku* (Rand McNally, 1958). This is the famous adventurer's account of Easter Island and the discoveries he made there in the 1950s. Good photos and discussion of re-erecting the huge megalithic heads that make the islands famous. My tattered copy predates ISBN numbers, but the work is easily found in most libraries and used book stores.

Prynn, Edward. *A Boy in Hobnailed Boots* (Tabb House, 1981, ISBN: 0-907018-07-6). Ed Prynn's boyhood in Cornwall. The story is continued in *No Problem* (Tabb House, 1982, ISBN: 0-907018-16-5). Together, these books give a unique look into the life of a Cornishman from World War Two until about 1980, just prior to Ed's megalithic phase. See my chapter 5. I await volume three of Ed's saga with keen anticipation. The books are available from Ed. See my sources section.

Secrets of Lost Empires: Reconstructing the Glories of Ages Past (Sterling, 1997). The 38-page first chapter, "Stonehenge," by Cynthia Page, recounts the experiences of a NOVA crew as they raised a full-scale model of the Great Trilithon at Stonehenge for their 1996 video (*Secrets of Lost Empires: Stonehenge*). Well illustrated in color. Other chapters deal with building a small pyramid, standing an obelisk, the Roman Colosseum, and the remarkable megalithic stonework of the ancient Incas. ISBN: 0-8069-9584-X.

If you enjoyed the book in your hand, you might like some of the following books by the same author, namely me. Available from Earthwood (see my sources section).

Roy, Rob. *Complete Book of Cordwood Masonry Housebuilding* (Sterling, 1992). How to build a low-cost, beautiful, energy-efficient home out of short logs unsuitable for making into lumber. The Earthwood Stone Circle made its debut in this book. ISBN: 0-8069-8590-9.

————. *The Complete Book of Underground Houses: How to Build a Low-Cost Home* (Sterling, 1994). Underground homes are cozy and warm, yet cool in the summer. They can be expensive, or, if the techniques in this book are followed, quite inexpensive for the owner-builder. ISBN: 0-8069-0728-2.

————. *The Sauna* (Chelsea Green, 1996). I've yet to meet a stone builder who wasn't also interested in sauna. This book gives sauna lore, health benefits, how to use the sauna, and, most importantly, how to build your own cordwood masonry sauna at low cost. ISBN: 0-930031-87-3.

————. *Mortgage-Free! Radical Strategies for Home Ownership* (Chelsea Green, 1998). Never, ever take out a mortgage to build a stone circle. This goes for your house, too. *Mortgage* means, literally, "death pledge." This book tells you how to steer clear of lending institutions. ISBN: 0-930031-98-9.

Index

CHELSEA GREEN

Sustainable living has many facets. Chelsea Green's celebration of the sustainable arts has led us to publish trend-setting books about organic gardening, solar electricity and renewable energy, innovative building techniques, regenerative forestry, local and bioregional democracy, and whole foods. The company's published works, while intensely practical, are also entertaining and inspirational, demonstrating that an ecological approach to life is consistent with producing beautiful, eloquent, and useful books, videos, and audio cassettes.

For more information about Chelsea Green, or to request a free catalog, call toll-free (800) 639–4099, or write to us at P.O. Box 428, White River Junction, Vermont 05001. Visit our Web site at www.chelseagreen.com.

Chelsea Green's titles include:

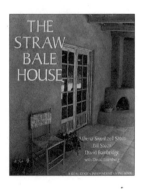

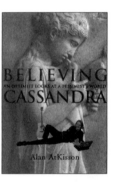

The New Independent Home

The Straw Bale House

Independent Builder: Designing & Building a House Your own Way

The Rammed Earth House

The Passive Solar House

The Sauna

Wind Power for Home & Business

Wind Energy Basics

The Solar Living Sourcebook

A Shelter Sketchbook

Mortgage-Free!

Hammer. Nail. Wood.

Four-Season Harvest

The Apple Grower

The Bread Builder

Keeping Food Fresh

Simple Food for the Good Life

The Flower Farmer

Passport to Gardening

The New Organic Grower

Solar Gardening

Straight-Ahead Organic

Good Spirits

The Contrary Farmer

The Contrary Farmer's Invitation to Gardening

Whole Foods Companion

Believing Cassandra

Gaviotas: A Village to Reinvent the World

The Man Who Planted Trees

Who Owns the Sun?

Global Spin: The Corporate Assault on Environmentalism

Seeing Nature

Hemp Horizons

Genetic Engineering, Food and Our Environment

Scott Nearing: The Making of a Home Steader

Loving and Leaving the Good Life

Wise Words for the Good Life